MACHINERY DYNAMICS

MACHINERY DYNAMICS

CE ZHANG
Department of Mechanical Engineering
Tianjin University
Tianjin, PR China

JIANMING YANG
Department of Mechanical Engineering
Memorial University of Newfoundland
St. John's, NL, Canada

ZONGYU CHANG
Department of Mechanical Engineering
Ocean University of China
Qingdao, PR China

ELSEVIER
ACADEMIC PRESS
An imprint of Elsevier
清華大學出版社
TSINGHUA UNIVERSITY PRESS

Academic Press is an imprint of Elsevier
125 London Wall, London EC2Y 5AS, United Kingdom
525 B Street, Suite 1650, San Diego, CA 92101, United States
50 Hampshire Street, 5th Floor, Cambridge, MA 02139, United States
The Boulevard, Langford Lane, Kidlington, Oxford OX5 1GB, United Kingdom

Library of Congress Cataloging-in-Publication Data
A catalog record for this book is available from the Library of Congress

British Library Cataloguing-in-Publication Data
A catalogue record for this book is available from the British Library

ISBN: 978-0-12-815785-5

For information on all Academic Press publications
visit our website at https://www.elsevier.com/books-and-journals

Publisher: Matthew Deans
Acquisitions Editor: Carrie Bolger
Editorial Project Manager: Naomi Robertson
Production Project Manager: Kamesh Ramajogi
Designer: Greg Harris

Typeset by VTeX

To our families for the unwavering love, support and encouragement.

Contents

List of figures

Biography

Ce Zhang

Ce Zhang is Professor Emeritus of the School of Mechanical Engineering at Tianjin University, PR China. He has more than 50 years experience in teaching and research in mechanical dynamics. He has authored (co-authored) seven books and more than 150 conference and journal papers.

Jianming Yang

Jianming Yang is an Associate Professor of Mechanical Engineering at Memorial University of Newfoundland, Canada. He has more than a total of 30 years experience in teaching at universities and practicing mechanical engineering in industry. His research focuses on nonlinear and random machinery dynamics.

Zongyu Chang

Zongyu Chang is a Professor of Mechanical Engineering at Ocean University of China. He has more than 20 years experience in teaching and research in machinery dynamics. His research focus is nonlinear machinery dynamics, with application to offshore systems, in particular.

Preface

This book has been planned with two main purposes: as a textbook of machinery dynamics for graduate students, and as a reference for researchers and practicing engineers. The following features are intentionally crafted to serve the two purposes:

- Vibration theory is enhanced by including a brief introduction to nonlinear vibration, in addition to traditional linear vibration theory. The arrangements comes with two considerations. First, elasto-dynamics is based on the vibration theory. Thus, students don't have to refer to a separate vibration text for adequate coverage of fundamental vibration theory. Second, attempts have been made in some universities to cluster vibration and machinery dynamics, which were traditionally two stand-alone courses in the undergraduate curriculum.
- The historical evolution of the subject, from static analysis through elasto-dynamic analysis, is clearly presented. This consideration helps students to establish an overall picture of the subject. The instructor, thus, can pick the materials in which he/she is interested for class presentation. At the same time, this provides abundant material outside the class coverage that is still available for the students to read.

Most English-language textbooks available in the market cover only the very basic rigid-body dynamics with a few of exceptions, such as the one by Dresig and Holzweißig (2010). They are adequate for undergraduate, but not for a graduate, course. This book is intended to fill this gap. In fact, efforts have been made for more than several years. A book written in Chinese by our leading author, Professor Zhang, and his associates was published in 2000 (1st edition) and 2008 (2nd edition) by China Higher Education Press. This book has been widely welcomed by many institutions in China, and it was recommended by the China Ministry of Education as a graduate textbook. The current book bears some resemblance to the counterpart in Chinese, but the materials are greatly expanded. We would like to express our sincere gratitude to the several contributors to the Chinese version; these include Dr. Yan'an Yao (Professor of Beijing Jiaotong University), Drs. Yuhu Yang, Yuehai Sun, and Shiyu Wang (Professors of Tianjin University), as well as Dr. Zhiyou Feng (Professor of Tianjin Polytechnic University). In addition, Dr. Yan'an Yao also wrote Section 14.4 of this English version.

This book also serves as a summary of the research work conducted by Professor Zhang and his former graduate students over more than 30 years. Professor Zhang started research on elasto-dynamics of linkages when he studied at Worcester Polytechnic Institute (WPI) in the US during 1980–1982. After returning to China, he continued the work in machinery dynamics, and he extended his research to include many other mechanisms, such as cam mechanisms, gear drives, intermittent mechanisms, and mechanical systems. More than 30 graduate students, including his other co-authors of this book, have been trained under Professor Zhang's supervision. Most of them are now faculty members in universities across China and Canada.

There are many people to be thanked for the realization of this book. In particular, the staff from Tsinghua University Press and Elsevier are greatly appreciated. Their professional work has provided the authors great help in writing this book.

Ce Zhang
Jianming Yang
Zongyu Chang
June 2021

PART 1

Rigid-body dynamics

The first two Industrial Revolutions introduced fundamental changes in driving power, materials, and production technologies. At the same time, higher requirements were put forward on machines. To improve productivity, machines were expected to run faster, be more powerful, and possess higher levels of automation. Higher precision and lighter weight have become necessary for the purpose of quality improvement and cost efficiency. These new requirements have required the support of new theories and new design and analysis tools. Machinery dynamics is a subject based on classical mechanics. The continuous pursuit of higher machine speeds greatly inspired the development of machinery dynamics. Newton was a pioneer of modern classical mechanics, who created the theory of Newtonian mechanics in the 17th century, laying the foundation for the dynamic analysis of machines. This is also the theoretical base of civil and many other engineering disciplines. In 1742, D'Alembert put forward the D'Alembert's principle, which incorporates Newton's second law of motion into the equilibrium equation of statics. As a result, a dynamic problem can be elegantly transformed into a static one. In 1765, Euler extended Newton's second law of motion for a point particle to rigid bodies, forming the theory of rigid-body dynamics. All these developments employs the manipulation of vector quantities, such as acceleration and force; thus, they are also called vectorial mechanics. In 1788, Lagrange established analytical mechanics that derives the dynamic equation of motion from the viewpoint of energy, including kinetic energy and potential energy. Energy is represented by scalars; thus, vectorial and graphic representation is avoided. In addition, the formulation of analytical mechanics does not involve the reaction forces of joints. These features make analytical mechanics more suitable to model complicated systems than the classical vectorial mechanics. It has been widely employed to derive the governing equation of motion in machinery dynamics.

Historically, force analysis of machines has gone through four different levels, namely, static analysis, kineto-static analysis, dynamic analysis, and kineto-elasticdynamic analysis (Erdman and Sandor, 1972). The first three are based on the assumption that rigid-body members, whereas the fourth takes the flexibility of the members into consideration (see Part 3). The three methods that are based on rigid-body assumption are outlined as follows:

1) Static analysis

In the early days, machine speed was generally low, and the links were generally over-designed having high stiffness. In this case, static analysis is adequate, and the rigid assumption is valid.

2) Kineto-static analysis

With the increase of machine speed, neglecting inertial loads may lead to unsatisfactory accuracy. With the D'Alembert's principle, the inertial loads can be added to the equilibrium equation, forming the so-called kineto-static analysis. In this method the input is assumed to maintain a constant speed.

3) Dynamic analysis

For some machines, such as crushers, stamping presses, and steel rolling mills, the working load changes dramatically over one cycle, resulting in large speed fluctuations in operation. Large speed fluctuation may not be desirable for some applications, making dynamic analysis necessary. In dynamic analysis, the assumption of constant input speed is abandoned. The interaction between the driver, such as motor or engines, and the mechanism is considered. For this reason, it is also referred to as dynamics of mechanical systems.

In internal combustion engines, the high-speed reciprocating motion of the piston causes severe vibration, which may worsen the machine performance or be transmitted to the base. The vibration is often controlled through two routes. One is through various vibration reduction and isolation measures, which are generally covered in vibration texts. The other is through balancing measures to directly control the excitation. Balancing techniques are generally covered in mechanism courses; they do not involve vibration theory.

CHAPTER 1

Kineto-static analysis

1.1 Introduction

If machines run at low speed, static analysis may be adequate. With inceased speed, however, inertial loads need to be properly accounted for in analysis and design. The easiest way to include the inertial loads in analysis is to use the D'Alembert's principle, which incorporates the inertial loads into static equilibrium equation. This technique is referred to as kineto-static analysis.

Static analysis is still in use in design. The dynamic effect in static analysis is generally accounted for by an estimated corrective dynamic factor. Kineto-static analysis proceeds one step beyond the static analysis, being necessary for machines operating at relatively high speeds.

To calculate the inertial loads, the acceleration is required. Thus, kinematic analysis is a prerequisite for kineto-static analysis. Kinematic analysis is based on the assumption that the driving link operates with a prescribed motion. More often, it is assumed to rotate at a constant angular speed. Obviously, kineto-static analysis does not consider the interaction between the driving power and the mechanism motion since it is an idealization of a real machine. It is commonly used in inverse dynamics to calculate the joint reaction forces, the required driving torque to maintain the prescribed motion, and the *shaking force* and *shaking moment*.

Kineto-static analysis can be conducted either graphically or analytically. The graphic method, generally covered in an undergraduate course, is straightforward. However, the analytic method is more conveniently implemented with computers.

This chapter focuses on the analytic technique. The chapter is organized as below:

- A short introduction of kineto-static analysis is presented in Section 1.1.
- Analysis of four-bar linkages are introduced in Sections 1.2 and 1.3. Specifically, Section 1.2 focuses on general four-bar linkages having only revolute joints, while Section 1.3 treats the slider-crank mechanism.
- Section 1.4 analyzes a planar cam mechanism.

Machinery Dynamics
https://doi.org/10.1016/B978-0-12-815785-5.00010-8

1.2 Analysis of planar linkages

1.2.1 Inertial force and inertial moment

We limit our discussion to planar mechanisms. The motion of a rigid body in a plane is represented by a combination of a translation of the center of mass and a rotation. As illustrated in Fig. 1.1, the center of mass is labeled as S. The position of point S and the orientation of the object can be described by two vectors, $\boldsymbol{S}_i$ and $\boldsymbol{\varphi}_i$, respectively. The translational and rotational accelerations are $\ddot{\boldsymbol{S}}_i$ and $\ddot{\boldsymbol{\varphi}}_i$. The inertial force and inertial moment are $-m_i\ddot{\boldsymbol{S}}_i$ and $-J_i\ddot{\boldsymbol{\varphi}}_i$, respectively. m_i and J_i are the mass and the moment of inertia around S. The sign $-$ simply indicates that the direction of the inertial force (or the inertial moment) is opposite to that of the accelerations. Two special cases are:

- the inertial moment is zero if a link translates along a straight line without rotation, and
- the inertial force is zero if a link has pure rotation around S without translation.

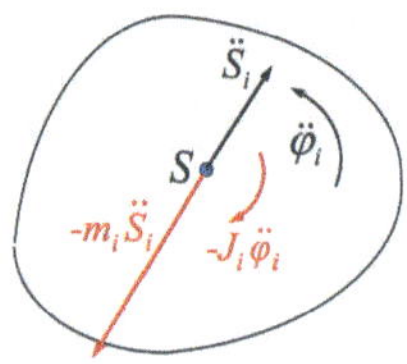

Figure 1.1 Inertial force and inertial moment.

1.2.2 Analysis of planar linkage

In the following, a four-bar linkage is used to illustrate the procedures of the kineto-static analysis. The method developed based on the four-bar mechanism, however, can be easily extended to more complicated linkages.

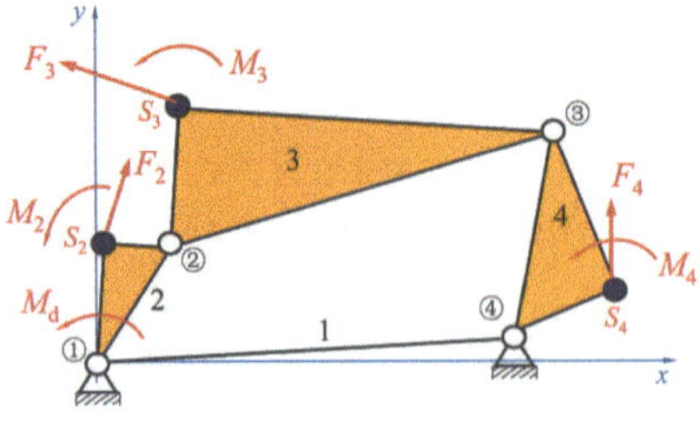

Figure 1.2 A four-bar linkage.

A four-bar mechanism is shown in Fig. 1.2. For clarity, some conventions are established as listed here:

- The links and joints are numbered in clockwise order.
- To distinguish the joint from the link, the joint number is put inside a circle.
- The fixed link is set as number 1.
- Friction is neglected.

S_i, $\boldsymbol{F}_i$, and $\boldsymbol{M}_i$ represent the center of mass, the external force, and the external moment applied to link i ($i = 2 - 4$). Double subscripts are used for the joint force; $\boldsymbol{F}_{ij}$ means the joint force applied by link i to link j. With these notations and conventions, the free body diagrams (FBD) of all moving links can be drawn as in Fig. 1.3.

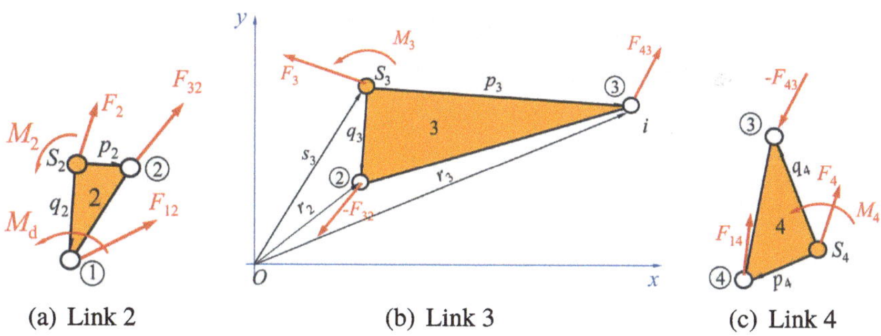

Figure 1.3 FBD of moving links.

According to the D'Alembert's principle, the equilibrium equations for a moving link i can be represented in a vectorial form as

$$\begin{cases} \boldsymbol{F}_{i\Sigma} - m_i\ddot{\boldsymbol{S}}_i = 0 \\ \boldsymbol{M}_{i\Sigma} - J_i\ddot{\boldsymbol{\varphi}}_i = 0, \end{cases} \tag{1.1}$$

where $\boldsymbol{F}_{i\Sigma}$ and $\boldsymbol{M}_{i\Sigma}$ are the resultant force and resultant moment, respectively, on the i-th link.

Applying these to each individual moving link, one has the following.

Link 2

$$\begin{cases} \boldsymbol{F}_{32} + \boldsymbol{F}_{12} + \boldsymbol{F}_2 = m_2\ddot{\boldsymbol{s}}_2 \\ (\boldsymbol{p}_2 \times \boldsymbol{F}_{32}) + (\boldsymbol{q}_2 \times \boldsymbol{F}_{12}) + \boldsymbol{M}_d + \boldsymbol{M}_2 = J_2\ddot{\boldsymbol{\varphi}}_2 \end{cases} \tag{1.2}$$

Link 3

$$\begin{cases} \boldsymbol{F}_{43} - \boldsymbol{F}_{32} + \boldsymbol{F}_3 = m_3\ddot{\boldsymbol{s}}_3 \\ (\boldsymbol{p}_3 \times \boldsymbol{F}_{43}) - (\boldsymbol{q}_3 \times \boldsymbol{F}_{32}) + \boldsymbol{M}_3 = J_3\ddot{\boldsymbol{\varphi}}_3 \end{cases} \tag{1.3}$$

Link 4

$$\begin{cases} \boldsymbol{F}_{14} - \boldsymbol{F}_{43} + \boldsymbol{F}_4 = m_4\ddot{\boldsymbol{s}}_4 \\ (\boldsymbol{p}_4 \times \boldsymbol{F}_{14}) - (\boldsymbol{q}_4 \times \boldsymbol{F}_{43}) + \boldsymbol{M}_4 = J_4\ddot{\boldsymbol{\varphi}}_4. \end{cases} \tag{1.4}$$

Two notes should be made regarding these equations.

- $\boldsymbol{M}_{\mathrm{d}}$ in Eq. (1.2) is the driving torque required to maintain the prescribed motion for the mechanism. In some books, this is termed *equilibrium torque*. It will be shown in Chapter 2 that $\boldsymbol{M}_{\mathrm{d}}$ is different from the actual driving torque transmitted to the driving link.
- $\boldsymbol{F}_{ij} = -\boldsymbol{F}_{ji}$ has been applied provided that they are action–reaction force pairs.

Several position vectors $\boldsymbol{S}_i$, $\boldsymbol{p}_i$, $\boldsymbol{q}_i$, $\boldsymbol{r}_i$, and $\boldsymbol{r}_{i-1}$ are also involved, which are shown in Fig. 1.3(b) for convenience. $\boldsymbol{p}_i$, and $\boldsymbol{q}_i$ can be easily obtained from the figure as:

$$\begin{cases} \boldsymbol{p}_i = \boldsymbol{r}_i - \boldsymbol{S}_i \\ \boldsymbol{q}_i = \boldsymbol{r}_{i-1} - \boldsymbol{S}_i. \end{cases} \tag{1.5}$$

Eqs. (1.2) through (1.4) can be transformed into scalar form, which is more convenient for computer coding.

Link 2

$$\begin{cases} F_{32}^{\mathrm{x}} + F_{12}^{\mathrm{x}} + F_2^{\mathrm{x}} = m_2\ddot{x}_{\mathrm{S2}} \\ F_{32}^{\mathrm{y}} + F_{12}^{\mathrm{y}} + F_2^{\mathrm{y}} = m_2\ddot{y}_{\mathrm{S2}} \\ (p_2^{\mathrm{x}}F_{32}^{\mathrm{y}} - p_2^{\mathrm{y}}F_{32}^{\mathrm{x}}) + (q_2^{\mathrm{x}}F_{12}^{\mathrm{y}} - q_2^{\mathrm{y}}F_{12}^{\mathrm{x}}) + M_{\mathrm{d}} + M_2 = J_2\ddot{\varphi}_2 \end{cases} \tag{1.6}$$

Link 3

$$\begin{cases} F_{43}^{\mathrm{x}} - F_{32}^{\mathrm{x}} + F_3^{\mathrm{x}} = m_3\ddot{x}_{\mathrm{S3}} \\ F_{43}^{\mathrm{y}} - F_{32}^{\mathrm{y}} + F_3^{\mathrm{y}} = m_3\ddot{y}_{\mathrm{S3}} \\ (p_3^{\mathrm{x}}F_{43}^{\mathrm{y}} - p_3^{\mathrm{y}}F_{43}^{\mathrm{x}}) - (q_3^{\mathrm{x}}F_{32}^{\mathrm{y}} - q_3^{\mathrm{y}}F_{32}^{\mathrm{x}}) + M_3 = J_3\ddot{\varphi}_3 \end{cases} \tag{1.7}$$

Link 4

$$\begin{cases} F_{14}^{x} - F_{43}^{x} + F_{4}^{x} = m_4 \ddot{x}_{S4} \\ F_{14}^{y} - F_{43}^{y} + F_{4}^{y} = m_4 \ddot{y}_{S4} \\ \left(p_4^{x} F_{14}^{y} - p_4^{y} F_{14}^{x}\right) - \left(q_{14}^{x} F_{43}^{y} - q_4^{y} F_{43}^{x}\right) + M_4 = J_4 \ddot{\varphi}_4. \end{cases} \tag{1.8}$$

The superscripts x and y represent the components of a vector along *thex* and y directions, respectively. Each moving link has three equations; thus, three moving links yield nine linear algebraic equations. They can be arranged into a more compact matrix form

$$\boldsymbol{AX} = \boldsymbol{B}_{\mathrm{I}} - \boldsymbol{B}. \tag{1.9}$$

$\boldsymbol{A}$ is a *nine* × *nine* matrix as

$$\boldsymbol{A} = \begin{bmatrix} 1 & 0 & 1 & 0 & 0 & 0 & 0 & 0 & 0 \\ 0 & 1 & 0 & 1 & 0 & 0 & 0 & 0 & 0 \\ -q_2^{y} & q_2^{x} & -p_2^{y} & p_2^{x} & 0 & 0 & 0 & 0 & 1 \\ 0 & 0 & -1 & 0 & 1 & 0 & 0 & 0 & 0 \\ 0 & 0 & 0 & -1 & 0 & 1 & 0 & 0 & 0 \\ 0 & 0 & q_3^{y} & -q_3^{x} & -p_3^{y} & p_3^{x} & 0 & 0 & 0 \\ 0 & 0 & 0 & 0 & -1 & 0 & 1 & 0 & 0 \\ 0 & 0 & 0 & 0 & 0 & -1 & 0 & 1 & 0 \\ 0 & 0 & 0 & 0 & q_4^{y} & -q_4^{x} & -p_4^{y} & p_4^{x} & 0 \end{bmatrix}. \tag{1.10}$$

$\boldsymbol{X}$ is the unknown vector.

$$\boldsymbol{X} = \begin{bmatrix} F_{12}^{x} & F_{12}^{y} & F_{32}^{x} & F_{32}^{y} & F_{43}^{x} & F_{43}^{y} & F_{14}^{x} & F_{14}^{y} & M_{\mathrm{d}} \end{bmatrix}^{\mathrm{T}}. \tag{1.11}$$

$\boldsymbol{B}$ and $\boldsymbol{B}_{\mathrm{I}}$ are the external load and the inertial load arrays, respectively. Their entries are as follows:

$$\boldsymbol{B} = \begin{bmatrix} F_2^{x} & F_2^{y} & M_2 & F_3^{x} & F_3^{y} & M_3 & F_4^{x} & F_4^{y} & M_4 \end{bmatrix}^{\mathrm{T}}, \tag{1.12}$$

$$\boldsymbol{B}_{\mathrm{I}} = \begin{bmatrix} m_2 \ddot{x}_{S2} & m_2 \ddot{y}_{S2} & J_2 \ddot{\varphi}_2 & m_3 \ddot{x}_{S3} & m_3 \ddot{y}_{S3} & J_3 \ddot{\varphi}_2 & m_4 \ddot{x}_{S4} & m_4 \ddot{y}_{S4} & J_4 \ddot{\varphi}_4 \end{bmatrix}^{\mathrm{T}}. \tag{1.13}$$

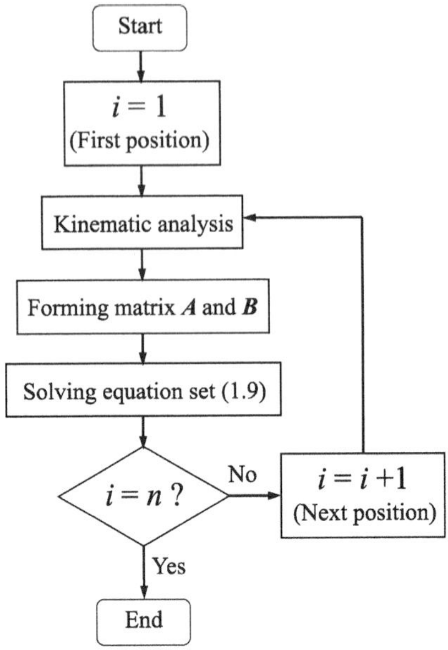

Figure 1.4 Flow chart of kineto-static analysis.

If the position vectors $\boldsymbol{p}$ and $\boldsymbol{q}$ are known, $\boldsymbol{X}$ can be solved for through Eq. (1.9). In the analysis, kinematic analysis should be conducted first to calculate $\boldsymbol{B}_{\mathrm{I}}$.

The procedures just cited are for one mechanism position. To obtain $\boldsymbol{X}$ in the whole motion cycle, the period of motion needs to be discretized into certain number (n) of positions. For each discrete position, Eq. (1.9) needs to be solved. The solution for all discrete positions gives the overall result. This may sound overwhelming for hand calculation, but is easily implemented on computers. Steps to solve with computers are summarized in a flowchart in Fig. 1.4.

If $\boldsymbol{B}_{\mathrm{I}}$ is set to be zero, this analysis would be degraded to static analysis.

According to the Gruebler's equation, the number of degrees of freedom (DOF) of a planar mechanism is calculated by

$$f = 3n - 2n_{\mathrm{L}} - n_{\mathrm{H}}, \tag{1.14}$$

where n_{L} and n_{H} are the number of lower pairs and higher pairs in the mechanism, respectively. n is the number of moving links in the mechanism.

For a general planar mechanism, f is the number of inputs (driving links) needed. In addition, each higher pair and lower pair introduce two and one reaction forces, respectively. Thus, the total number of unknowns, including reaction forces and equilibrium torques, in a mechanism is $2n_L + n_H + f = 3n$. On the other hand, each rigid moving link has three degrees of freedom, thus, three equations. The total number of equations available is $3n$ (n is the number of moving links). Obviously the number of equations equals the number of unknowns; thus, the equations are solvable.

1.2.3 Shaking force and shaking moment

The reaction forces at the joints on the machine frame are of special interest sometimes because they cause vibration to the frame, especially at high machine speeds.

The reaction forces on the frame joints create a shaking force $\boldsymbol{F}_s$ and a shaking moment $\boldsymbol{M}_s$. Taking a single-cylinder engine in Fig. 1.5 as an example, the core is a slider-crank mechanism, and the engine housing is the frame. For simplicity, we make the following assumptions:

1) the crank (link 2) rotates at a constant angular speed;
2) the center of mass of the crank is coincident with the joint O; and
3) the friction between the slider (link 4) and the track is neglected.

The center of mass of the crank is at point O; thus, there is no inertial force on it. The inertial loads of link 3 are represented by a force $\boldsymbol{F}_{I3}$ and a moment $\boldsymbol{M}_3$. The slider moves vertically so that an inertial force $\boldsymbol{F}_{I4}$ is formed. The frame applies a reaction force $\boldsymbol{F}_{12}$, which is resolved into F_{12}^x and F_{12}^y, through joint O and a horizontal force $\boldsymbol{F}_{14}$ through the sliding track on the mechanism. All these forces are shown in Fig. 1.5(a). If we look at the housing, however, it is subjected to a reaction force $\boldsymbol{F}_{21}$ at the revolute joint O and a reaction force $\boldsymbol{F}_{41}$ at the track from the mechanism as shown in Fig. 1.5(b). These reaction forces felt by the frame are transformed into a resultant shaking force $\boldsymbol{F}_S$ and a shaking moment $\boldsymbol{M}_S$ about joint O as shown in Fig. 1.5(c). Obviously, the location of point O has no effect on $\boldsymbol{F}_S$, but does influence $\boldsymbol{M}_s$.

For the mechanism in Fig. 1.2, the shaking force and shaking moment can be calculated as

$$\begin{cases} F_S^x = \sum_{i=2}^{4} (m_i \ddot{x}_{Si}) \\ F_S^y = \sum_{i=2}^{4} (m_i \ddot{y}_{Si}) \\ M_s = \sum_{i=2}^{4} \left[J_i \ddot{\varphi}_i + m_i \left(x_{Si} \ddot{y}_{Si} - y_{Si} \ddot{x}_{Si} \right) \right]. \end{cases} \tag{1.15}$$

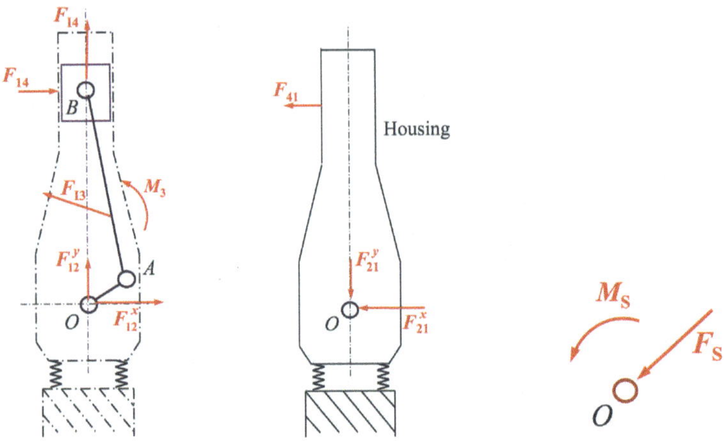

(a) Inertial and joint reaction forces (b) Joint reaction forces (c) Shaking force and shaking moment

Figure 1.5 Shaking force and shaking moment.

The shaking force and shaking moment can also be computed through Eq. (1.9). First, letting $\boldsymbol{B} = 0$, $\boldsymbol{X}$ can be solved for. From the reaction forces $\boldsymbol{F}_{21}$ and $\boldsymbol{F}_{41}$, the shaking force and shaking moment can be computed.

1.3 Analysis of the slider-crank mechanism

A variation of the general four-bar linkage is the slider-crank mechanism. Its analysis can be conducted by very similar procedures.

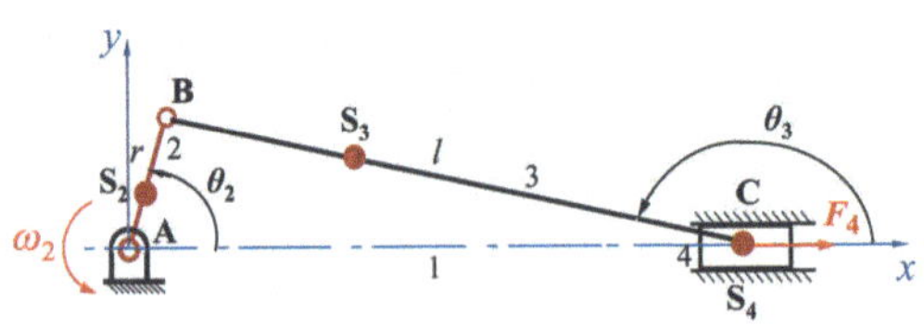

Figure 1.6 A slider-crank mechanism.

A slider-crank mechanism is shown in Fig. 1.6. The crank rotates anticlockwise, and a horizontal external force $\boldsymbol{F}_4$ is applied to the slider. For simplicity, the centers of mass S_2 and S_3 are assumed collinear with the two pin centers at ends. S_4 is coincident with joint C.

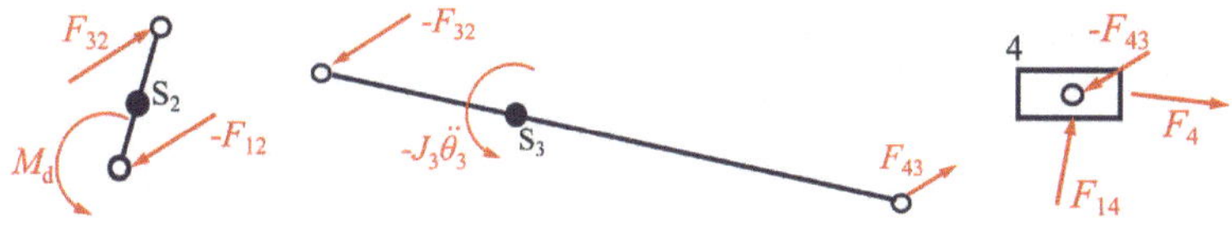

Figure 1.7 FBD of moving parts.

As in the general four-bar linkage, the FBD of the moving links are shown in Fig. 1.7. The equilibrium equations for the moving links are as

Link 2

$$\begin{cases} F_{32}^{x} + F_{12}^{x} = m_2 \ddot{x}_{S2} \\ F_{32}^{y} + F_{12}^{y} = m_2 \ddot{y}_{S2} \\ \left[F_{32}^{x}(\overline{S_2B})^{y} - F_{32}^{y}(\overline{S_2B})^{x}\right] + \left[F_{12}^{x}(\overline{S_2A})^{y} - F_{12}^{y}(\overline{S_2A})^{x}\right] + M_{d} = J_2\ddot{\theta}_2 \end{cases} \tag{1.16}$$

Link 3

$$\begin{cases} F_{43}^{x} - F_{32}^{x} = m_3 \ddot{x}_{S3} \\ F_{43}^{y} - F_{32}^{y} = m_3 \ddot{y}_{S3} \\ \left[F_{43}^{x}(\overline{S_3C})^{y} - F_{43}^{y}(\overline{S_3C})^{x}\right] - \left[F_{32}^{x}(\overline{S_3B})^{y} - F_{32}^{y}(\overline{S_3B})^{x}\right] = J_3\ddot{\theta}_3 \end{cases} \tag{1.17}$$

Link 4

$$\begin{cases} -F_{43}^{x} \pm \mu F_{14}^{y} + F_{4}^{x} = m_4 \ddot{x}_{S4} \\ F_{14}^{y} - F_{43}^{y} + F_{4}^{y} = 0. \end{cases} \tag{1.18}$$

Link 4 (the slider) has translation only in the horizontal direction, and all the forces on it intersect at joint C. Thus, two equations are adequate to describe the equilibrium. $\boldsymbol{F}_{14}$ is resolved into a component F_{14}^{y} and a component $F_{14}^{x} = \mu F_{14}^{y}$. The sign depends on the velocity because it is essentially the friction. Eqs. (1.16) through (1.18) are compacted into matrix form as

$$\boldsymbol{AX} = \boldsymbol{B}, \tag{1.19}$$

where

$$A = \begin{bmatrix} 1 & 0 & 1 & 0 & 0 & 0 & 0 & 0 \\ 0 & 1 & 0 & 1 & 0 & 0 & 0 & 0 \\ (\overline{S_2A})^{y} & -(\overline{S_2A})^{x} & (\overline{S_2B})^{y} & -(\overline{S_2B})^{x} & 0 & 0 & 0 & 1 \\ 0 & 0 & -1 & 0 & 1 & 0 & 0 & 0 \\ 0 & 0 & 0 & -1 & 0 & 1 & 0 & 0 \\ 0 & 0 & -(\overline{S_3B})^{y} & (\overline{S_3B})^{x} & (\overline{S_3C})^{y} & -(\overline{S_3C})^{x} & 0 & 0 \\ 0 & 0 & 0 & 0 & -1 & 0 & \pm\mu & 0 \\ 0 & 0 & 0 & 0 & 0 & -1 & 1 & 0 \end{bmatrix} \tag{1.20}$$

$$\mathbf{X} = \begin{bmatrix} F_{12}^{x} & F_{12}^{y} & F_{32}^{x} & F_{32}^{y} & F_{43}^{x} & F_{43}^{y} & F_{14}^{y} & M_{d} \end{bmatrix}^{T} \tag{1.21}$$

$$\boldsymbol{B} = \begin{bmatrix} m_2\ddot{x}_{S2} & m_2\ddot{y}_{S2} & J_2\ddot{\theta}_2 & m_3\ddot{x}_{S3} & m_3\ddot{y}_{S3} & J_3\ddot{\theta}_3 & m_4\ddot{x}_{C} - F_{4}^{x} & -F_{4}^{y} \end{bmatrix}^{T}. \tag{1.22}$$

Similarly, these procedures are for one mechanism position. To complete the overall analysis, the mechanism position are required to be discretized into different positions. For each position, repeat this process.

The shaking force and shaking moment about point A are calculated by

$$\begin{cases} F_{s}^{x} = F_{12}^{x} \pm \mu F_{14}^{y} \\ F_{s}^{y} = F_{12}^{y} + F_{14}^{y} \\ M_{s} = x_{c} F_{14}^{y}. \end{cases} \tag{1.23}$$

Example 1.1. A slider-crank mechanism has the following parameters, S_2 is coincident with pin A. $\omega_2 = 100$ rad/s, $r = 50.8$ mm, $l = 203$ mm, $\overline{BS_3} = 50.8$ mm, $m_3 = 1.36$ kg, $J_3 = 0.0102$ kg · m^2, and $m_4 = 0.907$ kg. Friction can be neglected. No external force is applied. Calculate the shaking force F_{S}^{x} and F_{S}^{y}, the shaking moment M_S about point A, and the required driving torque M_d.

Solution. $\mu = 0$, and $F_4 = 0$ are given. Divide the one revolution of the crank into 180 equal intervals and then solve Eq. (1.19). The results are given in Fig. 1.8.

1.4 Analysis of planar cam mechanisms

A planar cam-follower mechanism is shown in Fig. 1.9(a). The cam is assumed to rotate at a constant speed $\dot{\theta}$. A torque M_d drives the whole mechanism. The displacement of the follower s is vertical, and it is measured from its lowest position. It is a function of the cam's position θ. The

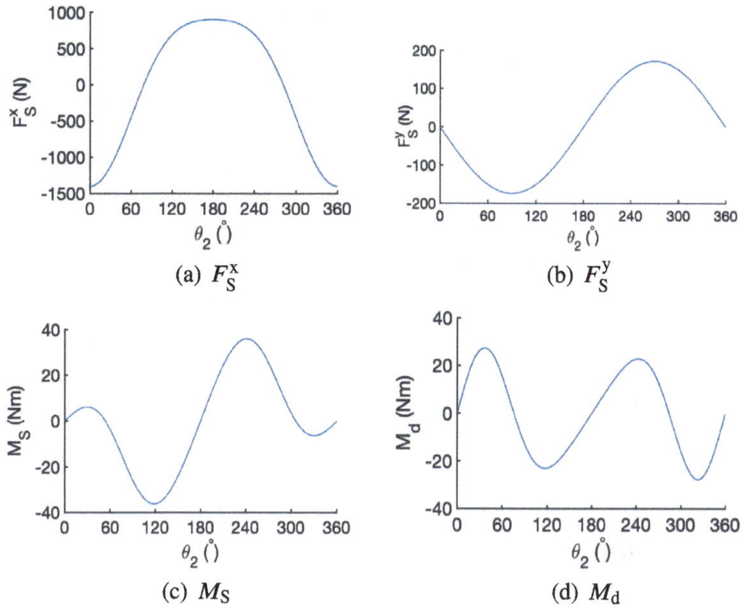

Figure 1.8 Simulation results.

follower is assumed to have a rise-dwell-fall-dwell (RDFD) motion cycle. To follow the numbering convention, the frame is numbered as link 1, whereas the cam and the follower are 2 and 3, respectively.

The FBD of the cam is shown in Fig. 1.9(b). The loads on it include a driving torque M_d and the reaction forces F_{12}^x and F_{12}^y, as well as a force applied through the roller $-F_{23}$.

The FBD of the follower is given in Fig. 1.9(c). The forces on it include the working load G, which is generally given as a function of θ; a vertical force $F_{p0} + ks$ applied by the closure spring (F_{p0} is the initial spring force and k is the spring stiffness); a horizontal force F_{13}^x from the track and a friction force fF_{13}^x (f is the friction coefficient); and an external moment M_3 and a force F_{23} that are exerted by the pin (α is the pressure angle).

The equilibrium equations of the cam and the follower are then represented by

Cam (link 2)

$$F_{12}^x - F_{23} \sin \alpha = 0 \tag{1.24}$$

$$F_{12}^y - F_{23} \cos \alpha = 0 \tag{1.25}$$

$$M_d - F_{23}(r_0 + s) \sin \alpha = 0. \tag{1.26}$$

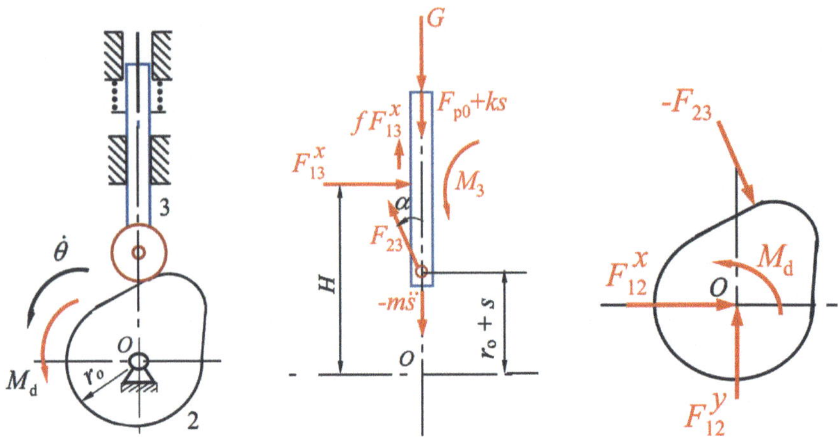

(a) A cam mechanism (b) FBD of the cam (c) FBD of the follower

Figure 1.9 Kineto-static analysis of the cam mechanism.

Follower (link 3)

$$-G-(F_0+ks)-\delta fF_{13}^x-m\ddot{s}+F_{23}\cos\alpha=0 \tag{1.27}$$

$$F_{13}^x-F_{23}\sin\alpha=0 \tag{1.28}$$

$$F_{23}(r_0+s)\sin\alpha-F_{13}^xH+M_3=0, \tag{1.29}$$

where δ is used to indicate the direction of the friction force. $\delta=1$ if $\dot{s}>0$, and $\delta=-1$ if $\dot{s}<0$.

The six linear algebraic equations from Eqs. (1.24) through (1.29) can be recast into matrix form as

$$\boldsymbol{AX}=\boldsymbol{B}, \tag{1.30}$$

where $\boldsymbol{X}$ is the unknown vector, whose entries are

$$\boldsymbol{X}=\begin{bmatrix}F_{13}^x & M_3 & F_{23} & F_{12}^x & F_{12}^y & M_d\end{bmatrix}^{\mathrm{T}}. \tag{1.31}$$

Eq. (1.30) can be solved with computers by similar procedures as in the previous section. For brevity, the details are not repeated here.

Now, let's take a close look at M_d. F_{23} is first solved from Eqs. (1.27) and (1.28).

$$F_{23}=\frac{G+F_{p0}+ks+m\ddot{s}}{\cos\alpha-\delta f\sin\alpha}. \tag{1.32}$$

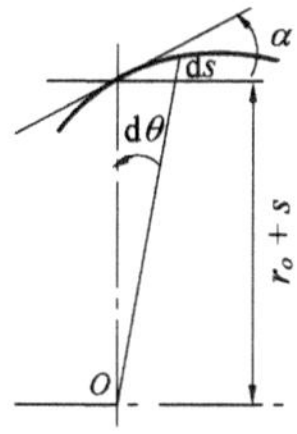

Figure 1.10 Geometry of the cam.

A closed-form representation of M_d can be obtained by substituting Eq. (1.32) into Eq. (1.26) as

$$M_d = (r_0 + s)\tan\alpha \frac{G + F_{p0} + ks + m\ddot{s}}{1 - \delta f \tan\alpha}. \tag{1.33}$$

$\tan\alpha$ can be derived from the geometry shown in Fig. 1.10 as follows:

$$\tan\alpha = \frac{ds}{(r_0 + s)d\theta} = \frac{\dot{s}}{(r_0 + s)\omega}. \tag{1.34}$$

Combining the two equations, the following can be obtained

$$M_d = \frac{\dot{s}}{\omega} \frac{G + F_{p0} + ks + m\ddot{s}}{1 - \delta f \tan\alpha}. \tag{1.35}$$

In many high-speed applications, the inertial load $m\ddot{s}$ is dominant, making $G + F_0 + ks$ negligible. If the friction is also neglected, the following approximation can be made:

$$M_d \approx \frac{m}{\omega} \dddot{s}s. \tag{1.36}$$

This relationship means that the value of $\dddot{s}s$ is proportional to M_d. Thus, for high-speed situations, a cam motion with a lower value of $\dddot{s}s$ is often desirable in terms of the driving torque required.

CHAPTER 2

Balancing of planar mechanisms and engine dynamics

2.1 Introduction

Mechanical parts moving at high speed cause significant *inertial forces* that have various harmful effects on the machine as outlined below:

- The inertial forces are generally periodic and are eventually transmitted to the machine frame in the form of a *shaking force* and *shaking moment*. The shaking forces and moments excite vibrations in the frame, degrade the machine's accuracy, and generate noise. In extreme cases, catastrophic failures or severe malfunction of the machine may occur.
- The inertial forces also cause the fluctuation of the driving torque (M_d) required to maintain the prescribed motion.
- The inertial forces add stress to links and joints of the machine, accelerating wear, and causing lubrication and fatigue problems.

There are two technical paths to deal with the harmful vibration of machines. One is through various vibration control techniques (Rao, 2010). This route is based on vibration theory to suppress, isolate, and dampen the harmful vibration. The other route is to reduce the shaking force and shaking moment through *balancing*. The direct objective of balancing is to reduce the vibration excitation, not the vibration response. Thus, it does not involve the vibration equation. In balancing theory, a basic assumption is that the machine has a prescribed motion, in most cases, at a constant input speed. For this reason, balancing is, in essence, on the level of kineto-static analysis.

Balancing can be classified as either *rotor balancing* or *mechanism balancing*. Rotor balancing is generally covered in an undergraduate mechanism course. This chapter deals with the more complicated mechanism balancing, focusing on planar mechanisms.

Systematic study of mechanism balancing started around the beginning of the 20th century, and the development has been closely related to internal combustion engines. One of the earliest publications in this field is probably by Fischer (1902) who proposed a method called *principal vectors*. One conclusion of this study is that the necessary and sufficient condition

Machinery Dynamics
https://doi.org/10.1016/B978-0-12-815785-5.00011-X

for balancing the shaking force is the fixation of the total center of mass of the moving links in the mechanism. Later on, the method of *static substitution of masses* was proposed. The method elegantly transforms a mechanism balancing problem into a simpler problem of balancing rotating links. To balance the slider-crank mechanism in an engine, a method based on *harmonic analysis* was developed during the 1960s by Emőd and Jurek (1967). In this method, the unbalanced force and moment are resolved into *Fourier series*, and the objective is to balance certain harmonics of the series. In the 1960s, a method for complete force balancing was proposed by Berkof and Lowen (1969). Several years later, Berkof (1973) developed a method for complete balancing of both the shaking force and the shaking moment. One dilemma of balancing is that a specific method may improve one or some of the criteria, but worsen some others simultaneously. For this reason, some optimization methods were proposed to combine optimization techniques and balancing theory with the objective to optimize multiple dynamic factors (Zhang et al., 1995).

Balancing can be conducted with various different objectives as outlined here:

- **Balancing of shaking force and shaking moment:** This is the most common case. The objective is to partially or completely remove the shaking force and shaking moment transmitted to the machine frame.
- **Balancing of driving torque required:** The inertial force causes the fluctuation of M_d. In this category, the objective is to control the range of fluctuations of M_d.
- **Balancing of joint loads:** The inertial force causes additional dynamic loads to the kinematic joints. The objective of this type of balancing is to minimize the dynamic loads on the joints.

In practice, balancing is conducted more often with the primary objective of reducing the shaking force and shaking moment, and simultaneously has some secondary objectives. Balancing techniques targeting at shaking force and shaking moment are further classified here:

- *Partial balancing* means the objective is to partially control one dynamic factor, most often the shaking force. To completely eliminate the shaking force and shaking moment is ideal in theory, but may be difficult to implement in practice. As a compromise, partial balancing becomes a trade-off option. Partial balancing has wide application, especially in the balancing of internal combustion engines.

- *Complete balancing* means to completely eliminate the shaking force, shaking moment, or both. In practice, elimination of the shaking force is more common. Thus, complete balancing may mean: 1) complete balancing of shaking force, or 2) complete balancing of both shaking force and shaking moment. Berkof (1973) was a pioneer in the development of complete balancing techniques.
- *Optimized balancing* combines the optimization techniques with balancing theory to optimize one or more dynamic factors. As stated earlier, either partial balancing or complete balancing has inherent shortcomings. Sometimes, optimized balancing becomes a realistic solution.

Balancing is often implemented through the following two measures:

- **Counterweights or counter moments of inertia:** Counterweights are used for balancing shaking forces, and counter moments of inertial are for shaking moments. Counterweights are relatively easier to add, thus, more commonly used in practice.
- **A symmetrical duplicate mechanism:** By adding a symmetrical mechanism to the original one, it is possible to cancel the shaking force and shaking moment completely. The downside of this method is that it makes the mechanism more complicated, and, for this reason, the us of this application is very limited. Applications are more likely to be adopted when the duplicated mechanism has also a functional benefit, such as in multiple-cylinder engines.

2.2 Equivalent masses

Balancing is generally based on kineto-static analysis. The general procedures for kineto-static analysis have been presented in Chapter 1. In practice, the analysis is often performed approximately for simplicity. For example, a continuous link is replaced by lumped masses connected with massless rods. This is especially useful when the accurate parameters of the links are not available yet in the early design stage. The lumped masses in this method are called *equivalent masses* and are usually chosen to be located at the joint center.

Let's look at an object of general shape as shown in Fig. 2.1. It has a mass m, and the center of mass is located at point S. The moment of inertia around S is J_S. According to the D'Alembert's principle, the inertial loads created by this link are as follows.

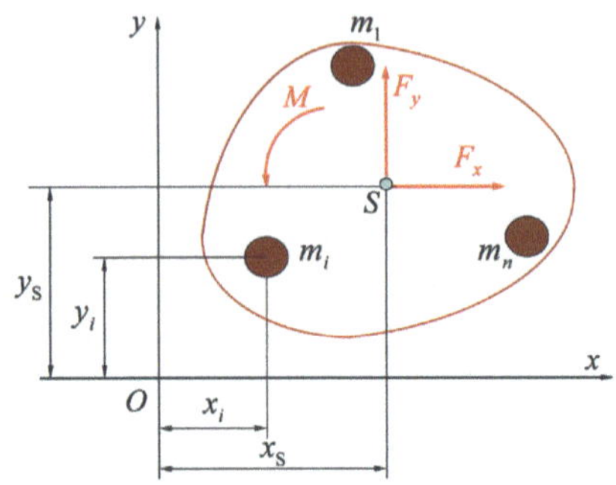

Figure 2.1 Substitute masses.

Inertial force:

$$\begin{cases} F_x = -m\ddot{x}_S \\ F_y = -m\ddot{y}_S. \end{cases} \tag{2.1}$$

Inertial moment:

$$M = -J_S\alpha, \tag{2.2}$$

where x_S and y_S are the coordinates of the center of mass, and α is the angular acceleration.

2.2.1 Equivalent criteria

If this object is to be replaced by n lumped masses, m_1, m_2, $\cdots$, m_n, two types of equivalent criteria, namely *static equivalency* and *dynamic equivalency*, are possible as discussed next.

2.2.1.1 Static equivalency

The conditions for static equivalency are

- The summation of the equivalent masses equals the mass of the object.
- The center of the lumped masses is coincident with that of the continuous object.

These can be represented mathematically as:

$$\sum_{i=1}^{n} m_i = m \tag{2.3}$$

$$\begin{cases} \sum_{i=1}^{n} m_i x_i = m x_S \\ \sum_{i=1}^{n} m_i y_i = m y_S, \end{cases} \tag{2.4}$$

where x_i, y_i are the coordinates of the lumped mass i as shown in Fig. 2.1. m is the mass of the continuous object.

2.2.1.2 Dynamic equivalency

To make the substitution dynamically equivalent, the moment of inertia of the lumped masses should be made equal to that of the continuous link, in addition to the two static equivalent conditions already stated. These can be defined as:

- The summation of the lumped masses equals the mass of the continuous object.
- The center of mass remains unchanged.
- The moment of inertia also remains the same.

The first two are already given in Eqs. (2.3) and (2.4), while the third is this:

$$\sum_{i=1}^{n} m_i (x_i^2 + y_i^2) = J_S + m(x_S^2 + y_S^2). \tag{2.5}$$

2.2.2 Real equivalent mass

If the mechanism link is replaced by two lumped masses, the values of the substitution masses will be real. This is the simplest case, and it is widely used in dynamic analysis of the slider-crank mechanism in internal combustion engines.

2.2.2.1 Static equivalency

A link AB shown in Fig. 2.2(a) has a mass of m. The center of mass S is located on the line AB. The moment of inertia about S is J_S. Two revolute joints A and B are located at the ends of the link.

For static equivalency, the joints A and B are chosen as the locations for the lumped masses. According to Eqs. (2.3) and (2.4), the following two equations should be satisfied:

$$\begin{cases} m_A + m_B = m \\ -m_A l_A + m_B l_B = 0. \end{cases} \tag{2.6}$$

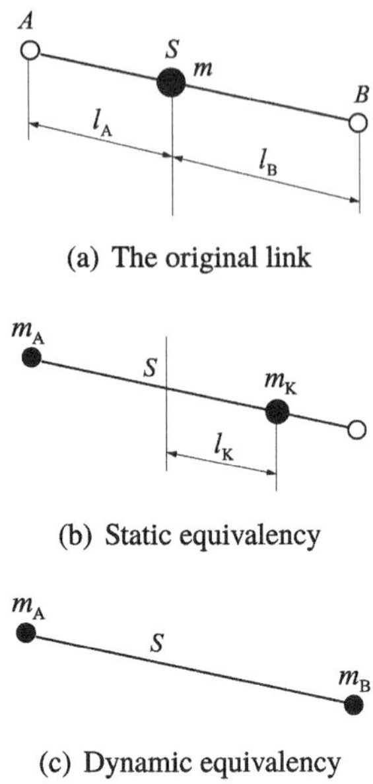

Figure 2.2 Real equivalent masses.

Then m_A and m_B can be solved as

$$\begin{cases} m_A = m\dfrac{l_B}{l_A + l_B} \\ m_B = m\dfrac{l_A}{l_A + l_B}. \end{cases} \tag{2.7}$$

The result is shown in Fig. 2.2(b).

2.2.2.2 Dynamic equivalency

If the link is substituted by two masses for dynamic equivalency, we assume one mass m_A is at point A and the other m_K at point K. Note that point K may not be point B. The following equations should be satisfied according to the conditions represented by Eqs. (2.3)–(2.5):

$$\begin{cases} m_A + m_K = m \\ -m_A l_A + m_K l_K = 0 \\ m_A l_A^2 + m_K l_K^2 = J_S. \end{cases} \tag{2.8}$$

From this we have

$$\begin{cases} l_K = \dfrac{J_S}{ml_A} \\ m_A = \dfrac{l_K}{l_A + l_K} m \\ m_K = \dfrac{l_A}{l_A + l_K} m. \end{cases} \tag{2.9}$$

In general, point K would not be coincident with the other revolute joint B. The result is schematically shown in Fig. 2.2(c).

When this is applied to the coupler of the slider-crank mechanism in an internal combustion engine, however, the common practice is to force $l_K \approx l_B$. Due to the special geometry of the coupler in the engine, this approximation is acceptable for accuracy. In this case, Eq. (2.9) becomes

$$\begin{cases} l_K \approx l_B \\ m_A \approx \dfrac{l_B}{l_A + l_B} m \\ m_B \approx \dfrac{l_A}{l_A + l_B} m. \end{cases} \tag{2.10}$$

The dynamic equivalency in this case is practically degraded to the static equivalency.

The method of real equivalent masses can be conveniently applied to the analysis of shaking-force balancing of a general four-bar mechanism shown in Fig. 2.3. The links 2 and 4 have pure rotation; thus, they can be designed intentionally to have their centers of mass at the pivot points A and D. The coupler, link 3, moves with both rotation and translation. It can be substituted by two masses, m_B and m_C, located at the revolute joints

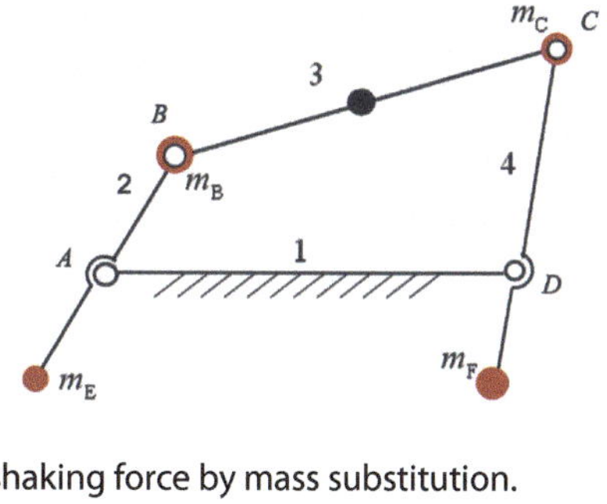

Figure 2.3 Balancing shaking force by mass substitution.

B and *C*. Then the force balancing of the mechanism is simplified to the problem of balancing the two lumped masses m_B and m_C. This can be easily achieved by adding two counter masses m_E and m_F opposite to m_B and m_C, respectively, as shown in the figure. Through this method, the shaking force can be completely removed, but a shaking moment remains.

2.3 Partial force balancing of slider-crank mechanisms

Slider-crank mechanisms, as shown in Fig. 2.4(a), are the core of an internal combustion engine. To achieve a continuous increase of engine speed has been the main motivation for the study of balancing techniques of this type of mechanism. A partial balance technique eliminating the main part of the shaking force has been widely used in the auto industry. This method is based on the equivalent-mass substitution introduced in the proceeding section and is especially useful in the initial stage of design when detailed design parameters are not available.

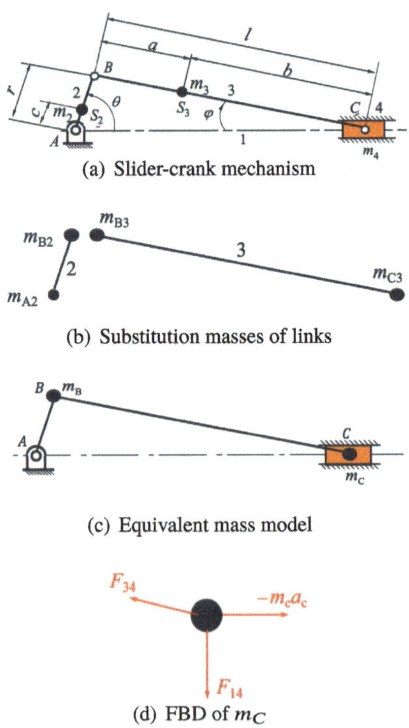

Figure 2.4 Equivalent-mass substitution of a slider-crank mechanism.

2.3.1 Dynamic analysis

In the following, the real equivalent-mass method is applied to the analysis of dynamics and the balancing of the slider-crank mechanism.

The crank, link 2, rotates around the revolute joint A at a constant speed. Thus, it can be substituted by a mass m_{A2} at A, and a mass m_{B2} at the other revolute joint B. m_{A2} and m_{B2} are calculated using the static equivalency criteria represented by Eq. (2.7). A is stationary; thus, only the mass m_{B2} is of interest:

$$m_{B2} = \frac{c}{r} m_2. \tag{2.11}$$

The subscript $B2$ represents the equivalent mass of link 2 at point B.

The coupler, link 3, has both translation and rotation. Literally, the criteria of dynamic equivalence, Eq. (2.8), should be used for the equivalent masses. However, static equivalency approximations represented by Eq. (2.10) are often used, which greatly simplifies the calculation:

$$\begin{cases} m_{B3} = m_3 \dfrac{b}{l} \\ m_{C3} = m_3 \dfrac{a}{l}. \end{cases} \tag{2.12}$$

The slider, link 4, moves in pure translation. There is no need to substitute its mass m_4.

After completing this process, a model represented by two lumped masses m_B and m_C is obtained:

$$\begin{cases} m_B = m_{B2} + m_{B3} \\ m_C = m_{C3} + m_4. \end{cases} \tag{2.13}$$

From Fig. 2.4(a) the following relationship can be established:

$$s = r\cos\theta + l\cos\varphi \tag{2.14}$$

$$\sin\varphi = \frac{r}{l}\sin\theta. \tag{2.15}$$

$\lambda = r/l$ is referred to as the *crank-conrod ratio* in an internal combustion engine. It is an important design parameter of the engine.

Taking advantage of the trigonometric identity, one can derive the following from Eq. (2.15):

$$\cos\varphi = \left(1-\sin^2\varphi\right)^{\frac{1}{2}} = \left(1-\lambda^2\sin^2\theta\right)^{\frac{1}{2}}$$
$$= 1 - \frac{1}{2}\left(\lambda^2\sin^2\theta\right) - \frac{1}{8}\left(\lambda^4\sin^4\theta\right) - \frac{1}{16}\left(\lambda^6\sin^6\theta\right) - \cdots$$
$$= \lambda\left(A_0 + \frac{1}{4}A_2\cos 2\theta - \frac{1}{16}A_4\cos 4\theta + \frac{1}{36}A_6\cos 6\theta + \cdots\right), \quad (2.16)$$

where A_i $(i = 0, 2, 4, 6)$ are constants defined as:

$$\begin{cases} A_0 = \dfrac{1}{\lambda} - \dfrac{1}{4}\lambda - \dfrac{3}{64}\lambda^3 - \dfrac{5}{286}\lambda^5 - \cdots \\ A_2 = \lambda + \dfrac{1}{4}\lambda^3 + \dfrac{15}{128}\lambda^5 + \cdots \\ A_4 = \dfrac{1}{4}\lambda^3 + \dfrac{3}{16}\lambda^5 + \cdots \\ A_6 = \dfrac{9}{128}\lambda^5. \end{cases} \quad (2.17)$$

To obtain the acceleration of point C, differentiating Eq. (2.14) twice with respect to time and taking advantage of Eq. (2.16) yields:

$$a_C = r\dot{\theta}^2\left(-\cos\theta - A_2\cos 2\theta + A_4\cos 4\theta - A_6\cos 6\theta + \cdots\right) +$$
$$r\ddot{\theta}\left(-\sin\theta - \frac{A_2}{2}\sin 2\theta + \frac{A_4}{4}\sin 4\theta - \frac{A_6}{6}\sin 6\theta + \cdots\right). \quad (2.18)$$

λ generally has a value between 0.16 and 0.40 in an internal combustion engine; thus, the terms containing λ^3 and higher orders can be neglected without significant loss of accuracy. In addition, the crank rotates at constant speed, meaning that $\ddot{\theta} = 0$. Taking these two factors into consideration, a_C can be approximated as:

$$a_C \approx -r\dot{\theta}^2\left(\cos\theta + \lambda\cos 2\theta\right). \quad (2.19)$$

The inertial force caused by m_B is $\boldsymbol{F}_{\mathrm{IB}} = -m_B\boldsymbol{a}_B$. Resolving this into components in x and y directions yields

$$\begin{cases} F_{\mathrm{IB}}^{\mathrm{x}} = -m_B r\dot{\theta}^2\cos\theta \\ F_{\mathrm{IB}}^{\mathrm{y}} = -m_B r\dot{\theta}^2\sin\theta. \end{cases} \quad (2.20)$$

The inertial force, only in the horizontal direction, caused by m_C is

$$F_{\mathrm{IC}}^{\mathrm{x}} = -m_C a_C = m_C r\dot{\theta}^2\left(\cos\theta + \lambda\cos 2\theta\right)$$
$$= m_C r\dot{\theta}^2\cos\theta + m_C r\dot{\theta}^2\lambda\cos 2\theta. \quad (2.21)$$

The first term on the righthand side related to $\cos\theta$ is called the *primary component*, and the second term related to $\cos 2\theta$ is the *secondary component*. In general, the primary component is much larger than the secondary one because λ is much less than *one*.

The resultant inertial force is the summation of these two:

$$\begin{cases} F_{\mathrm{I}}^{\mathrm{x}} = F_{\mathrm{IB}}^{\mathrm{x}} + F_{\mathrm{IC}}^{\mathrm{x}} = m_B r\dot{\theta}^2 \cos\theta + m_C r\dot{\theta}^2 \cos\theta + m_C r\dot{\theta}^2 \lambda \cos 2\theta \\ F_{\mathrm{I}}^{\mathrm{y}} = F_{\mathrm{IB}}^{\mathrm{y}} = m_B r\dot{\theta}^2 \sin\theta. \end{cases} \tag{2.22}$$

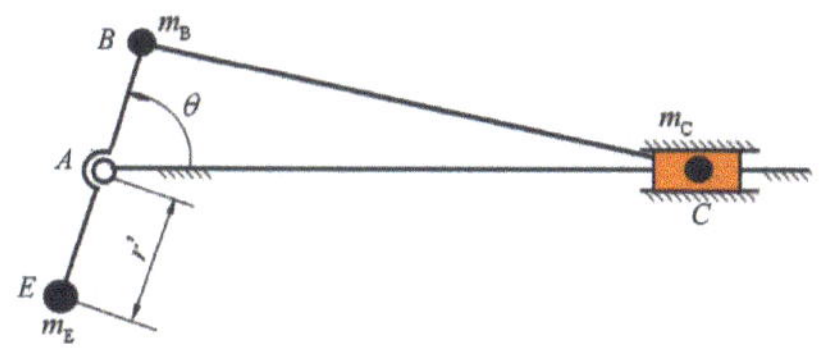

Figure 2.5 Balancing of a slider-crank mechanism.

The inertial force also causes a shaking moment. If the friction in the sliding track is neglected, the moment around point A (see Fig. 2.4(d)) is

$$M_A = s(-m_C \ddot{a}_C \tan\varphi). \tag{2.23}$$

Taking advantage of the relationship

$$\begin{cases} s = r\cos\theta + l\cos\varphi \approx r\cos\theta + l\left[1 - \dfrac{\lambda^2}{2}\sin^2\theta\right] \\ \tan\,\varphi = \dfrac{r\sin\theta}{l\cos\varphi} \approx \lambda\sin\theta\left(1 + \dfrac{\lambda^2}{2}\sin^2\theta\right), \end{cases}$$

M_A is represented as:

$$\begin{aligned} M_A &\approx -m_C r^2\dot{\theta}^2 \sin\theta\left(\frac{\lambda}{2} + \cos\theta + \frac{3\lambda}{2}\cos 2\theta\right) \\ &= \frac{1}{2} m_C r^2\dot{\theta}^2\left(\frac{\lambda}{2}\sin\theta - \sin 2\theta - \frac{3\lambda}{2}\sin 3\theta\right). \end{aligned} \tag{2.24}$$

To derive the relationship, the following trigonometric identities have been utilized:

$$\begin{cases} 2\sin\theta\cos 2\theta = \sin 3\theta - \sin\theta \\ 2\sin\theta\cos\theta = \sin 2\theta. \end{cases}$$

Obviously there is a third harmonic component in M_A.

2.3.2 Partial balancing of inertial forces

In this section, we will develop a way to partially balance the shaking force without considering the shaking moment of a slider-crank mechanism. After using the mass substitution presented in the previous section, we have a system with two lumped masses m_B and m_C at hand. The mass m_B has pure rotation around pivot A; thus, its inertial force can be easily balanced by adding a counterweight at a point, say point E, that is 180° from the vector $\boldsymbol{r}$ and at a distance of r' from A. This counterweight to balance the inertial force $\boldsymbol{F}_{\mathrm{IB}}$ is

$$m_{E1} = \frac{r}{r'} m_B. \tag{2.25}$$

The mass m_C translates horizontally. It is impossible to eliminate its inertial force by a counterweight because of the secondary harmonic component. However, the primary component can be balanced by a counterweight. Thus, a counterweight can only balance part of the inertial force.

If we add a counterweight m_{E2} at E, the two components of the inertial force caused by m_{E2} are

$$\begin{cases} F^{\mathrm{x}}_{\mathrm{IE}} = -m_{\mathrm{E2}} r' \dot{\theta}^2 \cos\theta \\ F^{\mathrm{y}}_{\mathrm{IE}} = -m_{\mathrm{E2}} r' \dot{\theta}^2 \sin\theta. \end{cases} \tag{2.26}$$

One can choose $m_{E2} r' = m_C r$ so that the primary component of $F^{\mathrm{x}}_{\mathrm{IC}}$ is completely balanced. However, the downside of this is that an additional $F^{\mathrm{y}}_{\mathrm{IE}}$ will be created. A compromise is to choose a smaller $m_{E2} r'$ so that the primary component is partially balanced, but the extra $F^{\mathrm{y}}_{\mathrm{IE}}$ is limited. The total counterweight at E is, thus, the sum of m_{E1} and m_{E2}:

$$m_E = m_{E1} + m_{E2} = \frac{r}{r'} \left(m_B + k m_C \right), \tag{2.27}$$

where k is a constant. The determination of its value depends on the specific balancing objective. Generally, k has a value between 1/3 and 2/3.

2.4 Complete balancing of planar mechanisms

2.4.1 Criteria for complete balancing

A four-bar linkage is shown in Fig. 2.6, in which the three moving links are numbered 2, 3, and 4. Link 1 is the stationary one. As usual, each moving link has a mass m_i and a moment of inertia J_i around its center of mass. x_i

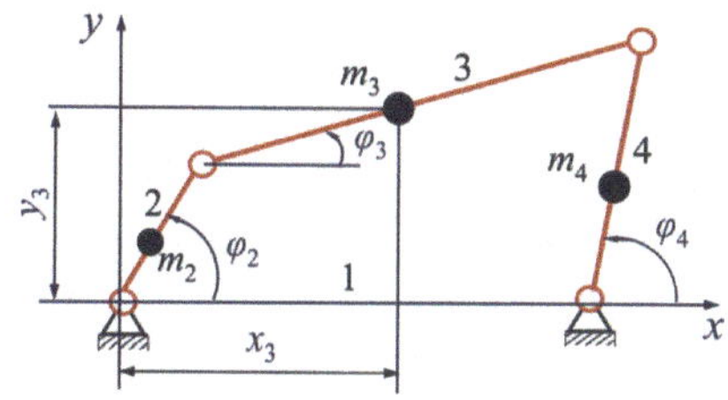

Figure 2.6 A planar four-bar linkage.

and y_i are the locations of the mass center of the link i, and φ_i is the position angle.

To completely eliminate the shaking force and shaking moment, the following conditions should be met:

$$\begin{cases} F_x = -\sum_{i=2}^{n} m_i \ddot{x}_i = 0 \\ F_y = -\sum_{i=2}^{n} m_i \ddot{y}_i = 0 \\ M_o = -\sum_{i=2}^{n} \left[m_i (x_i \ddot{y}_i - y_i \ddot{x}_i) + J_i \ddot{\varphi}_i \right] = 0, \end{cases} \tag{2.28}$$

where n is the number of links in the mechanism. In the case of Fig. 2.6 $n = 4$. On the other hand, the location of the center of mass of all the moving links in a mechanism is computed by

$$\begin{cases} x_S = \frac{1}{m} \sum_{i=2}^{n} m_i x_i \\ y_S = \frac{1}{m} \sum_{i=2}^{n} m_i y_i, \end{cases} \tag{2.29}$$

where m is the total mass of all moving links.

Comparing Eq. (2.29) to the first two parts of Eq. (2.28), one can conclude that, to completely balance the shaking force, the acceleration of the total mass center of the moving links in the mechanism has to be zero. To meet this condition, the total center of mass should be either 1) stationary or 2) moving along a straight line at a constant speed. Given that all machines have a periodic motion, state 2 is actually impossible for a real machine. In this case, state 1 becomes the only possibility for complete force balancing. Thus, the criteria for complete balancing of the shaking force can be also stated as *the total center of mass remaining stationary*. If we define $\sum_{i=2}^{n} m_i x_i$ and $\sum_{i=2}^{n} m_i y_i$ as moments of masses, the condition for complete force balancing can be also stated as *the total moment of mass of the mechanism remains constant.*

Now let's look at the shaking moment represented by the third part of Eq. (2.28). If we define the momentum around point O as

$$H_o = \sum_{i=2}^{n} \left[m_i(x_i\dot{y}_i - y_i\dot{x}_i) + J_i\dot{\varphi}_i \right], \tag{2.30}$$

the third part of Eq. (2.28) can be recast into the form

$$M_o = -\frac{dH_o}{dt} = 0. \tag{2.31}$$

This means that, to completely balance the shaking moment, *the total momentum of the mechanism remains constant.*

2.4.2 Complete balancing of shaking force through mass redistribution

Mass redistributions, using counterweights in particular, are the most common measures or balancing mechanisms. Several methods for complete balancing of shaking force through mass redistribution have been proposed. Among them the most representative is the method of *linearly independent vectors* proposed by Berkof and Lowen (1969). In the section following, this method will be briefly presented.

The procedural steps of the method are:

- To develop the equation for the global center of mass of the total mechanism.
- To simplify the equation in terms of only independent variables using the vector-loop equation.
- To keep the global center of mass stationary by ensuring that the time-dependent terms are zero.
- To maintain the mass redistribution scheme for the balancing based on the previous step.

These procedures are illustrated with more details for a four-bar linkage in Fig. 2.7.

Following the convention, the three moving links are numbered 2, 3 and 4, and the fixed link as 1. The centers of mass and the position angle of each of the three moving links are also shown in the figure. The location

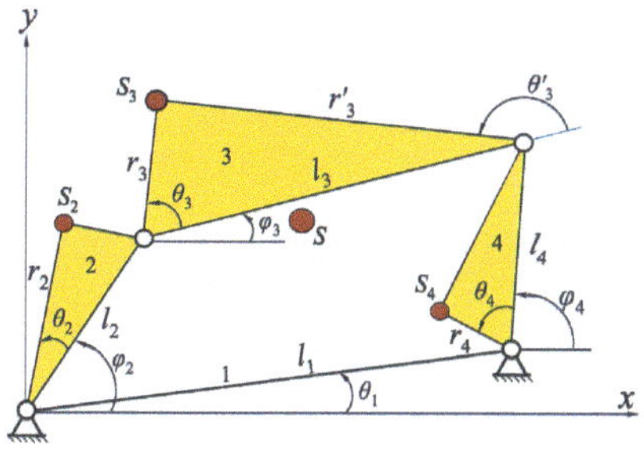

Figure 2.7 A four-bar linkage for the linearly independent vector method.

of global center of mass of the three links can be represented as:

$$\boldsymbol{r}_S = \frac{1}{m}\sum_{i=2}^{4} m_i \boldsymbol{r}_{si}, \tag{2.32}$$

where m and $\boldsymbol{r}_S$ are the total mass and the location vector of the global center of mass of the three moving links. m_i and $\boldsymbol{r}_{Si}$ are the mass and the location vector of the mass center of an individual link as shown in Fig. 2.7.

The location of the centers of mass of individual moving links can be represented as

$$\begin{cases} \boldsymbol{r}_{S2} = r_2 e^{i(\varphi_2+\theta_2)} \\ \boldsymbol{r}_{S3} = l_2 e^{i\varphi_2} + r_3 e^{i(\varphi_3+\theta_3)} \\ \boldsymbol{r}_{S4} = l_1 e^{i\theta_1} + r_4 e^{i(\varphi_4+\theta_4)}. \end{cases} \tag{2.33}$$

Substituting Eq. (2.33) into Eq. (2.32) gives

$$\boldsymbol{r}_S = \frac{1}{m}\left[\left(m_2 r_2 e^{i\theta_2} + m_3 l_2\right) e^{i\varphi_2} + \left(m_3 r_3 e^{i\theta_3}\right) e^{i\varphi_3} + \left(m_4 r_4 e^{i\theta_4}\right) e^{i\varphi_4} + m_4 l_1 e^{i\theta_1} \right]. \tag{2.34}$$

The vector-loop equation of the mechanism is

$$l_2 e^{i\varphi_2} + l_3 e^{i\varphi_3} - l_4 e^{i\varphi_4} - l_1 e^{i\theta_1} = 0. \tag{2.35}$$

From this relationship one obtains

$$e^{i\varphi_3} = \frac{l_1 e^{i\theta_1} - l_2 e^{i\varphi_2} + l_4 e^{i\varphi_4}}{l_3}. \tag{2.36}$$

Substituting Eq. (2.36) into (2.34) and rearranging the terms gives the following:

$$\begin{aligned} \boldsymbol{r}_S = & \frac{1}{m}\left[\left(m_2 r_2 e^{i\theta_2} + m_3 l_2 - m_3 r_3 \frac{l_2}{l_3} e^{i\theta_3}\right) e^{i\varphi_2} + \left(m_4 r_4 e^{i\theta_4} + m_3 r_3 \frac{l_4}{l_3} e^{i\theta_3}\right) e^{i\varphi_4} \right. \\ & \left. + \left(m_4 l_1 + m_3 l_3 \frac{l_1}{l_3} e^{i\theta_3}\right) e^{i\theta_1}\right]. \end{aligned} \tag{2.37}$$

φ_2 and φ_4 change with time. To make $\boldsymbol{r}_S$ a constant, the terms in the parentheses before $e^{i\varphi_2}$ and $e^{i\varphi_4}$ have to be zero. Thus, we have

$$\begin{cases} m_2 r_2 e^{i\theta_2} + m_3 l_2 - m_3 r_3 \frac{l_2}{l_3} e^{i\theta_3} = 0 \\ m_4 r_4 e^{i\theta_4} + m_3 r_3 \frac{l_4}{l_3} e^{i\theta_3} = 0. \end{cases} \tag{2.38}$$

These are the conditions for complete balancing of the shaking force.

If the parameters of link 3 are known, the following relationship can be obtained from Fig. 2.7:

$$r_3 e^{i\theta_3} = l_3 + r_3' e^{i\theta_3'}. \tag{2.39}$$

Substituting this relationship into Eq. (2.38) and rearranging the two equations give the following:

$$\begin{cases} m_2 r_2 e^{i\theta_2} = m_3 r_3' \frac{l_2}{l_3} e^{i\theta_3'} \\ m_4 r_4 e^{i\theta_4} = -m_3 r_3 \frac{l_4}{l_3} e^{i\theta_3}. \end{cases} \tag{2.40}$$

From this equation the following four relationships can be derived:

$$\begin{cases} m_2 r_2 = m_3 r_3' \frac{l_2}{l_3} \\ \theta_2 = \theta_3' \\ m_4 r_4 = m_3 r_3 \frac{l_4}{l_3} \\ \theta_4 = \theta_3 + \pi. \end{cases} \tag{2.41}$$

These are the conditions for links 2 and 4 to achieve complete balancing of the shaking force.

For a machine being designed, the stated conditions can be achieved through deliberate mass distribution of links 2 and 4. For an existing machine, the mass redistribution can be implemented by adding counter-weights. The calculation of the counterweights is presented next.

For an individual link i ($i = 2$ or 4), the following relationship exists:

$$m_i r_i e^{i\theta_i} = m_i^0 r_i^0 e^{i\theta_i^0} + m_i^* r_i^* e^{i\theta_i^*}, \tag{2.42}$$

where the superscripts ^{0} and * represent the existing and counterweight to be added. Solving for the counterweight, we have

$$m_i^* r_i^* e^{i\theta_i^*} = m_i r_i e^{i\theta_i} - m_i^0 r_i^0 e^{i\theta_i^0}. \tag{2.43}$$

From this the following two equations can be obtained:

$$\begin{cases} m_i^* r_i^* = \sqrt{(m_i r_i)^2 + (m_i^0 r_i^0)^2 - 2 m_i r_i m_i^0 r_i^0 \cos(\theta_i - \theta_i^0)} \\ \tan\theta_i^* = \dfrac{m_i r_i \sin\theta_i - m_i^0 r_i^0 \sin\theta_i^0}{m_i r_i \cos\theta_i - m_i^0 r_i^0 \cos\theta_i^0} \end{cases} \quad (i = 2, 4), \tag{2.44}$$

where $m_i = m_i^0 + m_i^*$ is the total of the original and the added masses.

Example 2.1. A four-bar linkage has the parameters given in Table 2.1. Use the method of linearly independent vectors just introduced to calculate the counterweights needed on links 2 and 4 for complete force balancing.

Solution. To meet the conditions of complete balancing, the parameters of links 2 and 4 are calculated according to Eq. (2.37):

$$m_2 r_2 = m_3 r_3' \frac{l_2}{l_3} = (0.125)(75.6)\frac{50}{150} = 3.15 \text{ kg·mm} \tag{a}$$

$$\theta_2 = \theta_3' = 164.1^\circ \tag{b}$$

Table 2.1 Parameters of an unbalanced four-bar linkage.

Link No.	2	3	4	1
l_i (mm)	50	150	75	140
r_3 (mm)		80		
r_i^0 (mm)	25		40	
θ_i^0 (°)	0	15	0	0
r_3' (mm)		75.6	0	
θ_3' (°)		164.1		
m_i (kg)	0.046	0.125	0.054	

$$m_4r_4 = m_3r_3\frac{l_4}{l_3} = (0.125)(80)\frac{75}{150} = 5 \text{ kg·mm} \tag{c}$$

$$\theta_4 = \theta_3 + \pi = 15 + 180 = 195°. \tag{d}$$

The product of *mr* for the original links 2 and 4 are as follows:

$$m_2^0r_2^0 = (0.046)(25) = 1.15 \text{ kg·mm} \tag{e}$$

$$m_4^0r_4^0 = (0.054)(40) = 2.16 \text{ kg·mm.} \tag{f}$$

The parameters of the counterweights on links 2 and 4 can be calculated by using Eq. (2.44):

$$\begin{aligned} m_2^*r_2^* &= \sqrt{(m_2r_2)^2 + (m_2^0r_2^0)^2 - 2m_2r_2m_2^0r_2^0\cos(\theta_2 - \theta_2^0)} \\ &= \sqrt{3.15^2 + 1.15^2 - (2)(3.15)(1.15)\cos(164.1° - 0°)} = 4.27 \text{ kg·mm} \end{aligned} \tag{g}$$

$$\begin{aligned} \tan\theta_2^* &= \arctan\left(\frac{m_2r_2\sin\theta_2 - m_2^0r_2^0\sin\theta_2^0}{m_2r_2\cos\theta_2 - m_2^0r_2^0\cos\theta_2^0}\right) \\ &= \arctan\left(\frac{3.15\sin 164.1° - 1.15\sin 0°}{3.15\cos 164.1° - 1.15\cos 0°}\right) = 168.3° \end{aligned} \tag{h}$$

$$\begin{aligned} m_4^*r_4^* &= \sqrt{(m_4r_4)^2 + (m_4^0r_4^0)^2 - 2m_4r_4m_4^0r_4^0\cos(\theta_4 - \theta_4^0)} \\ &= \sqrt{5^2 + 2.16^2 - (2)(5)(2.16)\cos(195° - 0°)} = 7.11 \text{ kg·mm} \end{aligned} \tag{i}$$

$$\begin{aligned} \tan\theta_4^* &= \arctan\left(\frac{m_4r_4\sin\theta_4 - m_4^0r_4^0\sin\theta_4^0}{m_4r_4\cos\theta_4 - m_4^0r_4^0\cos\theta_4^0}\right) \\ &= \arctan\left(\frac{5\sin 195° - 2.16\sin 0°}{5\cos 195° - 2.16\cos 0°}\right) = 190.5°. \end{aligned} \tag{j}$$

The mechanism with added counterweights on links 2 and 4 is schematically illustrated in Fig. 2.8.

The method of linearly independent vectors can be applied to more complicated linkages with multiple loops. An example of a six-bar linkage was given by Berkof and Lowen (1969). However, the procedure is much more complicated than a four-bar linkage. To limit this work's total page length, the detail is omitted here.

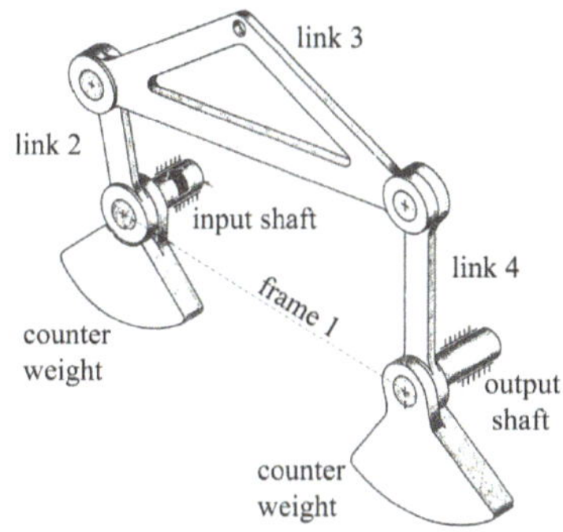

Figure 2.8 Mechanism with added counterweights.

2.4.3 Complete balancing of shaking force and shaking moment

Both shaking force and shaking moment cause vibration of the machine frame. In the previous section, the method for balancing the shaking force was presented, while the shaking moment remained untouched, while taking the four-bar mechanism shown in Fig. 2.9 as an example. After force balancing, although the resultant joint force $\boldsymbol{F}_{21} + \boldsymbol{F}_{41} = 0$, $\boldsymbol{F}_{21}$ and $\boldsymbol{F}_{41}$ are equal in magnitude, they are opposite in direction, forming a shaking moment. This moment also causes vibration of the machine.

Ideally, both the shaking force and shaking moment should be balanced. Several methods (Arakelian et al., 2000; Esat and Bahai, 1999; Kochev, 2000) have been proposed to this end. The one developed by Berkof (1973) is relatively clear in concept. This method is based on the concepts of dynamic mass substitution and *a physical pendulum*. The procedural steps of the method are presented subsequently.

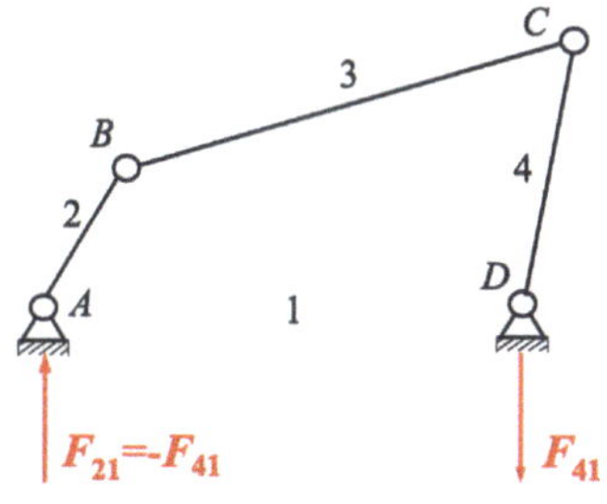

Figure 2.9 A four-bar linkage after force balancing.

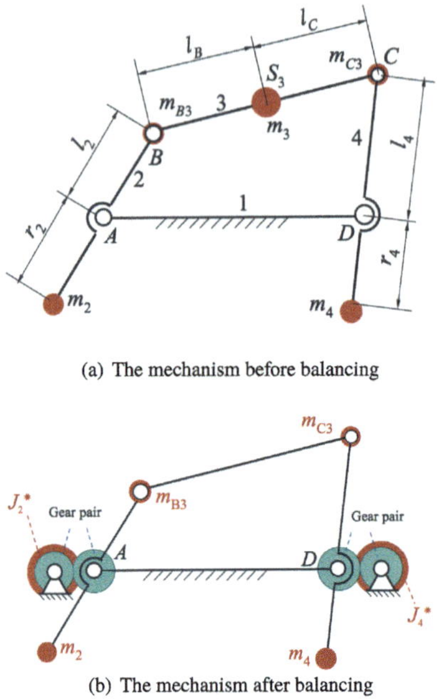

Figure 2.10 Complete balancing of force and moment.

A four-bar linkage shown in Fig. 2.10 is used to illustrate the procedures. For the sake of simplicity, links 2 and 4 are assumed to be already force balanced. Link 3 has a general planar motion, thus it is the key in the balancing. If link 3 can be dynamically substituted by two lumped masses located on the two revolute joints B and C, it is called a physical pendulum.

The two equivalent masses on B and C of link 3 are calculated as

$$\begin{cases} m_{B3} = \dfrac{l_C}{l_B + l_C} m_3 \\ m_{C3} = \dfrac{l_B}{l_B + l_C} m_3. \end{cases} \tag{2.45}$$

Then force balancing can be achieved by adding a counterweight $m_2 r_2$ and $m_4 r_4$ on links 2 and 4 respectively. $m_2 r_2$ and $m_4 r_4$ are calculated by

$$\begin{cases} m_2 r_2 = m_{B3} l_2 = \dfrac{l_C l_2}{l_3} m_3 \\ m_4 r_4 = m_{C3} l_4 = \dfrac{l_B l_4}{l_3} m_3, \end{cases} \tag{2.46}$$

where $l_3 = l_B + l_C$ has been used.

To this point, the inertial force is eliminated by m_2 and m_4. Now let's look at the shaking moment, which can be represented as (Norton, 2004)

$$M_S = \sum_{i=2}^{4} A_i \alpha_i, \tag{2.47}$$

where α_i is the angular acceleration of the link i ($i = 2$–4) and A_i are the coefficients

$$\begin{cases} A_2 = -m_2\left(k_2^2 + r_2^2 + l_2 r_2\right) \\ A_3 = -m_3\left(k_3^2 + l_B^2 - l_3 l_B\right) \\ A_4 = -m_4\left(k_4^2 + r_4^2 + l_2 r_2\right), \end{cases} \tag{2.48}$$

where $k_i = \sqrt{J_i/m_i}$ is the *radius of gyration*.

To balance the shaking moment, two geared rotating counter moments of inertia J_2^* and J_4^* as shown in Fig. 2.10(b) are added through gear trains of 1 : 1 speed ratio. The total moment after adding these rotating counter moments becomes

$$M_S = \sum_{i=2}^{4} A_i \alpha_i + J_2^* \alpha_2 + J_4^* \alpha_4. \tag{2.49}$$

To make M_S zero, the following conditions should be met:

$$\begin{cases} J_2^* = -A_2 \\ J_4^* = -A_4 \\ A_3 = 0. \end{cases} \tag{2.50}$$

From Eq. (2.45) we have

$$k_3^2 = -l_B(l_B - l_3) = l_B l_C. \tag{2.51}$$

Combining Eqs. (2.46), (2.48), (2.50), and (2.51), the following conditions to be met for complete balancing of shaking force and shaking

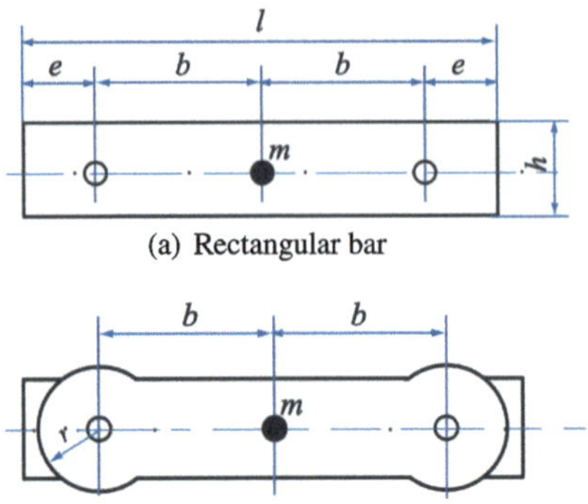

Figure 2.11 Two designs of link 3.

moment to be obtained are

$$\begin{cases} m_2 r_2 = m_{B3} l_2 = \dfrac{l_C l_2}{l_3} m_3 \\ m_4 r_4 = m_{C3} l_4 = \dfrac{l_B l_4}{l_3} m_3 \\ k_3^2 = l_B l_C \\ J_2^* = m_2 \left(k_2^2 + r_2^2 + l_2 r_2\right) \\ J_4^* = m_4 \left(k_4^2 + r_4^2 + l_2 r_2\right). \end{cases} \tag{2.52}$$

The first two equations are for shaking-force balancing; the third is the condition for physical pendulum, while the latter two define the moments of inertia required for the balancing. It should be pointed out that, if the crank (link 2) rotates at a constant angular speed, then M_S on link 2 is zero. Thus, only J_4^* is needed.

Berkof (1973) discussed two possible cases of a physical pendulum. One is the rectangular bar, and the other is the dogbone bar as shown in Fig. 2.11.

For the case of Fig. 2.11(a) this condition for physical pendulum is

$$\frac{e}{h} = \frac{1}{2}\sqrt{3\left(\frac{2b}{h}\right)^2 - 1} - \frac{b}{h}. \tag{2.53}$$

For the case of Fig. 2.11(b) the condition can be represented as

$$A\left(\frac{e}{h}\right)^3 + B\left(\frac{e}{h}\right)^2 + C\left(\frac{e}{h}\right) + D = 0, \tag{2.54}$$

where the coefficients are defined as

$$\begin{cases} A = 8 \\ B = 12\left(\frac{2b}{c}\right) + 24 \\ C = 24\left(\frac{2b}{c}\right) + 26 \\ D = -2\left(\frac{2b}{c}\right)^3 + 13\left(\frac{2b}{c}\right) + 12\pi - 10. \end{cases}$$

Obviously, complete balancing of both force and moment is much more complicated than only balancing the shaking force.

2.4.4 Complete balancing through duplicated mechanism

Complete balancing can be also achieved by adding a duplicated mechanism that is symmetrical to the original one. Due to the symmetry, both the inertial forces and the inertial moment of the moving links are canceled out. Three typical examples are illustrated in Fig. 2.12 (Arakelian and Briot, 2015). This balancing technique has found application in fields such as multiple-cylinder engines and some agricultural machines. It is, however, not as widely accepted as the method of traditional counterweight and mass redistribution.

2.4.5 Limitation of complete balancing

From the theoretical point of view, complete balancing is more desirable than partial balancing. However, some technical limitations of complete balancing should not be neglected. These can be outlined as:

- Complete balancing is only possible for some mechanisms. For example, complete force balancing of a planar mechanism requires that there exists a contour to the ground from every point on the linkage by way of revolute joints only (Esat and Bahai, 1999; Tepper and Lowen, 1972). If there are prismatic joints in a contour, complete force balancing is impossible by only the addition of counterweights.
- Complete balancing always makes the mechanism either heavier or more structurally complicated.
- Complete balancing of both force and moment requires not only counterweights, but also counter moment of inertia (see Section 2.4.3).

Due to these limitations, partial balancing and optimized balancing have wider application than complete balancing in practice.

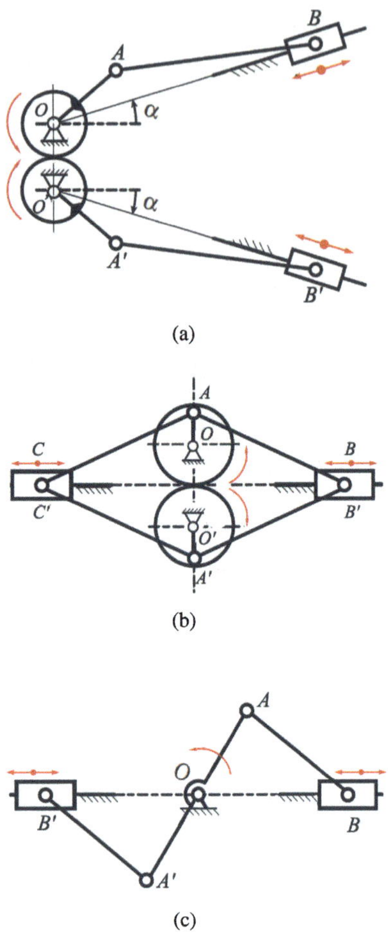

Figure 2.12 Complete balancing by adding a duplicated mechanism.

2.5 Optimized balancing

While a balancing technique improves one criterion, at the same time, it often worsens others. For example, adding counterweights reduces or eliminates the shaking force, but it may increase joint loads. Likewise, adding rotational moment of inertia cancels the shaking moment, but increases the required input torque. To achieve a compromise between multiple criteria, optimized balancing, which combines the optimum techniques and balancing theory, becomes an option. In recent years, great advances have been made in this direction, and many optimization algorithms have been proposed. Readers interested in this topic are advised to consult the review paper by Arakelian and Briot (2015).

In the following section, the work by Zhang et al. (1995) is briefly introduced. This work treats a planar mechanism as a mass-spring vibration system having three DOF as shown in Fig. 2.13. The vibration response, rather than the shaking force and shaking moment, was taken as the objective function.

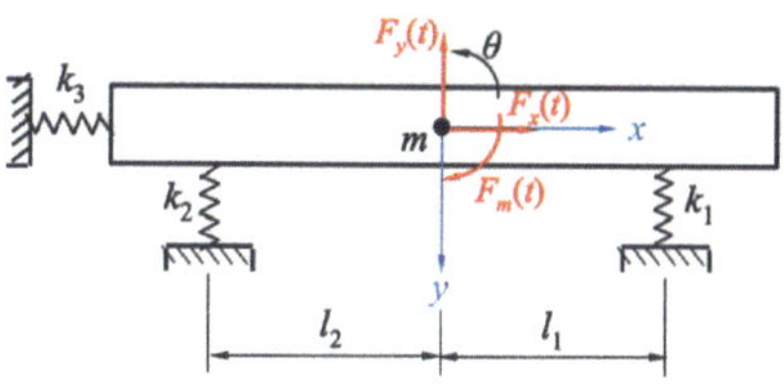

Figure 2.13 Mass-spring model.

The shaking forces in x and y directions are represented as F_x and F_y, respectively, and the shaking moment is M. Then the vibration of the machine under the shaking forces and shaking moment can be written as:

$$\begin{cases} m\ddot{x} + k_3 x = F_x \\ m\ddot{y} + (k_1 + k_2)\,y - (k_2 l_2 - k_1 l_1)\,\theta = F_y \\ J_m\ddot{\theta} - (k_2 l_2 - k_1 l_1)\,y + (k_1 l_1^2 + k_2 l_2^2) = M, \end{cases} \tag{2.55}$$

where m and J_m are the equivalent mass and moment of inertia about point m. k_i ($i = 1$ to 3) are the stiffness of springs as shown.

The response ranges of the x, y, and θ are represented as:

$$\begin{cases} R_x = x_{max} - x_{min} \\ R_y = y_{max} - y_{min} \\ R_\theta = \theta_{max} - \theta_{min}. \end{cases} \tag{2.56}$$

The objective function is constructed as a weighted combination of the response ranges:

$$f(\boldsymbol{X}) = w_x R_x + w_y R_y + w_\theta R_\theta, \tag{2.57}$$

where w_x, w_y, and w_θ are the weight coefficients that can be chosen by the designer based on the objective of balancing. $\boldsymbol{X}$ is the variable vector whose components include the amount of counterweights and their locations.

The objective function can be optimized through numerical algorithms that can be found in any well-written textbook on optimization.

2.6 Dynamics of engines

The core of a modern engine is a slider-crank mechanism. A single-cylinder internal combustion engine contains only one slider-crank mechanism, for which the dynamic analysis and partial force balancing were already discussed in Section 2.4. Here is a summary of the conclusions:

- The shaking force and shaking moment formed in a single-cylinder engine are calculated by Eqs. (2.22) and (2.24), respectively.
- By adding a counterweight to the crank, the shaking force can only be partially removed due to the existence of the secondary harmonic component, and the shaking moment remains unchanged.

Most modern engines have multiple cylinders. Through a proper configuration of the cylinders, it is possible to make the inertial forces and the shaking moment cancel out. In the following section, the dynamics of multiple-cylinder engines are introduced.

There are different configurations of multiple-cylinder engines, including inline engines, V engines, opposed engines, etc. The inline layout means all the cylinders are located in one common plane and on one side of the crankshaft. In the V engines, two banks of cylinders are arranged at some intervening V angle. Each bank is an inline version. Examples of a four-cylinder inline engine and four-cylinder V engine are shown in Figs. 2.14 and 2.15(a), respectively. Opposed engines are a special case of the V type having an angle of 180° (Norton, 2004).

2.6.1 Inline engines

To analyze the dynamics of an inline engine, the cylinders are numbered in sequence as in Fig. 2.14. Each slider-crank mechanism's position is represented by an angle with the y axis, θ_i. Generally the first mechanism is taken as the reference, and we have $\theta_1 = \theta$. The phase difference between the i-th mechanism and the first mechanism ϕ_i. For the case shown in Fig. 2.14, we have $\phi_1 = 0$, $\phi_2 = \pi/2$, $\phi_3 = \pi$ and $\phi_4 = 3\pi/2$. For clarity, ϕ_i is not shown in the figure.

2.6.1.1 *Shaking force* F_S

In the auto industry, it is common practice to balance the crankshaft in advance; thus, the inertial force caused by m_B in each individual mechanism

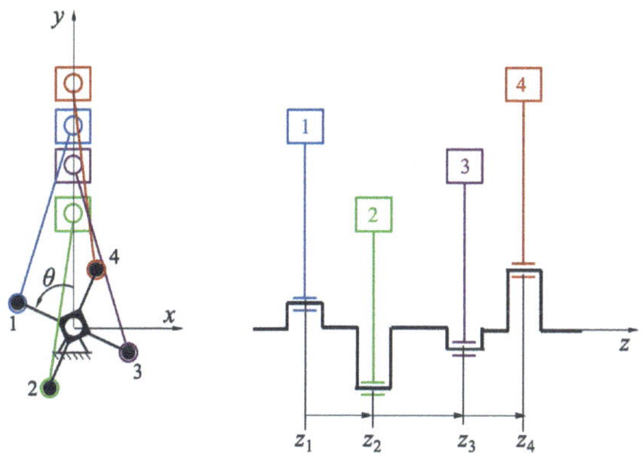

Figure 2.14 A four-cylinder inline engine.

(see Fig. 2.5) is eliminated. The inertial force caused by the reciprocating mass m_C remains and is calculated by Eq. (2.21). For the case of multiple cylinders, the total inertial force is the summation of that from all the individual mechanisms

$$F_S = mr\dot{\theta}^2 \sum_{i=1}^{n} (\cos\theta_i + \lambda \cos 2\theta_i), \tag{2.58}$$

where m is the mass of the reciprocating piston. Substituting $\theta_i = \theta + \phi_i$ into the previous equation gives

$$F_S = mr\dot{\theta}^2 \sum_{i=1}^{n} [\cos(\theta + \phi_i) + \lambda \cos 2(\theta + \phi_i)]. \tag{2.59}$$

Taking advantage of the trigonometric identity

$$\cos(A + B) = \cos A \cos B - \sin A \sin B, \tag{2.60}$$

the following is obtained:

$$F_S = mr\dot{\theta}^2 \left[\left(\cos\theta \sum_{i=1}^{n} \cos\phi_i - \sin\theta \sum_{i=1}^{n} \sin\phi_i \right) + \lambda \left(\cos 2\theta \sum_{i=1}^{n} \cos 2\phi_i - \sin 2\theta \sum_{i=1}^{n} \sin 2\phi_i \right) \right]. \tag{2.61}$$

To make it zero, the following conditions must be met:

$$\begin{cases} \sum_{i=1}^{n} \cos \phi_i = 0 \\ \sum_{i=1}^{n} \sin \phi_i = 0 \\ \sum_{i=1}^{n} \cos 2\phi_i = 0 \\ \sum_{i=1}^{n} \sin 2\phi_i = 0. \end{cases} \tag{2.62}$$

For the particular case in Fig. 2.14 having the parameters: $n = 4$, $\phi_1 = 0$; $\phi_2 = \pi/2$; $\phi_3 = \pi$, and $\phi_4 = 3\pi/2$, it is easy to verify that all the conditions of Eq. (2.62) are met.

2.6.1.2 Shaking moment M_z

The joint forces within each individual mechanism form a shaking moment in the z direction, which is calculated by Eq. (2.24). Similarly, the total moment along the z direction is the sum caused by all mechanisms:

$$\begin{aligned} M_z &= \frac{1}{2} m r \dot{\theta}^2 \sum_{i=1}^{n} \left(\frac{\lambda}{2} \sin \theta_i - \sin 2\theta_i - \frac{3\lambda}{2} \sin 3\theta_i \right) \\ &= \frac{1}{2} m r \dot{\theta}^2 \left[\frac{\lambda}{2} \sin (\theta + \phi_i) - \sin 2 (\theta + \phi_i) - \frac{3\lambda}{2} \sin 3 (\theta + \phi_i) \right] \\ &= \frac{1}{2} m r \dot{\theta}^2 \left[\frac{\lambda}{2} \left(\sin \theta \sum_{i=1}^{n} \cos \phi_i + \cos \theta \sum_{i=1}^{n} \sin \phi_i \right) - \right. \\ & \left(\sin 2\theta \sum_{i=1}^{n} \cos 2\phi_i + \cos 2\theta \sum_{i=1}^{n} \sin 2\phi_i \right) - \\ & \left. \frac{3\lambda}{2} \left(\sin 3\theta \sum_{i=1}^{n} \cos 3\phi_i + \cos 3\theta \sum_{i=1}^{n} \sin 3\phi_i \right) \right]. \end{aligned} \tag{2.63}$$

To arrive at these formulations, the following identity has been used:

$$\sin(A + B) = \sin A \cos B + \cos A \sin B. \tag{2.64}$$

To make it zero, the following relationships must be met:

$$\begin{cases} \sum_{i=1}^{n} \cos\phi_i = 0 \\ \sum_{i=1}^{n} \sin\phi_i = 0 \\ \sum_{i=1}^{n} \cos 2\phi_i = 0 \\ \sum_{i=1}^{n} \sin 2\phi_i = 0 \\ \sum_{i=1}^{n} \cos 3\phi_i = 0 \\ \sum_{i=1}^{n} \sin 3\phi_i = 0. \end{cases} \tag{2.65}$$

It is easy to check that for, the case of $n = 4$, $\phi_1 = 0$, $\phi_2 = \pi/2$, $\phi_3 = \pi$, and $\phi_4 = 3\pi/2$, all the conditions are met.

2.6.1.3 Shaking moment M_x

In a multiple-cylinder engine, the cylinders have to be separated at a distance z_i in the axial direction to avoid interference as shown in Fig. 2.14. In this case, the inertial force in y direction will cause a rocking moment along the x direction. If we take the plane where the first cylinder's center line is located as the point to calculate the moment, the total rocking moment M_x can be calculated as:

$$\begin{aligned} M_x = \sum_{i=1}^{n} z_i F_{Si} &= mr\dot{\theta}^2 \sum_{i=1}^{n} z_i \left[\cos(\theta + \phi_i) + \lambda \cos 2(\theta + \phi_i)\right] \\ &= mr\dot{\theta}^2 \sum_{i=1}^{n} \left[\left(\cos\theta \sum_{i=1}^{n} z_i \cos\phi_i - \sin\theta \sum_{i=1}^{n} z_i \sin\phi_i \right) + \right. \\ &\left. \lambda \left(\cos 2\theta \sum_{i=1}^{n} z_i \cos 2\phi_i - \sin 2\theta \sum_{i=1}^{n} z_i \sin 2\phi_i \right) \right]. \end{aligned} \tag{2.66}$$

To make it zero, the following conditions have to be met.

$$\begin{cases} \sum_{i=1}^{n} z_i \cos \phi_i = 0 \\ \sum_{i=1}^{n} z_i \sin \phi_i = 0 \\ \sum_{i=1}^{n} z_i \cos 2\phi_i = 0 \\ \sum_{i=1}^{n} z_i \sin 2\phi_i = 0. \end{cases} \tag{2.67}$$

For the case $n = 4$, if we assume $z_1 = 0$, $z_2 = 1$, $z_3 = 2$ and $z_4 = 3$, the following is obtained:

$$\begin{cases} \sum_{i=1}^{4} z_i \cos \phi_i = -2 \\ \sum_{i=1}^{4} z_i \sin \phi_i = -2 \\ \sum_{i=1}^{4} z_i \cos 2\phi_i = -2 \\ \sum_{i=1}^{4} z_i \sin 2\phi_i = 0. \end{cases} \tag{2.68}$$

From this analysis, we can conclude that the four-cylinder inline engine with $\phi_1 = 0$; $\phi_2 = \pi/2$; $\phi_3 = \pi$, and $\phi_4 = 3\pi/2$ performs well in terms of the shaking force F_S and shaking moment M_z. However, in terms of the shaking moment M_x, the cancelation conditions are not met, meaning that some balancing measures are still required.

2.6.2 V engines

In V engines, the mechanisms are arranged into two banks that form an angle of 2γ as shown in Fig. 2.15(a). Each bank, the left or right shown in Figs. 2.15(b) and 2.15(c), can be regarded as an inline engine.

2.6.2.1 *Shaking force* F_S

Without loss of generality, we number the cylinders in sequence and assume the odd numbers are on the left bank and the even numbers are on the

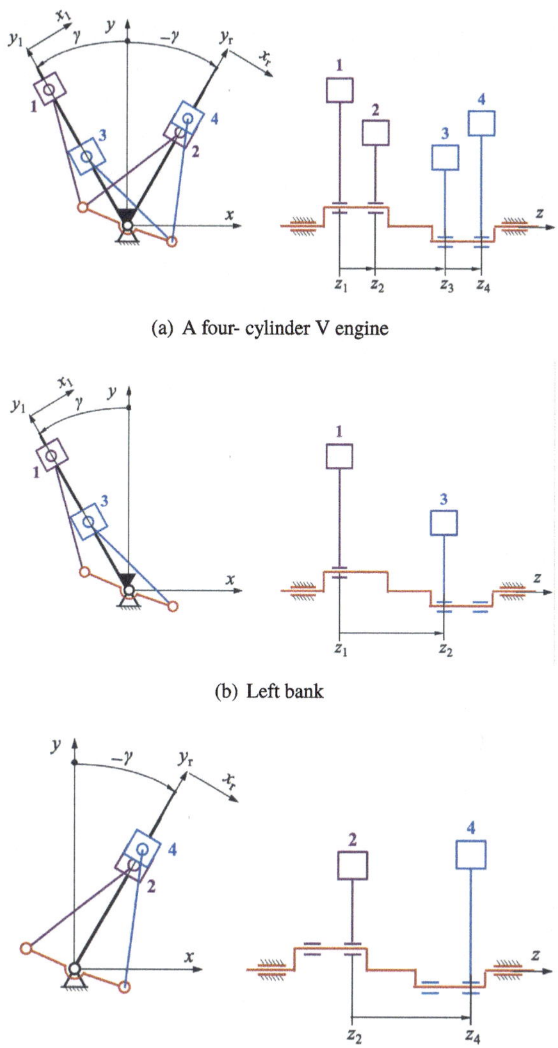

(a) A four- cylinder V engine

(b) Left bank

(c) Right bank

Figure 2.15 V engine.

right. Then the shaking force on the left and right bank can be written as

$$\begin{cases} F_{Sl} = mr\dot{\theta}^2 \displaystyle\sum_{i=1,3,\cdots}^{n-1} (\cos\theta_i + \lambda\cos 2\theta_i) \\ F_{Sr} = mr\dot{\theta}^2 \displaystyle\sum_{i=2,4,\cdots}^{n} (\cos\theta_i + \lambda\cos 2\theta_i)\,. \end{cases} \tag{2.69}$$

The subscript r and l represent the right and left banks, respectively. $\boldsymbol{F}_{\mathrm{Sr}}$ and $\boldsymbol{F}_{\mathrm{Sl}}$ are in the y_{r} and y_{l} direction. It should be noted that the crank position θ_i here is defined with reference to the center line of the respective cylinder. To transform it to a common reference y, the relationship between θ_i and θ is represented as follows:

$$\theta_i = \begin{cases} \theta + \phi_i - \gamma & \text{for left bank} \\ \theta + \phi_i + \gamma & \text{for right bank,} \end{cases} \tag{2.70}$$

where ϕ_i is the phase shift as previously defined. Taking advantage of the identities Eqs. (2.60) and (2.64), Eq. (2.69) can be written as

$$\begin{cases} F_{\mathrm{Sl}} = & mr\dot{\theta}^2 \left[\cos(\theta - \gamma) \sum\limits_{i=1,3,\cdots}^{n-1} \cos\phi_i - \right. \\ & \sin(\theta - \gamma) \sum\limits_{i=1,3,\cdots}^{n-1} \sin\phi_i + \\ & \lambda \cos 2(\theta - \gamma) \sum\limits_{i=1,3,\cdots}^{n-1} \cos 2\phi_i - \\ & \left. \lambda \sin 2(\theta - \gamma) \sum\limits_{i=1,3,\cdots}^{n-1} \sin 2\phi_i \right] \\ F_{\mathrm{Sr}} = & mr\dot{\theta}^2 \left[\cos(\theta + \gamma) \sum\limits_{i=2,4,\cdots}^{n} \cos\phi_i - \right. \\ & \sin(\theta + \gamma) \sum\limits_{i=2,4,\cdots}^{n} \sin\phi_i + \\ & \lambda \cos 2(\theta + \gamma) \sum\limits_{i=2,4,\cdots}^{n} \cos 2\phi_i - \\ & \left. \lambda \sin 2(\theta + \gamma) \sum\limits_{i=2,4,\cdots}^{n} \sin 2\phi_i \right]. \end{cases} \tag{2.71}$$

To transform them into the x and y direction, the following is obtained:

$$\begin{cases} F_{\mathrm{Sx}} = (F_{\mathrm{Sl}} + F_{\mathrm{Sr}}) \cos\gamma \\ F_{\mathrm{Sy}} = (F_{\mathrm{Sl}} - F_{\mathrm{Sr}}) \sin\gamma. \end{cases} \tag{2.72}$$

2.6.2.2 Shaking moment M_z

As in the inline case, each of the left and right bank forms a shaking moment in the direction of z. The moment for each bank can be calculated by Eq. (2.63) as follows:

$$\begin{cases} M_{zl} = \dfrac{1}{2}mr\dot{\theta}^2\left[\dfrac{\lambda}{2}\sin(\theta-\gamma)\sum_{i=1,3,\cdots,}^{n-1}\cos\phi_i + \dfrac{\lambda}{2}\cos(\theta-\gamma)\sum_{i=1,3,\cdots,}^{n-1}\sin\phi_i - \sin 2(\theta-\gamma)\sum_{i=1,3,\cdots,}^{n-1}\cos 2\phi_i - \cos 2(\theta-\gamma)\sum_{i=1,3,\cdots,}^{n-1}\sin 2\phi_i - \dfrac{3\lambda}{2}\sin 3(\theta-\gamma)\sum_{i=1,3,\cdots,}^{n-1}\cos 3\phi_i - \dfrac{3\lambda}{2}\cos 3(\theta-\gamma)\sum_{i=1,3,\cdots,}^{n-1}\sin 3\phi_i\right] \\ M_{zr} = \dfrac{1}{2}mr\dot{\theta}^2\left[\dfrac{\lambda}{2}\sin(\theta+\gamma)\sum_{i=2,4,\cdots,}^{n}\cos\phi_i + \dfrac{\lambda}{2}\cos(\theta+\gamma)\sum_{i=2,4,\cdots,}^{n}\sin\phi_i - \sin 2(\theta+\gamma)\sum_{i=2,4,\cdots,}^{n}\cos 2\phi_i - \cos 2(\theta+\gamma)\sum_{i=2,4,\cdots,}^{n}\sin 2\phi_i - \dfrac{3\lambda}{2}\sin 3(\theta+\gamma)\sum_{i=2,4,\cdots,}^{n}\cos 3\phi_i - \dfrac{3\lambda}{2}\cos 3(\theta+\gamma)\sum_{i=2,4,\cdots,}^{n}\sin 3\phi_i\right]. \end{cases} \tag{2.73}$$

The total moment M_z is

$$M_z = M_{zl} + M_{zr}. \tag{2.74}$$

2.6.2.3 Shaking moment M_x

Referring to the inline engine, each of the left and the right bank creates a moment along the x_l and x_r direction, respectively. They are calculated by

Eq. (2.66):

$$\begin{cases} M_{\mathrm{xl}} = & mr\dot{\theta}^2\left[\cos(\theta-\gamma)\left(\sum_{i=1,3,\cdots,}^{n-1} z_i\cos\phi_i\right)-\right. \\ & \sin(\theta-\gamma)\left(\sum_{i=1,3,\cdots,}^{n-1} z_i\sin\phi_i\right)+ \\ & \lambda\cos 2(\theta-\gamma)\left(\sum_{i=1,3,\cdots,}^{n-1} z_i\cos 2\phi_i\right)- \\ & \left.\lambda\sin 2(\theta-\gamma)\left(\sum_{i=1,3,\cdots,}^{n-1} z_i\sin 2\phi_i\right)\right] \\ M_{\mathrm{xr}} = & mr\dot{\theta}^2\left[\cos(\theta+\gamma)\left(\sum_{i=2,4,\cdots,}^{n} z_i\cos\phi_i\right)-\right. \\ & \sin(\theta+\gamma)\left(\sum_{i=2,4,\cdots,}^{n} z_i\sin\phi_i\right)+ \\ & \lambda\cos 2(\theta+\gamma)\left(\sum_{i=2,4,\cdots,}^{n} z_i\cos 2\phi_i\right)- \\ & \left.\lambda\sin 2(\theta+\gamma)\left(\sum_{i=2,4,\cdots,}^{n} z_i\sin 2\phi_i\right)\right]. \end{cases} \tag{2.75}$$

Transforming them into x and y components yields the following:

$$\begin{cases} M_{\mathrm{x}} = (M_{\mathrm{xl}} - M_{\mathrm{xr}})\sin\gamma \\ M_{\mathrm{y}} = (-M_{\mathrm{xl}} - M_{\mathrm{xr}})\cos\gamma. \end{cases} \tag{2.76}$$

One can obtain the conditions for the zero shaking force and shaking moments of one individual bank with procedures similar to those of inline engines. This will give the specific choice of ϕ_i. This is adequate, but not necessary. For cases in which the zero condition is not met for an individual bank, zero shaking force and shaking moments for the whole engine are still possible by properly choosing γ. Specifically, letting Eqs. (2.72), (2.74) and (2.76) equal zero gives the options for γ. Obviously, the designer has more options in designing well balanced V engines.

For the shaking force and shaking moments that are not canceled out through the described procedure, some balancing measures can be applied to remove the unbalanced quantities. Due to the need for brevity, this is beyond the scope of this book. Readers who are interested in this topic can refer to the book by Norton (2004).

CHAPTER 3

Dynamics of single-DOF machines

3.1 Introduction

There are two types of dynamic analysis for rigid machines, namely, *inverse dynamics* and *direct dynamics*. In inverse dynamics, the objective is to find the required driving torque and the reaction forces at the joints to maintain a prescribed motion. In direct dynamics, the objective is to find the actual motion of the machine under given driving force and external loads. Clearly, Chapter 1 addresses inverse dynamics. Direct dynamics is the topic of Chapters 3 and 4. This chapter focuses on machines having a single DOF, while Chapter 4 is dedicated to machines having multiple DOFs. It is noted that the scope of research in direct dynamics includes not only the mechanism itself, but also the driving input. In this book the combination of the functional mechanism and the driving input is referred to as the *mechanical system*.

The operation of a machine can be generally divided into three stages, including start-up, steady running, and stop-down as shown in Fig. 3.1. During the start-up, the machine is accelerated to the speed to be maintained during the expected work. In the stead-running stage, the speed of the machine generally fluctuates around the working speed. Speed variation in this stage causes a set of detrimental effects, such as causing inertial forces in joints and vibration of the machine. Thus, variation should be limited. In the stop-down stage, however, large deceleration is expected, so that the machine is able to come to a stop within a required length of time.

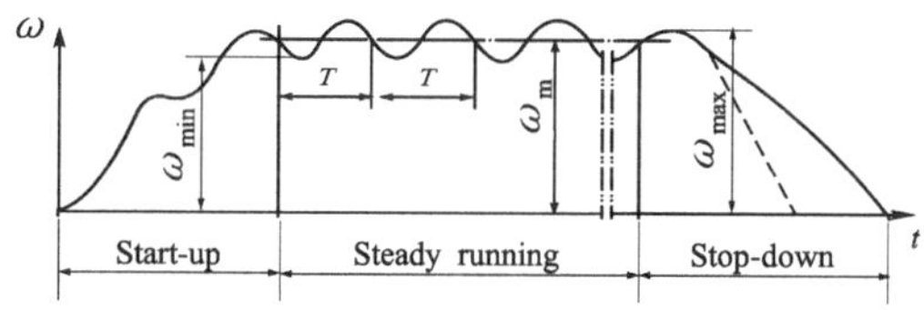

Figure 3.1 Stages of machine operation.

The chapter is organized as below:

- Section 3.2 focuses on the classification of possible forces that are applied to a machine.

Machinery Dynamics
https://doi.org/10.1016/B978-0-12-815785-5.00012-1

- The characteristics of electric motors are discussed in Section 3.3.
- Section 3.4 introduces the derivation of the dynamics of single-DOF machines using the Lagrange's equation, while Section 3.5 is dedicated to the solution of the dynamic equations.
- Section 3.6 introduces fast algorithms to find the steady-state operational speed. Speed smoothness is discussed in Section 3.7.

3.2 Forces on machines

3.2.1 Classification of forces

The forces, including torques, exerted on a machine can be generally categorized amoning these types:

- **Driving force or torque**: The driving force normally takes the form of a torque and is generally provided by a motor or engine. Power flows into the machine through the driving force.
- **Working resistance**: Depending on the type of machine, this force (or torque) may assume a different form. It consumes energy, thus doing negative work.
- **Weight**: The weight of a moving machine component acts when moving downward so it consumes energy only when moving upward. Over a cycle, the net work is zero. When a machine runs at high speed, inertial forces generally overtake the static weight, becoming the dominant load. Thus, component weight can be neglected without significant loss of accuracy in analysis. For some heavy-duty machines operating at low speeds, the weight nevertheless plays a considerable role in dynamic analysis.
- **Friction**: Friction consumes energy and is regarded as detrimental in most cases. However, friction has limited effect on dynamics except in several special cases.

The working resistance, depending on the types of machines, may

- be constant, such as in cranes and rolling mills,
- vary with machine position, such as in reciprocating compressors and pumps,
- vary with machine speed, such as in centrifugal compressors and pumps, or
- depend on time, such as in rock crushers and ball mills.

The driving force may also be a constant, vary with machine position, or vary with machine speed. For example, if the driving force derives from the weight of an object, it is a constant; if the driving force originates from a spring, the force depends on the position; while if the driver is a motor, then the driving force varies with the machine speed.

3.3 Characteristics of induction motors

Electric motors, three-phase induction motors, in particular, are the most widely used drivers in machines. In the following, a brief introduction is presented to the characteristics of induction motors. For other types of drivers, such as engines and DC motors, readers can consult relevant manuals or specialized books.

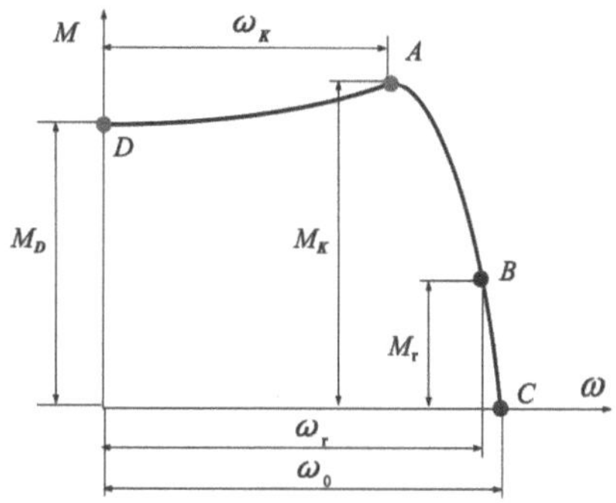

Figure 3.2 Characteristic curve of a three-phase induction motor.

A typical characteristic curve of an induction motor is shown in Fig. 3.2, in which the abscissa and the ordinate represent the angular speed and the output torque, respectively. The curve can be roughly divided into two segments *AD* and *AC* which are discussed subsequently:

- Segment *AD*: In this part, the motor is unstable. For a general machine, an increase of the external load would lead to a lowering of speed. However, in this section, a lower speed corresponds to a lower driving torque. Thus, there is not possible for the lowering of motor torque to establish an equilibrium with the increased working load. For this reason, section *AD* is not suitable for machine operation.
- Segment *AC*: In this segment, the driving torque decreases if the speed increases. On the other hand, an increased machine speed is always accompanied by lowered resistance of the working load. Thus, it is possible for the driving torque and the working resistance to develop and maintain a balance. Motors are expected to work over this segment. The nominal working status of a motor corresponds to a point, such as *B*, within this segment.

Generally the following information can be obtained from the nameplate of a motor:

- Rated power P_r (kW), rated torque M_r (N·m), and rated speed n_r (rpm).
- Synchronous speed n_0.
- Breakdown torque (M_K) (generally given by a ratio $\lambda = M_K/M_r$).
- Pull-up torque (M_D), (generally given by a ratio $\lambda_1 = M_D/M_r$).

The relationships among these parameters (in the SI unit) can be summarized as:

$$\begin{cases} M_r = 9550\dfrac{P_r}{n_r} \\ \omega_r = \dfrac{\pi}{30}n_r \\ \omega_0 = \dfrac{\pi}{30}n_0 \\ M_K = \lambda M_r \\ M_D = \lambda_1 M_r \\ \omega_K \approx \omega_0 - (\omega_0 - \omega_r)\left(\lambda + \sqrt{\lambda^2 - 1}\right). \end{cases} \tag{3.1}$$

It is common practice to approximate the characteristic curve with straight lines or polynomials. For example, AD can be approximated by a straight line passing the two points A and D. The working part AC can be represented by a parabola passing through the three points A, B, and C. If accuracy is not a concern, BC can be approximated by a straight line passing the two points B and C. In the following, the procedures for approximation are presented.

To approximate the segment ABC with a parabola, the following general equation is assumed:

$$M = a + b\omega + c\omega^2, \tag{3.2}$$

where M and ω are the output torque and the angular speed of the motor. Plugging the three point coordinates into the equation, we obtain

$$\begin{cases} a + b\omega_0 + c\omega_0^2 = 0 \\ a + b\omega_r + c\omega_r^2 = M_r \\ a + b\omega_K + c\omega_K^2 = M_K. \end{cases} \tag{3.3}$$

Then, the constants a, b, and c can be determined by solving the equations:

$$\begin{cases} a = \omega_0 \omega_K \alpha + \omega_0 \omega_K \beta \\ b = -(\omega_0 + \omega_K)\alpha - (\omega_0 + \omega_K)\beta \\ c = \alpha + \beta, \end{cases} \tag{3.4}$$

where

$$\begin{cases} \alpha = \dfrac{M_r}{(\omega_r - \omega_0)(\omega_r - \omega_K)} \\ \beta = \dfrac{M_K}{(\omega_K - \omega_0)(\omega_K - \omega_r)}. \end{cases} \tag{3.5}$$

To approximate the segment of BC with a straight line, one can assume the equation

$$M = a + b\omega. \tag{3.6}$$

Plugging in the coordinates of points B and C, we obtain

$$\begin{cases} a = \dfrac{-\omega_0 M_r}{\omega_r - \omega_0} \\ b = \dfrac{M_r}{\omega_r - \omega_0}. \end{cases} \tag{3.7}$$

Example 3.1. A motor of model Y90L-4 is used in a hoisting machine. It has the following parameters: $P_r = 1.5$ kW, $n_0 = 1500$ rpm, and $n_r = 1410$ rpm. Find the approximate straight line to represent the segment BC.

Solution. The equation for a straight line takes the form of Eq. (3.6). First, the following parameters are calculated with Eq. (3.1):

$$\begin{cases} \omega_r = \dfrac{\pi}{30} n_r = \dfrac{\pi}{30} 1410 = 147.65 \text{ rad/s} \\ \omega_0 = \dfrac{\pi}{30} n_0 = \dfrac{\pi}{30} 1500 = 157.08 \text{ rad/s} \\ M_r = 9550 \dfrac{P_r}{n_r} = 9550 \dfrac{1.5}{1410} = 10.16 \text{ N·m}, \end{cases} \tag{a}$$

where ω_r, M_r, and ω_0 are the rated speed, rated output torque, and the synchronous speed, respectively.

With Eq. (3.7), one obtains $a = 169.23$ and $b = -1.0773$. Thus, the straight-line representation of the segment is

$$M = 169.23 - 1.0773\omega. \tag{b}$$

Example 3.2. A Y100L2-4 motor has the following parameters: $P_r = 3$ kW, $n_0 = 1500$ rpm, $n_r = 1420$ rpm, $\lambda = 2.2$. Find a parabolic approximation of the segment ABC.

Solution. The form of the parabolic equation is Eq. (3.2). As in the first example, the following parameters are calculated first with Eq. (3.1).

$$\begin{cases} \omega_r = \dfrac{\pi}{30} n_r = \dfrac{\pi}{30} 1480 = 147.70 \text{ rad/s} \\ \omega_0 = \dfrac{\pi}{30} n_0 = \dfrac{\pi}{30} 1500 = 157.08 \text{ rad/s} \\ M_r = 9550 \dfrac{P_r}{n_r} = 9550 \dfrac{3}{1420} = 20.176 \text{ N·m} \\ \omega_K = \omega_0 - (\omega_0 - \omega_r)\left(\lambda + \sqrt{\lambda^2 - 1}\right) = 122.23 \text{ rad/s} \\ M_K = \lambda M_r = 2.2 \times 20.176 = 44.39 \text{ N·m}. \end{cases} \tag{a}$$

α and β are calculated with Eq. (3.5).

$$\begin{cases} \alpha = \dfrac{M_r}{(\omega_r - \omega_0)(\omega_r - \omega_K)} = -0.091 \\ \beta = \dfrac{M_K}{(\omega_K - \omega_0)(\omega_K - \omega_r)} = 0.04812. \end{cases} \tag{b}$$

From Eq. (3.4), we have

$$\begin{cases} a = \omega_0 \omega_K \alpha + \omega_0 \omega_K \beta = -622.9 \\ b = -(\omega_0 + \omega_K)\alpha - (\omega_0 + \omega_K)\beta = 10.70 \\ c = \alpha + \beta = -0.04286. \end{cases} \tag{c}$$

The parabolic representation is then obtained as

$$M = -622.9 + 10.698\omega - 0.04286\omega^2. \tag{d}$$

3.4 Dynamics of single-DOF systems

3.4.1 Lagrange's equation

In inverse dynamics, Newton's second law and D'Alembert's principle are the tools to derive the equations. In the process, the reaction forces of joints, which are not necessary to be known in direct dynamics, have to be included. Moreover, because equations have to be formulated for every moving link; the number of equations for a machine is often rather high. For these two reasons, Newton's law is not often used in direct dynamics. The *Lagrange's equation*, however, provides a much more elegant approach for direct dynamics. The general form of the equation is

$$\frac{\mathrm{d}}{\mathrm{d}t}\left(\frac{\partial E_k}{\partial \dot{q}_i}\right)-\frac{\partial E_k}{\partial q_i}+\frac{\partial E_{\mathrm{p}}}{\partial q_i}=Q_i \quad (i=1,\cdots,n), \tag{3.8}$$

where E_k and E_{p} are the kinetic energy and potential energy, respectively, and q_i is the *generalized coordinate*, which is defined as the independent coordinate to fully describe the position of the mechanism. Q_i is the *generalized force*, and n is the number of the generalized coordinates, also called the degrees of freedom (DOF).

For a single-DOF mechanism, only one equation is needed. Compared with the Newton's law, the Lagrange's formulation has the following features:

- The method is developed on the basis of energy, involving only scalar calculation. The complicated vector calculation in the Newton's method is avoided.
- The number of equations equals the number of the generalized coordinates, which, in general cases, is much smaller than the number of equations derived from Newton's method.
- It does not involve the reaction forces of joints, greatly simplifying calculation and the derivation process.

Due to these unique features, the Lagrange's equation is widely adopted in the study of direct dynamics of complicated mechanical systems.

3.4.2 Governing equation of single-DOF systems

Machines having a single DOF need only one generalized coordinate to describe their motion. For example, in a slider-crank mechanism, the angular position of the crank is often chosen as the generalized coordinate. In the following, the procedures to derive the governing equation of a planar mechanism are presented.

3.4.2.1 Kinetic energy

The kinetic energy of a link in a planar mechanism is represented as

$$E_{\mathrm{k}i}=\frac{1}{2}m_i v_{\mathrm{S}i}^2+\frac{1}{2}J_{\mathrm{S}i}\omega_i^2, \tag{3.9}$$

where m_i and J_{Si} are the mass and moment of inertia about the center of mass of the link. v_{Si} and ω_i are the translational velocity of the mass center and the angular velocity, respectively. The subscript i indicates the i-th link.

The total kinetic energy of the mechanism is, thus, the summation of all the moving links:

$$E_k = \sum_{i=1}^{l} E_{ki} = \sum_{i=1}^{l} \left(\frac{1}{2} m_i v_{Si}^2 + \frac{1}{2} J_{Si} \omega_i^2 \right), \tag{3.10}$$

where l is the number of moving links in the mechanism.

If the driving link is purely rotational, the angular position of the driving link is often chosen as the generalized coordinate q. The total kinetic energy of the mechanism can be transformed to the driving link (also called an equivalent link) in the following form:

$$E_k = \frac{1}{2} J_e \dot{q}^2, \tag{3.11}$$

where J_e is an equivalent moment of inertia of the whole mechanism transformed to the driving link. Comparison of Eq. (3.11) with Eq. (3.10) gives

$$J_e = \sum_{i=1}^{l} \left[m_i \left(\frac{v_{Si}}{\dot{q}} \right)^2 + J_{Si} \left(\frac{\omega_i}{\dot{q}} \right)^2 \right], \tag{3.12}$$

where q and $\dot{q}$, the generalized coordinate and generalized velocity, are usually functions of the mechanism position (or time). $v_{Si}/\dot{q}$ and $\omega_i/\dot{q}$ are often called *generalized velocity ratios*; they depend on the position and geometry of the mechanism.

3.4.2.2 Potential energy

The potential energy of a link in machines may comprise two forms, elastic and gravitational. If the link is rigid, gravitational potential becomes the only form. Generally, the potential energy of a link is much smaller than the kinetic energy. Thus, it is common practice to neglect the potential energy in the derivation.

3.4.2.3 Generalized force

If F_k $(k = 1, 2, \cdots, m)$ and M_j $(j = 1, 2, \cdots, n)$ are the external forces and moments applied on the mechanism, the *virtual work* by these forces and moments are calculated as

$$\delta W = \sum_{k=1}^{m} (F_k \delta s_k \cos\alpha_k) + \sum_{j=1}^{n} (\pm M_j \delta\varphi_j), \tag{3.13}$$

where δs_k and $\delta\varphi_j$ are the virtual displacement at the point of F_k and the virtual angular displacement of the link of M_j.

The virtual work can also be calculated from the equivalent link as

$$\delta W = Q\delta q, \tag{3.14}$$

where δq is the virtual displacement of the generalized coordinate. The virtual work calculated from Eqs. (3.14) and (3.13) should be equal to each other; then we obtain

$$\begin{aligned} Q &= \sum_{k=1}^{m} \left(\frac{F_k \delta s_k \cos\alpha_K}{\delta q} \right) + \sum_{j=1}^{n} \left(\frac{\pm M_j \delta\varphi_j}{\delta q} \right) \\ &= \sum_{k=1}^{m} \left(\frac{F_k v_k \cos\alpha_K}{\dot{q}} \right) + \sum_{j=1}^{n} \left(\frac{\pm M_j \omega_j}{\dot{q}} \right). \end{aligned} \tag{3.15}$$

For the case of a rotating driving link, the generalized force is an equivalent torque on the driving link, which depends on the generalized velocity ratios $v_k/\dot{q}$ and $\omega_j/\dot{q}$. That is $M_e = Q$.

From Eq. (3.11) we can obtain

$$\frac{\partial E_k}{\partial q} = \frac{1}{2} \frac{\partial J_e}{\partial q} \dot{q}^2 \tag{3.16}$$

$$\frac{\partial E_k}{\partial \dot{q}} = J_e \dot{q} \tag{3.17}$$

$$\frac{d}{dt} \left(\frac{\partial E_k}{\partial \dot{q}} \right) = J_e \ddot{q} + \frac{\partial J_e}{\partial q} \dot{q}^2. \tag{3.18}$$

J_e is not a constant; it changes with q. Thus, the chain rule is utilized in deriving Eq. (3.18).

As stated before, the gravitational potential energy is generally neglected. Then we have

$$\frac{\partial E_p}{\partial q} = 0. \tag{3.19}$$

Substituting Eqs. (3.15) through (3.19) into Eq. (3.8) gives the governing equation of motion for the single DOF machine:

$$J_e\ddot{q} + \frac{1}{2}\frac{\partial J_e}{\partial q}\dot{q}^2 = M_e. \tag{3.20}$$

3.4.3 Equivalent model

Now let's revisit some of the concepts used in the derivation of Eq. (3.20) to get more insight into the problem. These include:

Equivalent link: the link chosen to describe the motion of the mechanism.

Equivalent moment of inertia J_e: the moment of inertia of the whole machine transformed to the equivalent link.

Equivalent torque M_e: a quantity transformed to the equivalent link representing all the external forces and moments.

The key to obtaining J_e and M_e is that the total kinetic energy and the work done by all external forces and moments equal those represented by the equivalent link.

With J_e and M_e, the dynamics of a machine having a rotating equivalent link can be investigated with an equivalent model as shown in Fig. 3.3(a).

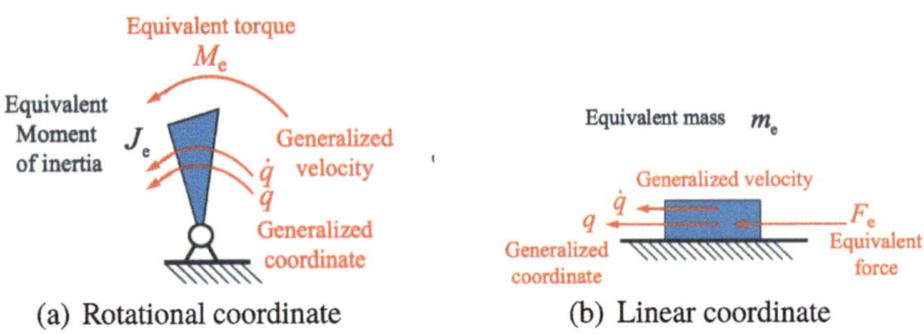

Figure 3.3 Equivalent model.

If the driving link (or the equivalent link) of the machine has a linear motion (linear generalized coordinate), an equivalent model shown in Fig. 3.3(b) can be obtained in a similar way. In this model, an equivalent mass m_e and an equivalent force F_e correspond to the equivalent moment of inertia J_e and the equivalent moment M_e in the rotational driving link. The governing equation of motion in this case takes the following form:

$$m_e\ddot{q} + \frac{1}{2}\frac{\partial m_e}{\partial q}\dot{q}^2 = F_e. \tag{3.21}$$

In summary J_e (or m_e) and M_e (or F_e) depend on the generalized speed ratios (Eqs. (3.12) and (3.15)). Thus, they can be calculated without knowing the actual motion of the mechanism.

Example 3.3. A hoisting machine is shown in Fig. 3.4. The power is delivered by an electric motor through a planetary gear train with a small difference of tooth numbers. The parameters are given next.

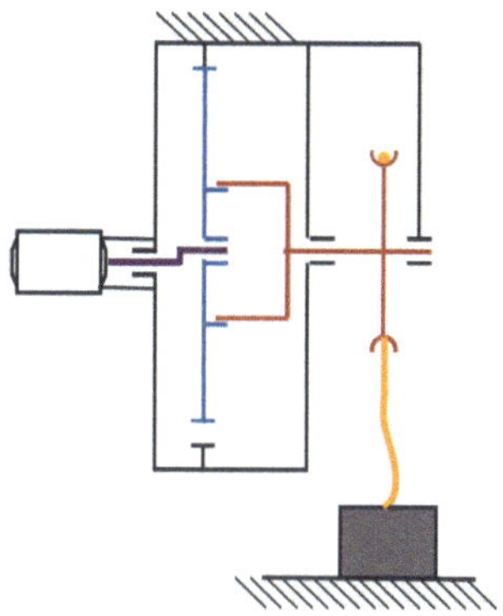

Figure 3.4 A hoisting machine.

The motor is the model of Y90L-4 as given in Example 3.1. The total moment of inertia of the motor rotor and the crankshaft of the gear set is $J_1 = 0.00715$ kg·m^2; the total moment of inertia of the planet gear, the sprocket and, the output shaft is $J_2 = 0.15$ kg·m^2. The planet gear weighs $G_1 = 20$ N; the diameters of the planet center trajectory and the pitch circle of the sprocket are $d = 5.0$ mm and $D = 200$ mm, respectively; the weight of the object to be lifted is $G = 4000$ N; the speed ratio of the gear set is $i = 40$. Taking the rotation of the motor shaft as the generalized coordinate, find the equivalent torque and the equivalent moment of inertia before and after lifting the object.

Solution. The external loads applied to the system include the motor's driving torque (M_d) and the torque caused by the weight of the object to be lifted. In Example 3.1, we already found an approximate form of the motor's driving torque:

$$M_d = 169.23 - 1.0773\omega \text{ N·m.} \tag{a}$$

It is directly applied to the equivalent part; thus, no transformation is needed.

The weight of the object forms a torque, which can be transformed as an equivalent torque M_r with Eq. (3.15)

$$M_r = \frac{Gv\cos\alpha}{\dot{q}} = -\frac{GD}{2i} = -\frac{4000 \times 0.2}{40} = -10 \text{ N·m}. \tag{b}$$

Then, the equivalent moment is the summation of the two:

$$M_e = M_d + M_r = 169.23 - 1.0773\omega - 10 = 159.23 - 1.0773\omega \text{ N·m}. \tag{c}$$

The equivalent moment of inertia before the object (J_e) is lifted off the ground has three contributions, including: 1) J_1, which does not need transformation; 2) J_2, which needs to be transformed to the equivalent part according to Eq. (3.12); and 3) the planet gear weight G_1, which forms a moment of inertia, and which needs to be transformed into the equivalent part. The total equivalent moment of inertia J_e is the sum of the three contributions as follows:

$$J_e = J_1 + \frac{J_2}{i^2} + \frac{G_1 d^2}{4g} = 0.00715 + \frac{0.15}{40^2} + \frac{20 \times 0.005^2}{9.81 \times 4} = 0.007256 \text{ kg·m}^2. \tag{d}$$

The equivalent moment of inertia after the object is lifted, designated as J_e', contains the contribution from the weight of the object in addition to J_e. The moment of inertial of the object's weight about the center line of the sprocket, J_4, is

$$J_4 = \frac{GD^2}{4g} = \frac{4000 \times 0.2^2}{9,81 \times 4} = 4.0775 \text{ kg·m}^2. \tag{e}$$

The sprocket rotates at rates different from the motor shaft; thus, it needs to be transformed. The total J_e' would be

$$J_e' = J_e + \frac{J_4}{i^2} = 0.007106 + \frac{4.0775}{40^2} = 0.009654 \text{ kg·m}^2. \tag{f}$$

3.4.4 Equation of energy form

Eqs. (3.20) and (3.21) are in the form of moment and force. The equation can also be represented in the form of energy as developed in this subsection.

The virtual work W done by the generalized force is transformed into kinetic energy E_k.

$$W = \Delta E_k. \tag{3.22}$$

If the equivalent link rotates from φ_1 to φ_2, the angular speed would change from ω_1 to ω_2. According to Eq. (3.22), we have

$$\frac{1}{2}J_{e2}\omega_2^2 - \frac{1}{2}J_{e1}\omega_1^2 = \int_{\varphi_1}^{\varphi_2} M_e \mathrm{d}\phi, \tag{3.23}$$

where J_{e1} and J_{e2} are the equivalent moments of inertia at φ_1 and φ_2, respectively.

In the case of a linear generalized coordinate, a similar equation as this can be obtained:

$$\frac{1}{2}m_{e2}v_2^2 - \frac{1}{2}m_{e1}v_1^2 = \int_{s_1}^{s_2} F_e \mathrm{d}s, \tag{3.24}$$

where m_{e1} and m_{e2} are the equivalent masses at position s_1 and s_2, respectively.

Eqs. (3.23) and (3.24) are the governing equations of motion in the form of energy for a rotational and a linear generalized coordinate, respectively.

For a mechanism having constant speed ratio, the second term in Eq. (3.20) is zero; the equation becomes

$$J_e\ddot{q} = M_e. \tag{3.25}$$

This is the Newton–Euler equation for the rotation of a rigid body.

The second term in Eq. (3.20) is caused by the change of the equivalent moment of inertia. Thus, Eq. (3.20) is, in essence, the Newton–Euler equation for varying inertial systems.

3.4.5 Discussion

The equivalent moment of inertial J_e (or mass m_e) varies with the mechanism position. It is generally not easy to represent it in a closed form, especially for mechanisms with time-changing speed ratios, $\omega_i/\dot{q}$ and $v_{Si}/\dot{q}$, such as linkages and cams. In addition, $\mathrm{d}J_e/\mathrm{d}q$ and $\mathrm{d}m_e/\mathrm{d}q$ are also difficult to express in closed form. In most cases, these quantities must be calculated numerically, rewriting Eq. (3.12) in the following form by decomposing v_{Si}^2

into $v_{Six}^2 + v_{Siy}^2$:

$$J_e = \sum_{i=1}^{l}\left[m_i \frac{\left(v_{Six}^2 + v_{Siy}^2\right)}{\dot{q}^2} + J_{Si}\frac{\omega_i^2}{\dot{q}^2}\right]. \tag{3.26}$$

Differentiating this with respect to q gives

$$\frac{dJ_e}{dq} = \frac{dJ_e}{dt}\frac{dt}{dq} = \frac{2}{\dot{q}^3}\sum_{i=1}^{l}\left[m_i\left(v_{Six}a_{Six} + v_{Siy}a_{Siy}\right) + J_i\omega_i\alpha_i\right]. \tag{3.27}$$

With Eqs. (3.26) and (3.27), J_e and dJ_e/dq can be calculated numerically. It is worth pointing out that both J_e and dJ_e/dq have nothing to do with the actual motion parameters of the machine. Rather they depend on speed ratios $v_{Si}/\dot{q}$ and $\omega_i/\dot{q}$. Thus, $\dot{q}$ can be chosen as any value in the calculation. For simplicity it is often set as unity.

Example 3.4. A slider-crank mechanism shown in Fig. 3.5 has the following parameters. The length of links 2 and 3 are $l_2 = 0.2$ m, $l_3 = 0.5$ m;

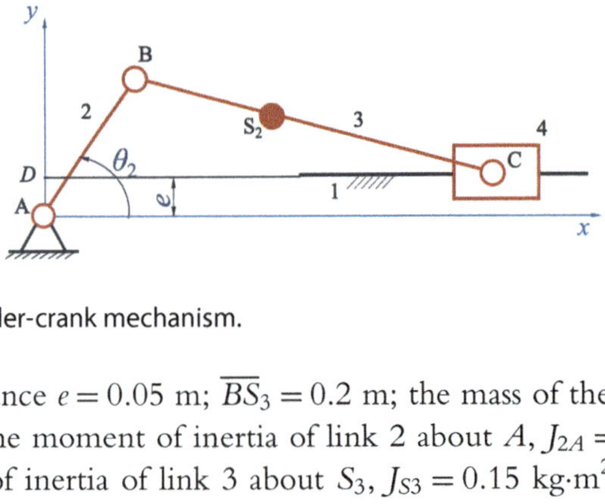

Figure 3.5 A slider-crank mechanism.

the offset distance $e = 0.05$ m; $\overline{BS_3} = 0.2$ m; the mass of the slider (link 4) $m_4 = 10$ kg; the moment of inertia of link 2 about A, $J_{2A} = 3$ kg·m^2; and the moment of inertia of link 3 about S_3, $J_{S3} = 0.15$ kg·m^2. Numerically calculate J_e and dJ_e/dq.

Solution. According to Eqs. (3.26) and (3.27), J_e and dJ_e/dq for this problem can be expressed as

$$J_e = J_{2A} + J_{S3}\frac{\omega_3^2}{\omega_2^2} + m_3\frac{v_{S3x}^2 + v_{S3y}^2}{\omega_2^2} + m_4\frac{v_C^2}{\omega_2^2} \tag{a}$$

$$\frac{dJ_e}{dq} = \frac{2}{\omega_2^3}\left[J_{S3}\omega_3\alpha_3 + m_3\left(v_{S3x}a_{S3x} + v_{S3y}a_{S3y}\right) + m_4 v_C a_C\right]. \tag{b}$$

All the kinematic parameters on the right-hand side of the equation can be obtained through kinematic analysis by assuming an arbitrary input ω_2. For example, we assume $\omega_2 = 1$ rad/s. Then, one cycle of the crank motion is divided into 300 steps, and J_e and dJ_e/dq are calculated with the cited equations for each step. The results are as given in Fig. 3.6.

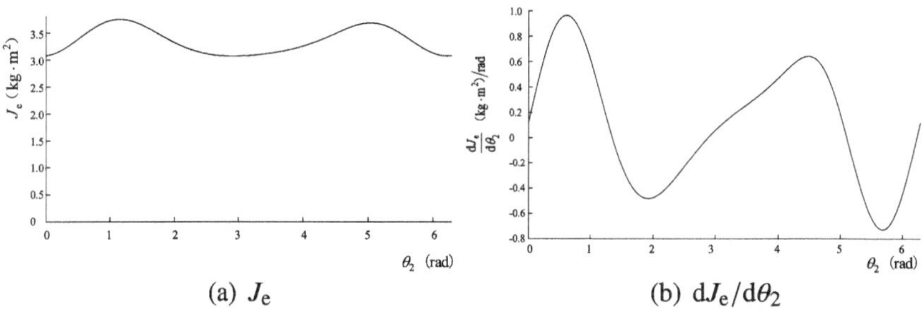

Figure 3.6 J_e and $dJ_e/d\theta_2$ of the slider-crank mechanism.

3.5 Solution of equation

Solving the governing equations, either the moment (force) form or the energy form, gives the actual motion of the equivalent part. Depending on the form of the equation, it can be solved in various ways. Closed form solutions are only possible for very simple cases. Numerical methods are more often used for solving practical problems. In the following, several cases are presented, focusing on numerical methods.

3.5.1 Case 1: M_e being function of q

When the driving torque and the resistant force of the mechanism depend on only the position of the mechanism, the equivalent moment M_e is a function of q. For example, the equivalent moment transformed to the crank from the gas pressure of an internal combustion engine belongs to this category.

In this case, the energy form, Eq. (3.23), is more convenient. For two adjacent position φ_0 and φ, the equation is

$$\frac{1}{2}J_e\omega^2 - \frac{1}{2}J_{e0}\omega_0^2 = \int_{\varphi_0}^{\varphi} M_e(\varphi)d\varphi, \tag{3.28}$$

where J_e and ω are the equivalent moment of inertia and the angular velocity at position φ. The subscript e and 0 refer to equivalent and initial, respectively. $M_e(\varphi)$ is the equivalent moment. To emphasize its dependence on φ, $M_e(\varphi)$ is deliberately used.

In Eq. (3.28) the right-hand side is actually the work done by $M_e(\varphi)$ during the period between φ and φ_0. If this work is designated as $W(\varphi)$, ω can be solved easily as

$$\omega = \left[\frac{J_{e0}\omega_0^2 + 2W(\varphi)}{J_e(\varphi)}\right]^{\frac{1}{2}}. \tag{3.29}$$

If $M_e(\varphi)$ is given in closed form and integrable, a closed form ω can be obtained. Otherwise, the integration can be evaluated numerically with proper methods, such as the trapezoidal and Simpson methods (Chapra and Canale, 2011). Taking the trapezoidal method as an example, the procedure of calculation is illustrated in Fig. 3.7. First, the integration interval is discretized into multiple smaller intervals of equi-length $\Delta\varphi$. The essence of this method is to approximate the area of each segment under the dashed curve with the one under the straight solid line.

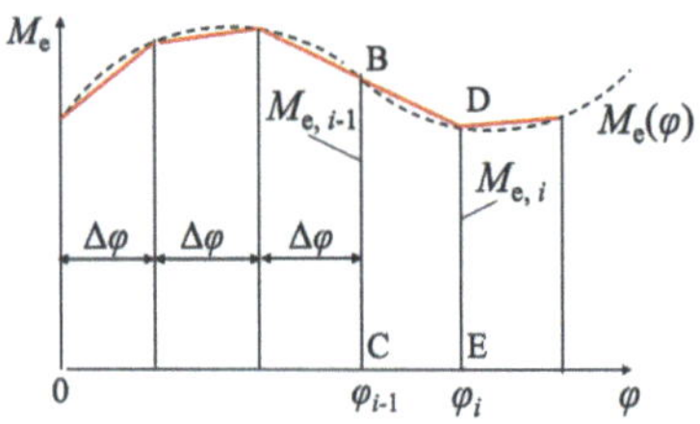

Figure 3.7 Trapezoidal method of numerical integration.

Once ω has been obtained, the time to a specific position can be computed as

$$t = t_0 + \int_{\varphi_0}^{\varphi} \frac{d\varphi}{\omega}. \tag{3.30}$$

Example 3.5. In the slider-crank mechanism of Example 3.4, the relationship between the equivalent moment M_e and the angular position θ_2 is given in Table 3.1. The initial conditions are at $t = 0$, $\theta_2 = 0°$, and $\omega_0 = 62$ rad/s. Find the angular velocity ω of the crank.

Solution. According to Table 3.1, the angular difference between two adjacent positions of the crank is

$$\Delta\theta = \frac{\pi}{180°} \times 10° = 0.1745 \text{ rad}. \tag{a}$$

Table 3.1 Equivalent moment and crank angle.

θ_2 (°)	0	10	20	30	40	50	60	70	80
M_e (N·m)	720	540	360	180	0	−240	−480	−720	−840
θ_2 (°)	90	100	110	120	130	140	150	160	170
M_e (N·m)	−900	−840	−720	−480	−240	0	180	360	480
θ_2 (°)	180	190	200	210	220	230	240	250	260
M_e (N·m)	540	420	240	0	−180	−360	−480	−600	−480
θ_2 (°)	270	280	290	300	310	320	330	340	350
M_e (N·m)	−360	−180	0	240	480	720	840	960	840

Then it is straightforward to derive the following iteration based on the trapezoidal integration scheme:

$$W_0 = 0 \tag{b}$$

$$W_i = W_{i-1} + \frac{\Delta\theta}{2}\left[(M_e)_{i-1} + (M_e)_i\right], \tag{c}$$

where W_0 and W_i are the virtual work done by M_e at the start and the i-th discretized position (see Fig. 3.7).

Then, the value of ω at each discrete position i can be obtained by Eq. (3.29). The results are graphically shown in Fig. 3.8.

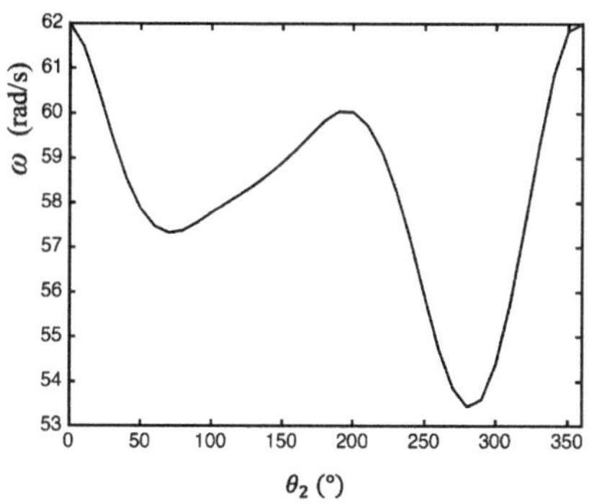

Figure 3.8 ω versus θ_2.

3.5.2 Case 2: constants J_e and M_e depending on ω

For machines with a constant speed ratio, the equivalent moment of inertia is also constant. If electric motors are used as the drivers, then the driving torque depends only on the angular speed ω. In this case, if the resistant force also depends only on ω, the equivalent moment M_e will only depend

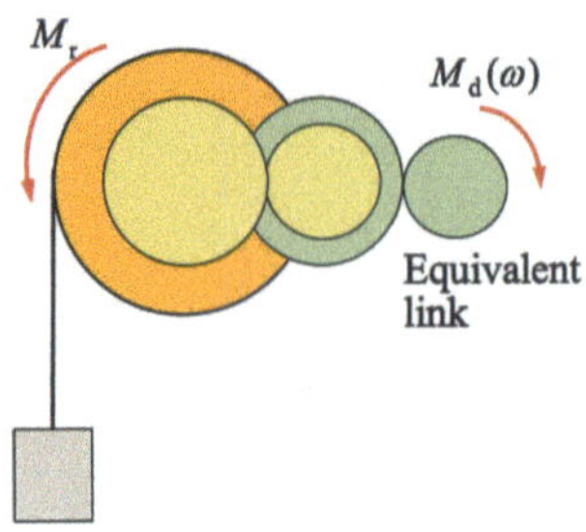

Figure 3.9 Lifting device.

on ω. The lifting device shown in Fig. 3.9 is such an example. Other examples include pumps, fans, and blowers driven by electric motors.

For such cases, closed-form solutions can be obtained if the M_e is integrable. Otherwise, numerical calculation can be used.

3.5.2.1 Closed-form solution

In this case, Eq. (3.20) is more convenient. Given that J_e is constant, Eq. (3.20) can be simplified as

$$M_e(\omega) = J_e \frac{d\omega}{dt}. \tag{3.31}$$

It is straightforward to obtain the time t as

$$t = t_0 + J_e \int_{\omega_0}^{\omega} \frac{d\omega}{M_e(\omega)}. \tag{3.32}$$

In some cases M_e can be represented as a linear or quadratic function of ω. For the linear case $M_e = a + b\omega$, the integration of Eq. (3.32) leads to

$$t = t_0 + \frac{J_e}{b} \ln \frac{a + b\omega}{a + b\omega_0}. \tag{3.33}$$

If a quadratic function $M_e = a + b\omega + c\omega^2$ is used, the integration of Eq. (3.32) gives

$$t = \begin{cases} t_0 + \dfrac{2J_e}{\sqrt{4ac - b^2}}(A_1 - A_2) & \text{if } b^2 - 4ac < 0 \\ t_0 + \dfrac{2J_e}{\sqrt{b^2 - 4ac}} \ln \dfrac{A_3}{A_4} & \text{if } b^2 - 4ac > 0, \end{cases} \tag{3.34}$$

where

$$A_1 = \arctan\left(\frac{2c\omega + b}{\sqrt{4ac - b^2}}\right)$$
$$A_2 = \arctan\left(\frac{2c\omega_0 + b}{\sqrt{4ac - b^2}}\right)$$
$$A_3 = \left(2c\omega + b - \sqrt{b^2 - 4ac}\right)\left(2c\omega_0 + b + \sqrt{b^2 - 4ac}\right)$$
$$A_4 = \left(2c\omega + b + \sqrt{b^2 - 4ac}\right)\left(2c\omega_0 + b - \sqrt{b^2 - 4ac}\right).$$

If $b^2 - 4ac < 0$, the equation $a + b\omega + c\omega^2 = 0$ has no real roots, indicating that no stable angular velocity exists in the first case of Eq. (3.34). Thus, this case is only for the braking stage of a machine.

In the start-up stage, ω gradually picks up. When ω reaches a certain value, the equivalent moment $M_e(\omega) = 0$. In this case, if J_e is constant, ω will stabilize on a certain value.

Eq. (3.34) can be shortened to $t = f(\omega)$. The inverse function $\omega = g(t)$ defines the dependence of ω on time t.

If we prefer to represent ω as a function of time t, Eq. (3.31) is first recast into the following form

$$M_e(\omega) = J_e\omega\frac{d\omega}{d\theta}. \tag{3.35}$$

Separating the variables and then integrating leads to

$$\theta = \theta_0 + J_e\int_{\omega_0}^{\omega}\frac{\omega\, d\omega}{M_e(\omega)}. \tag{3.36}$$

If a closed-form $M_e(\omega)$ is available, plugging into the equation and completing the integration give θ. For example if $M_e(\omega) = a + b\omega$, θ is in the form

$$\theta = \theta_0 + \frac{J_e}{b}\left(\omega - \omega_0 - \frac{a}{b}\ln\frac{a + b\omega}{a + b\omega_0}\right). \tag{3.37}$$

However, if $M_e(\omega)$ is given in a tabular form, the integration has to be performed numerically.

Example 3.6. In the hoisting machine of Example 3.3, the motor speed before the rope is tightened up is $\omega_0 = 157.08$ rad/s. Find the motor speed during the lifting process.

Solution. In Example 3.3, the equivalent moment of inertia before and after lifting the object off the ground is $J_e = 0.007106$ kg·m^2 and $J'_e = 0.009654$ kg·m^2, respectively. The equivalent moment is $M_e = 159.23 - 1.0773\omega$.

At the instant when the rope is tightened up, the equivalent moment of inertia abruptly increases because of the weight of the object, leading to a decease in speed. However, the kinetic energy before and after the tightening-up instant remains the same. Then, we have

$$\frac{1}{2}J_e\omega_0^2 = \frac{1}{2}J'_e(\omega'_0)^2. \qquad \text{(a)}$$

ω'_0 can be solved as

$$\omega'_0 = \sqrt{\frac{J_e}{J'_e}}\omega_0 = \sqrt{\frac{0.007106}{0.009654}} \times 157.08 = 134.77 \text{ rad/s}. \qquad \text{(b)}$$

The equivalent moment can be approximated as a linear function of ω. If the initial condition is $t_0 = 0$ and $\omega = \omega'_0$, we have the following according to Eq. (3.33)

$$t = \frac{J'_e}{b} \ln \frac{a + b\omega}{a + b\omega'_0}. \qquad \text{(c)}$$

Then, we obtain

$$\omega = \frac{1}{b}\left[e^{qt}\left(a + b\omega'_0\right) - a\right] = -13.041e^{-111.591t} + 147.8, \qquad \text{(d)}$$

where $q = b/J'_e$.

This relationship between ω and time t is drawn in Fig. 3.10.

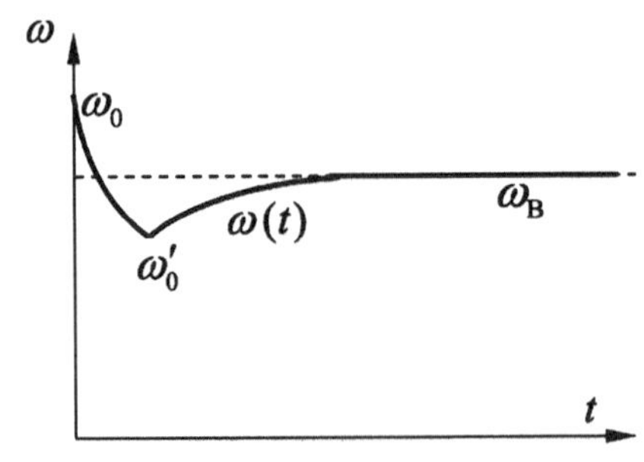

Figure 3.10 $\omega(t)$.

It can be seen that ω drops from ω_0 to ω_0' with the lifting of the weight. However, the drop is not vertical because of the elastic deformation of the rope. Instead, it is occurs within a short period of time. After this, it exponentially approaches a limit ω_B, which is $-a/b = 147.81$ rad/s.

The time from ω_0' to ω_B is called the transition time. Theoretically, this time is infinite. In practice, however, the transition is regarded as completed when ω reaches a threshold, such as $0.99\omega_B$. In this case, the transition time is computed by

$$t' = \frac{J_e'}{b} \ln \frac{a + 0.99b\omega_B}{a + b\omega_0'} = 0.0195 \text{ s.} \tag{e}$$

Generally, the transition time is very short.

3.5.2.2 Numerical solution

When $M_e(\omega)$ is too complicated for integration or given as discrete values in a tabular form, numerical calculation becomes necessary.

Eq. (3.31) can be rewritten in the form

$$\frac{d\omega}{dt} = \frac{M_e(\omega)}{J_e}. \tag{3.38}$$

This is a first-degree ordinary differential equation (ODE). It can easily be solved numerically from an initial condition, $t = t_0$, $\omega = \omega_0$ with many schemes, among which the most common are the Runge–Kutta methods (Chapra and Canale, 2011).

3.5.3 Case 3: M_e being a function of q and $\dot{q}$

For some machines, the speed ratio is not a constant. The contribution to M_e from the external forces will, thus, not be a constant even if the external forces may be constant. In addition, the driving torque of a motor depends on the angular speed ω. In this case, M_e may be represented as a function of both the generalized coordinate q and the generalized velocity $\dot{q}$. Numerical methods can be used to solve the equation in this case. Two methods will be discussed in the following.

3.5.3.1 General method

If the input is a rotation, then $q=\theta$, and $\dot{q}=\omega$. Eq. (3.20) will be in the form

$$J_e(\theta)\frac{d\omega}{dt}+\frac{1}{2}J_e'(\theta)\omega^2=M_e(\theta,\omega), \tag{3.39}$$

where $J_e'=dJ_e(\theta)/d\theta$.

Before solving the equation, the following transformation is first made

$$\frac{d\omega}{dt}=\frac{d\omega}{d\theta}\frac{d\theta}{dt}=\frac{d\omega}{d\theta}\omega. \tag{3.40}$$

Substituting this into Eq. (3.39) and rearranging the terms give

$$\frac{d\omega}{d\theta}=\frac{M_e(\theta,\omega)-0.5J_e'(\theta)\omega^2}{J_e(\theta)\omega}. \tag{3.41}$$

This is a first-order ODE; θ is the independent variable. It can be solved with various Runge–Kutta methods if initial condition $\theta=\theta_0$ and $\omega=\omega_0$ are given. The process can be found in any numerical-methods textbook; thus, the detail is not given here. One special case requiring special attention is for the zero initial condition $\theta=0$, $\omega=0$. In this case, the cited numerical method fails because of the zero denominator in Eq. (3.41). This issue can be easily solved by flipping over Eq. (3.41) into the following form:

$$\frac{d\theta}{d\omega}=\frac{J_e(\theta)\omega}{M_e(\theta,\omega)-0.5J_e'(\theta)\omega^2}. \tag{3.42}$$

Solve this ODE first for some steps until $\theta=\theta_1$, $\omega=\omega_1$, which can be used as the initial conditions for Eq. (3.41); then, the normal numerical procedures for Eq. (3.41) can be started.

3.5.3.2 Fast method

In some cases, fast solution methods are available. For example, if θ and ω in M_e can be separated, the equation can be solved with a fast algorithm. In this case, M_e can be represented as a summation of M_d and M_r. M_d and M_r are the equivalent moment from the driving and resistance, respectively. If the driver is a motor, then M_d can be approximated as

$$M_d=a+b\omega+c\omega^2. \tag{3.43}$$

The equivalent resistance torque M_r can be divided into two parts. One relies on θ, $M_r'(\theta)$, and the other depends on ω. It is then represented as

$$M_r = M_r'(\theta) + b'\omega + c'\omega^2. \tag{3.44}$$

It is more convenient to solve the energy-form equation, Eq. (3.23), in this case.

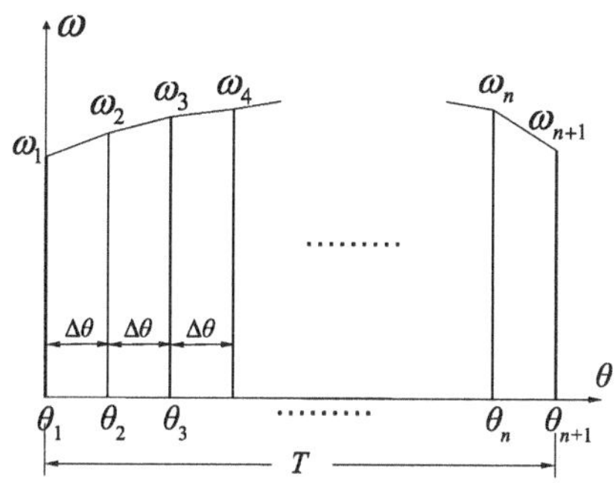

Figure 3.11 Fast algorithm.

First, θ is discretized into n intervals with a width of $\Delta\theta$ as shown in Fig. 3.11. Applying Eq. (3.23) to the interval between θ_i and θ_{i+1} gives

$$\frac{1}{2}J_{e(i+1)}\omega_{i+1}^2 - \frac{1}{2}J_{ei}\omega_i^2 = \int_{\theta_i}^{\theta_{i+1}} (M_d + M_r)\, d\theta, \tag{3.45}$$

where ω_i and J_{ei} are the angular speed and equivalent moment of inertia at position θ_i. With the trapezoidal integration scheme, the previous equation can be computed as

$$\frac{1}{2}J_{e(i+1)}\omega_{i+1}^2 - \frac{1}{2}J_{ei}\omega_i^2 = \frac{\Delta\theta}{2}\left[M_d(\theta_i) + M_r(\theta_i) + M_d(\theta_{i+1}) + M_r(\theta_{i+1})\right]. \tag{3.46}$$

Taking advantage of Eqs. (3.43) and (3.44), Eq. (3.46) can be represented as

$$A_i\omega_{i+1}^2 + B_i\omega_{i+1} + C_i = 0, \tag{3.47}$$

where

$$\begin{cases} A_i = J_{e(i+1)} - \Delta\theta\left(c + c'\right) \\ B_i = -\Delta\theta\left(b + b'\right) \\ C_i = -\Delta\theta\left[M_r'(\theta_i) + M_r'(\theta_{i+1}) + 2a + \left(b + b'\right)\omega_i + \left(c + c'\right)\omega_i^2\right] - J_{ei}\omega_i^2. \end{cases} \tag{3.48}$$

Starting from ω_1, ω_i can be computed using the iterative relationship of Eqs. (3.47) and (3.48).

3.6 Machine speed in the steady stage

As discussed at the beginning of this chapter, a machine generally goes through the start-up, steady-running and stop-down stages. In most cases, the speed, especially the speed fluctuation, in the steady stage is of more interest than in the other stages. In the section following, two methods will be introduced to calculate the speed fluctuation during the steady-running stage

$$\omega(\theta) = \omega(\theta + T), \tag{3.49}$$

where T is the angle of one cycle, generally 2π.

The dynamic equation can be solved numerically with the zero initial condition $\omega = 0$. With modern computers, the time to attain the steady status is generally rather quick. If the computational time is a concern, however, it can be shortened through the following two methods.

3.6.1 Estimated initial conditions

During the steady-running stage, the machine speed fluctuates periodically. The speed in the steady-running stage can be obtained from the numerical integration from zero initial conditions presented in the previous section. However, if we are concerning only in the speed in the steady-running stage, the numerical process can start from estimated non-zero initial conditions. Thus, the integration time to the steady stage can be significantly shortened.

The machine-speed periodicity can be expressed as

$$\omega(\theta) = \omega(\theta - T), \tag{3.50}$$

where T is the period, generally 2π.

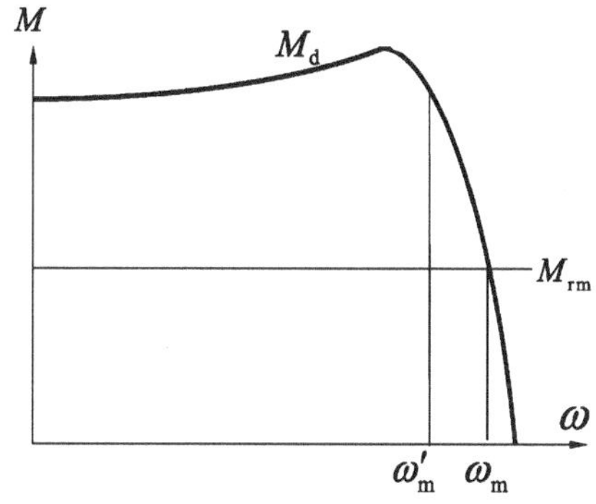

Figure 3.12 Estimation of ω_{m}.

If the equivalent driving and resistant torques are represented by $M_{\mathrm{d}}(\omega)$ and $M_{\mathrm{r}}(\omega)$, respectively, the average resistant torque in one period is

$$M_{\mathrm{rm}} = \frac{1}{T}\int_0^T M_{\mathrm{r}}(\theta)\,\mathrm{d}\theta. \tag{3.51}$$

The average speed in one cycle, ω_{m}, will be the value at the intersection of M_{d} and M_{rm}, as illustrated in Fig. 3.12. To quickly converge to the steady-state result, the numerical process can be started from $\omega_0 = \omega_{\mathrm{m}}$. The value can be regarded as steady when the following criterion is met:

$$\frac{|\omega(\theta) - \omega(\theta - T)|}{\omega(\theta)} \leq \epsilon. \tag{3.52}$$

ϵ is small number, the selection of which is a trade-off between computational time and accuracy, generally being between 0.001 and 0.0001. The speed of convergence heavily depends on the degree of self-regulation of the machine. Self-regulation means the feature by which a machine adjusts its speed to an external load disturbance. This is illustrated in Fig. 3.13.

Fig. 3.13(a) illustrates the types of self-regulating machines. The driving torque M_{d} and the resistant torque M_{r} intersect at point A, which represents the current operation status. If somehow the resistance M_{r} drops from the solid line to the dashed line, the machine speed ω will be increased because $M_{\mathrm{d}} > M_{\mathrm{r}}$. This increased ω will lead to a decreased M_{d}. Then, a new point of balance B will be formed. The case in Fig. 3.13(b) is an illustration of nonself-regulating machines. In this case, a drop in M_{r} also leads to an increase of ω. However, the driving torque M_{d} goes up with the increased ω, making the machine speed ω grow further. If no protective

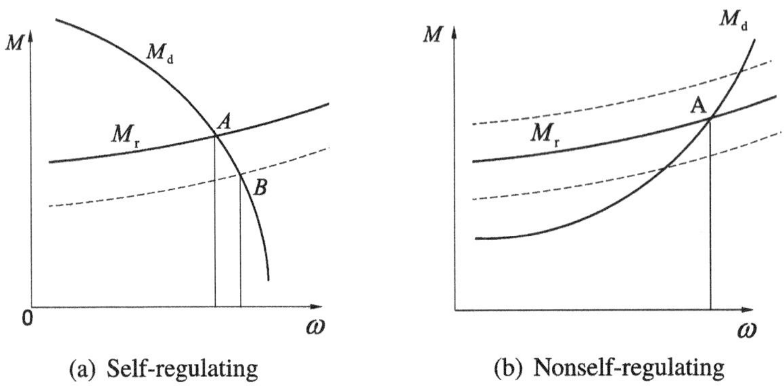

Figure 3.13 Self-regulating and nonself-regulating.

measures are taken, the continuous increase of ω may result in the breakage of the machine. In practice, speed governors, such as the centrifugal governor invented by J. Watt, are often used in nonself-regulating machines to limit ω.

3.6.2 Roots of equations

The convergence of numerical integration highly depends on the degree of self-regulation. For machines without a good self-regulating feature, such as machines driven by internal combustion engines, the numerical integration may converge very slowly. In this case, the following method can be used.

Given that the steady speed ω is periodic, one motion cycle of the machine can be divided into n small intervals with equal width. The values of the position angle and angular speed at the discrete points are represented as θ_i and ω_i, respectively $(i = 1, 2, \cdots, n+1)$. The periodicity of ω can be represented as

$$\Delta\omega = \omega_{n+1} - \omega_1 = 0. \tag{3.53}$$

The value of $\Delta\omega$ can be obtained through numerical integration over one motion cycle, starting from an arbitrary initial ω_1. The root of Eq. (3.53) will be the steady speed. Many numerical schemes are available for finding the root, such as various bracketing methods (Chapra and Canale, 2011).

Alternatively, the root can also be approximated by that of a cubic polynomial as shown in Fig. 3.14. To construct the approximate cubic

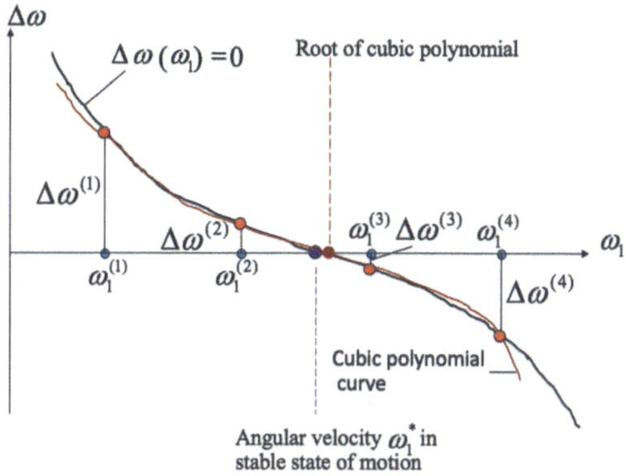

Figure 3.14 Root of polynomial.

polynomial, four points are needed. These can be constructed as

$$\begin{cases} \omega_1^{(1)} = \omega_a \\ \omega_1^{(2)} = \omega_a + \dfrac{1}{3}(\omega_b - \omega_a) \\ \omega_1^{(3)} = \omega_a + \dfrac{2}{3}(\omega_b - \omega_a) \\ \omega_1^{(4)} = \omega_b, \end{cases} \tag{3.54}$$

where ω_b and ω_a are the two ends of a bracket with opposed sign of $\Delta\omega$. Then, the cubic polynomial has the following form:

$$\Delta\omega(\omega_1) = c_0 + c_1\omega_1 + c_2\omega_1^2 + c_3\omega_1^3. \tag{3.55}$$

The coefficient c_i are computed by solving the following equations:

$$\begin{cases} c_0 + c_1\omega_1^{(1)} + c_2\left(\omega_1^{(1)}\right)^2 + c_3\left(\omega_1^{(1)}\right)^3 = \Delta\omega^{(1)} \\ c_0 + c_1\omega_1^{(2)} + c_2\left(\omega_1^{(2)}\right)^2 + c_3\left(\omega_1^{(2)}\right)^3 = \Delta\omega^{(2)} \\ c_0 + c_1\omega_1^{(3)} + c_2\left(\omega_1^{(3)}\right)^2 + c_3\left(\omega_1^{(3)}\right)^3 = \Delta\omega^{(3)} \\ c_0 + c_1\omega_1^{(4)} + c_2\left(\omega_1^{(4)}\right)^2 + c_3\left(\omega_1^{(4)}\right)^3 = \Delta\omega^{(4)}. \end{cases} \tag{3.56}$$

An approximation of the steady angular speed then can be obtained by solving the following equation:

$$\Delta\omega(\omega_1) = c_0 + c_1\omega_1 + c_2\omega_1^2 + c_3\omega_1^3 = 0, \tag{3.57}$$

where $\Delta\omega(\omega_1)$ merely indicates the dependency of $\Delta\omega$ on ω_1. If the approximation meets a prescribed accuracy criterion, stop and take it as the solution. If not, narrow the bracket and repeat the procedures until a satisfactory solution is found.

3.7 Smoothening velocity fluctuation

3.7.1 Traditional flywheel

As indicated in previous section, the speed of a machine during steady stage fluctuates periodically. The degree of speed fluctuation is commonly represented by a coefficient δ, termed the coefficient of speed fluctuation, which is defined as (Mata et al., 2016)

$$\delta = \frac{\omega_{max} - \omega_{min}}{\omega_m}, \tag{3.58}$$

where ω_{max}, ω_{min} and ω_m represent the maximum, the minimum, and the mean speeds, respectively.

In practice, the mean speed is often replaced by the median speed because of the resulting ease of computation:

$$\omega_m = \frac{1}{2}\left(\omega_{max} + \omega_{min}\right). \tag{3.59}$$

Substituting Eq. (3.59) into Eq. (3.58) gives another form of δ:

$$\delta = \frac{2(\omega_{max} - \omega_{min})}{\omega_{max} + \omega_{min}}. \tag{3.60}$$

Too large a speed fluctuation has a series of negative effects on machine performance and/or safety. For example, speed fluctuation of a generator affects the voltage quality; unstable speed in machine tools degrades the machining quality. The recommended values of δ for various machines can be found in relevant manuals and textbooks (Mata et al., 2016).

The speed variation of a machine within one cycle can be illustrated as in Fig. 3.15. In one cycle, the work done by the driving torque (M_d) equals that by the resistant torque (M_r). However, the instant values of M_d and M_r vary with time or machine position as shown in Fig. 3.15(a). In some sections, such as the one between 0 and 1, M_d is greater than M_r; thus, the machine gains energy and speeds up as shown in Fig. 3.15(c). In some other sections, for instance, the one between 1 and 2, M_d is lower than M_r; the machine loses energy (negative work), and the speed drops down.

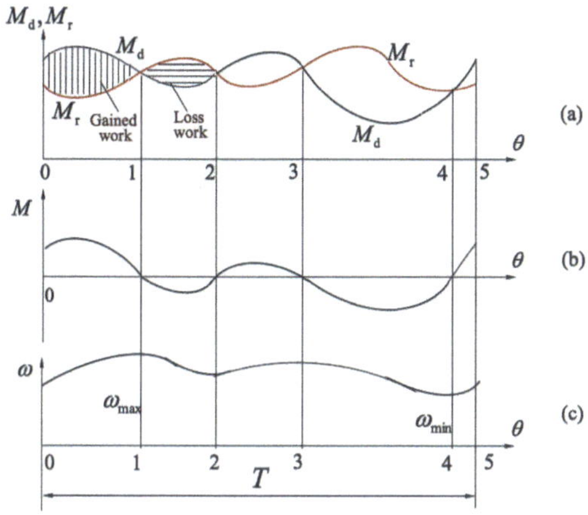

Figure 3.15 Machine-speed fluctuation.

To smoothen the speed variation, the most common method is to install a flywheel. The principle of flywheel is explained next.

If we look at a cycle between θ_0 and θ, the change of kinetic energy stored in the machine equals the work done by the equivalent torque, that is,

$$\Delta E_k = \int_{\theta_0}^{\theta} (M_d - M_r)\, d\theta. \tag{3.61}$$

If a flywheel having a moment of inertia J_f is installed, the change in kinetic energy is

$$\Delta E_k = \frac{1}{2}(J + J_f)\,\omega^2 - \frac{1}{2}(J_0 + J_f)\,\omega_0^2, \tag{3.62}$$

where J and J_0 are the equivalent moments of inertia at θ and θ_0, respectively. ω and ω_0 are the equivalent angular speeds at positions θ, and θ_0.

Generally, the flywheel's moment of inertia J_f is much larger than J and J_0. So, the difference between J and J_0 can be neglected. Thus, J_0 can be approximated by J here. Eq. (3.62), then, becomes

$$\Delta E_k = \frac{1}{2}(J + J_f)\left(\omega^2 - \omega_0^2\right). \tag{3.63}$$

Obviously, for a certain ΔE_k, the significant increase of moment of inertia by adding a flywheel leads to lower speed variation. In theory, a flywheel is a reservoir of kinetic energy that absorbs the increased kinetic energy when $M_d > M_r$, and releases it when $M_d < M_r$.

The other benefit of a flywheel is that it can flatten out the required driving torque, M_d, and decrease the size of the motor or engine required. Taking a mechanical press in Fig. 3.16 as an example, the typical working-load cycles in stamping operation are shown in Fig. 3.17(a). In the majority of the times, the resistant load on the machine is only the rather small friction. The stamping time requiring large driving torque comprises a very small part of a working cycle.

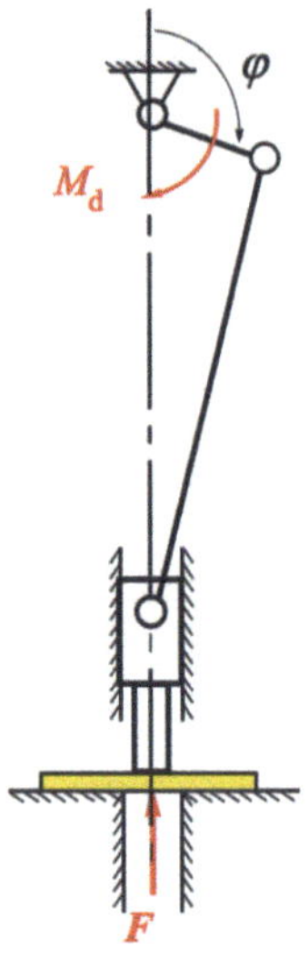

Figure 3.16 Stamping press.

If the selection is made based on the very large stamping load, the motor would be oversized for the nonstamping operation. This is unreasonable from either the technical or economic viewpoint. If a flywheel is used, however, the flywheel will store energy during nonstamping operation and release it when stamping. As shown in Fig. 3.17(b), the equivalent driving torque is determined on the basis that the gained work during nonstamping equals the lost work during the short stamping period within one cycle. This greatly decreases the size of the motor required. The machine speed, in this case, experiences large fluctuations. However, speed variation is not a big concern for stamping.

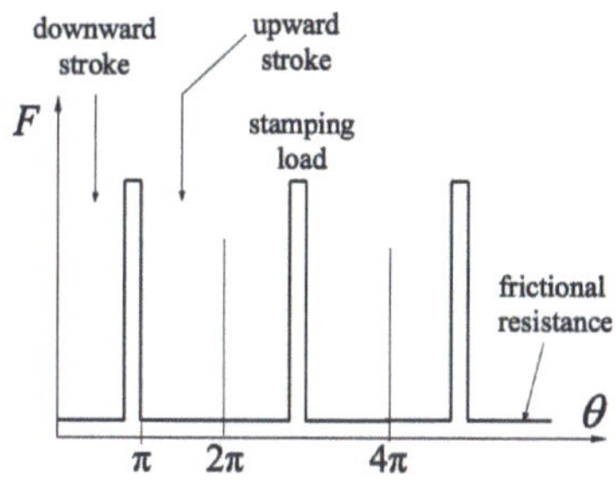

(a) Load

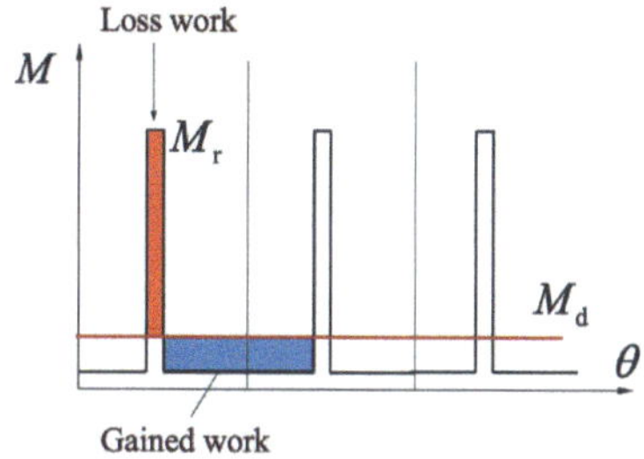

(b) Torque

Figure 3.17 Loads and torques of the press.

3.7.2 Sizing flywheels

Several approximate formulations are available for sizing a flywheel based on permissible δ (Norton, 2004). There are also some numerical methods (Mata et al., 2016).

For a given δ, one can also use the methods introduced in Section 3.5 of this book to obtain the stable machine speed, and then check if the requirement on δ is met. If not, modify the moment of inertia of the flywheel and repeat the procedures until the requirement on δ is satisfied. The procedure of this method can be summarized thus:

- For a given permissible coefficient of speed fluctuation $[\delta]$, estimate a value of moment of inertial $J_{\mathrm{f}}^{(0)}$.
- Use the methods in Section 3.5 to calculate the machine steady operation speed.
- Compute δ (Eq. (3.60)), and check if $|\delta - [\delta]| < \epsilon$.
- If yes, stop.
- If not, modify J_{f} as $J_{\mathrm{f}}^{(i+1)} = (J_{\mathrm{f}}^{(i)} + J_0)\delta/[\delta] - J_0$ and repeat the previous step until a satisfactory J_{f} is found.

Here, ϵ is a given small number controlling the accuracy, and J_0 is the average moment of inertia of the moving parts of the machine before installing the flywheel.

3.7.3 An innovative mini-flywheel

A flywheel introduced here can control the speed fluctuation of a machine within a given range. However, it can not completely eliminate the speed fluctuation. In addition, a small $[\delta]$ often leads to very high J_f, which translates to high cost and other consequences. Huang and Shi (1990) proposed a unique flywheel concept, named mini-flywheel, which has the following two outstanding superiorities over the traditional flywheel:

- having the potential to completely eliminate the speed fluctuation;
- having dimensions much smaller than traditional flywheels. A case study indicates that the moment of inertia of this type of flywheel is only about 1/30 of the traditional one (Huang and Shi, 1990).

The flywheel is connected to a mechanism through either a dyad (Fig. 3.18(a)) or a noncircular gear pair (Fig. 3.18(b)). The angular speed of

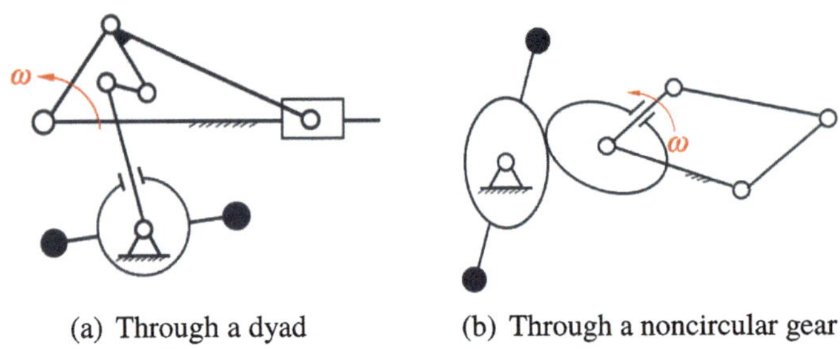

Figure 3.18 Mini-flywheel.

the flywheel varies and so does the kinetic energy. Through careful design of the speed ratio of the dyad or the noncircular gears, the kinetic energy of the flywheel could be made to exactly compensate for the gain or loss of work. One obvious difference between this type of flywheel and the traditional ones is that it smooths out the machine speed through the varying speed of the flywheel, while the traditional ones achieve the same objective through adding to the moment of inertia.

CHAPTER 4

Dynamics of machines with multiple DOFs

4.1 Introduction

In various automatic machines and robots, multi-inputs are common, meaning that these machines have multiple DOFs. In the previous chapter, the dynamic analysis of machines of single DOF using Lagrange's equation was introduced. For such a machine, the whole machine can be represented by an equivalent link. In this chapter, dynamic modeling of more general multiple-DOF machines with Lagrange's equation will be presented with a focus on machines having two DOFs.

The chapter is organized as follows:

- In Section 4.2, the modeling of two DOF machines is introduced. With a five-bar linkage as an example, the steps to develop the equation of motion with the Lagrange's equation are presented in detail.
- Using the procedures introduced in a previous section, the dynamics of a two-link manipulator are modeled and analyzed in Section 4.3.
- Section 4.4 offers a brief introduction to and review of the dynamics of robotic manipulators.

4.2 Dynamics of machines of two DOFs

Typical machines with two DOFs include five-bar linkages and planetary differential drives. The methods and procedures developed for two-DOF machines can be easily generalized to machines having more DOFs. In view of this, a five-bar mechanism, as shown in Fig. 4.1, is used as an example to illustrate the detailed procedures.

As shown in the figure, there are two inputs, link 2 and 4. Thus, the generalized coordinates can be chosen as $q_1 = \varphi_2$ and $q_2 = \varphi_5$. Lagrange's equation for a two-DOF machine is

$$\frac{\mathrm{d}}{\mathrm{d}t}\left(\frac{\partial E_{\mathrm{k}}}{\partial \dot{q}_i}\right) - \frac{\partial E_{\mathrm{k}}}{\partial q_i} + \frac{\partial E_{\mathrm{p}}}{\partial q_i} = Q_i \quad (i = 1, 2). \tag{4.1}$$

Machinery Dynamics
https://doi.org/10.1016/B978-0-12-815785-5.00013-3

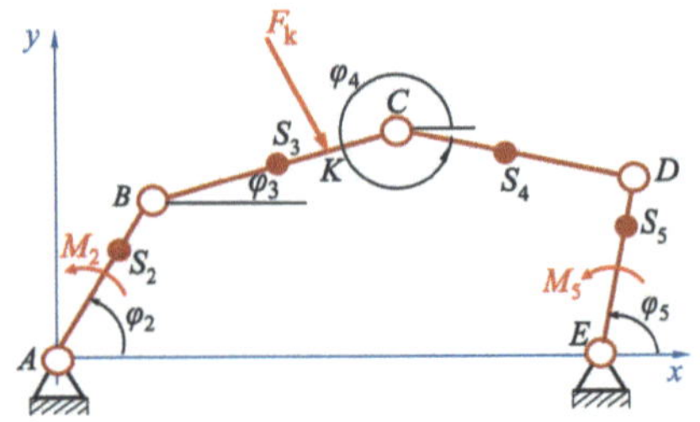

Figure 4.1 A five-bar linkage.

In the following, the gravitational potential is neglected. Thus, the potential energy can be dropped because no elastic elements are involved.

4.2.1 Kinematics

To calculate the kinetic energy, kinematic analysis should be first conducted. For a link moving within a plane, its position can be described by the coordinates of the link's center of mass, x_{Sj}, y_{Sj}, and the position angle φ_j of the link. They all depend on the generalized coordinates q_1 and q_2. Without loss of generality, we represent them as

$$\begin{cases} \varphi_j = \varphi_j(q_1, q_2) \\ x_{Sj} = x_{Sj}(q_1, q_2) \\ y_{Sj} = y_{Sj}(q_1, q_2). \end{cases} \tag{4.2}$$

The subscripts j and S denote the link number and the center of mass, respectively. The variables can be seen clearly in Fig. 4.1.

Differentiating Eq. (4.2) with respect to time gives the velocities:

$$\begin{cases} \dot{\varphi}_j = \dfrac{\partial \varphi_j}{\partial q_1}\dot{q}_1 + \dfrac{\partial \varphi_j}{\partial q_2}\dot{q}_2 \\ \dot{x}_{Sj} = \dfrac{\partial x_{Sj}}{\partial q_1}\dot{q}_1 + \dfrac{\partial x_{Sj}}{\partial q_2}\dot{q}_2 \\ \dot{y}_{Sj} = \dfrac{\partial y_{Sj}}{\partial q_1}\dot{q}_1 + \dfrac{\partial y_{Sj}}{\partial q_2}\dot{q}_2. \end{cases} \tag{4.3}$$

The linear velocity of S_j is then calculated as:

$$v_{Sj} = \sqrt{\dot{x}_{Sj}^2 + \dot{y}_{Sj}^2}. \tag{4.4}$$

4.2.2 Kinetic energy

With the angular and linear velocities of each individual moving link known, the total kinetic energy of the machine is calculated as

$$E_k = \sum_{j=2}^{l} \frac{1}{2}\left(m_j v_{Sj}^2 + J_j \dot{\varphi}_j^2\right), \tag{4.5}$$

where m_j and J_j are the mass and moment of inertia about S_j of the link, respectively. Taking advantage of Eqs. (4.2) through (4.4), one has

$$\begin{aligned} E_k = &\sum_{j=2}^{l} \frac{1}{2}\left\{ m_j\left[\left(\frac{\partial x_{Sj}}{\partial q_1}\right)^2 + \left(\frac{\partial y_{Sj}}{\partial q_1}\right)^2\right] + J_j\left(\frac{\partial \varphi_j}{\partial q_1}\right)^2\right\}\dot{q}_1^2 + \\ &\sum_{j=2}^{l} \frac{1}{2}\left\{ m_j\left[\left(\frac{\partial x_{Sj}}{\partial q_2}\right)^2 + \left(\frac{\partial y_{Sj}}{\partial q_2}\right)^2\right] + J_j\left(\frac{\partial \varphi_j}{\partial q_2}\right)^2\right\}\dot{q}_2^2 + \\ &\sum_{j=2}^{l} \frac{1}{2}\left\{ m_j\left[\left(\frac{\partial x_{Sj}}{\partial q_1}\frac{\partial x_{Sj}}{\partial q_2}\right) + \left(\frac{\partial y_{Sj}}{\partial q_1}\frac{\partial y_{Sj}}{\partial q_2}\right)\right] + J_j\left(\frac{\partial \varphi_j}{\partial q_1}\frac{\partial \varphi_j}{\partial q_2}\right)\right\}\dot{q}_1\dot{q}_2. \end{aligned} \tag{4.6}$$

This can be shortened to

$$E_k = \frac{1}{2}J_{11}\dot{q}_1^2 + J_{12}\dot{q}_1\dot{q}_2 + \frac{1}{2}J_{22}\dot{q}_2^2, \tag{4.7}$$

where

$$\begin{cases} J_{11} = \sum_{j=2}^{l}\left[m_j\left(\frac{\partial x_{Sj}}{\partial q_1}\right)^2 + m_j\left(\frac{\partial y_{Sj}}{\partial q_1}\right)^2 + J_j\left(\frac{\partial \varphi_j}{\partial q_1}\right)^2\right] \\ J_{22} = \sum_{j=2}^{l}\left[m_j\left(\frac{\partial x_{Sj}}{\partial q_2}\right)^2 + m_j\left(\frac{\partial y_{Sj}}{\partial q_2}\right)^2 + J_j\left(\frac{\partial \varphi_j}{\partial q_2}\right)^2\right] \\ J_{12} = \sum_{j=2}^{l}\left[m_j\left(\frac{\partial x_{Sj}}{\partial q_1}\frac{\partial x_{Sj}}{\partial q_2}\right) + m_j\left(\frac{\partial y_{Sj}}{\partial q_1}\frac{\partial y_{Sj}}{\partial q_2}\right) + J_j\left(\frac{\partial \varphi_j}{\partial q_1}\frac{\partial \varphi_j}{\partial q_2}\right)\right]. \end{cases} \tag{4.8}$$

J_{11}, J_{22}, and J_{12} constitute the unit of moment of inertia in the case where q_j are rotation angles; thus, they are known as equivalent moments of inertia (or reduced moments of inertia). Obviously they are functions of the mechanism's geometry, mass, moment of inertia, and the generalized coordinates q_j, but independent of the generalized velocities. In addition, it is seen from Eq. (4.8) that they depend on both generalized coordinates, q_1

and q_2, meaning it is impossible to translate a multiple-DOF machine to only one equivalent link as was done in the previous chapter.

4.2.3 Generalized forces

The determination of generalized forces is based on the virtual work principle as stated:

$$\delta W = Q_1 \delta q_1 + Q_2 \delta q_2, \tag{4.9}$$

where δq_1 and δq_2 are the virtual displacements and δW is the virtual work done by the generalized forces along the two virtual displacements. The virtual work δW can also be calculated from the external forces and moments as:

$$\delta W = \sum_{k=1}^{m} \left(F_k \delta s_k \cos \alpha_k \right) + \sum_{j=1}^{n} \left(\pm M_j \delta \varphi_j \right), \tag{4.10}$$

where δs_k and $\delta \varphi_j$ are the translational and rotational virtual displacements at the point of F_k and M_j. α_k is the angle between F_k and δs_k. δs_k and $\delta \varphi_j$ can be represented as:

$$\begin{cases} \delta s_k = \dfrac{\partial s_k}{\partial q_1} \delta q_1 + \dfrac{\partial s_k}{\partial q_2} \delta q_2 \\ \delta \varphi_j = \dfrac{\partial \varphi_j}{\partial q_1} \delta q_1 + \dfrac{\partial \varphi_j}{\partial q_2} \delta q_2. \end{cases} \tag{4.11}$$

Substitution of Eq. (4.11) into Eq. (4.10) gives

$$\begin{aligned} \delta W = & \sum_{k=1}^{m} \left[F_k \left(\frac{\partial s_k}{\partial q_1} \delta q_1 + \frac{\partial s_k}{\partial q_2} \delta q_2 \right) \cos \alpha_k \right] + \\ & \sum_{j=1}^{n} \left[\pm M_j \left(\frac{\partial \varphi_j}{\partial q_1} \delta q_1 + \frac{\partial \varphi_j}{\partial q_2} \delta q_2 \right) \right]. \end{aligned} \tag{4.12}$$

Comparing Eq. (4.12) and Eq. (4.9) yields

$$\begin{cases} Q_1 = \displaystyle\sum_{k=1}^{m} \left[F_k \frac{\partial s_k}{\partial q_1} \cos \alpha_k \right] + \sum_{j=1}^{n} \left[\pm M_j \frac{\partial \varphi_j}{\partial q_1} \right] \\ Q2 = \displaystyle\sum_{k=1}^{m} \left[F_k \frac{\partial s_k}{\partial q_2} \cos \alpha_k \right] + \sum_{j=1}^{n} \left[\pm M_j \frac{\partial \varphi_j}{\partial q_2} \right]. \end{cases} \tag{4.13}$$

These are the equations used to calculate the generalized forces.

4.2.4 Governing equation of motion

With the equations of E_k, Q_1, and Q_2, we can have the following:

$$\begin{cases} \dfrac{\partial E_k}{\partial q_1} = \dfrac{1}{2}\dot{q}_1^2\dfrac{\partial J_{11}}{\partial q_1} + \dot{q}_1\dot{q}_2\dfrac{\partial J_{12}}{\partial q_1} + \dfrac{1}{2}\dot{q}_2^2\dfrac{\partial J_{22}}{\partial q_1} \\ \dfrac{\partial E_k}{\partial q_2} = \dfrac{1}{2}\dot{q}_2^2\dfrac{\partial J_{22}}{\partial q_2} + \dot{q}_1\dot{q}_2\dfrac{\partial J_{12}}{\partial q_2} + \dfrac{1}{2}\dot{q}_1^2\dfrac{\partial J_{11}}{\partial q_2} \\ \dfrac{\partial E_k}{\partial \dot{q}_1} = J_{11}\dot{q}_1 + J_{12}\dot{q}_2 \\ \dfrac{\partial E_k}{\partial \dot{q}_2} = J_{22}\dot{q}_2 + J_{12}\dot{q}_1 \\ \dfrac{d}{dt}\left(\dfrac{\partial E_k}{\partial \dot{q}_1}\right) = J_{11}\ddot{q}_1 + J_{12}\ddot{q}_2 + \left(\dfrac{\partial J_{11}}{\partial q_1}\right)\dot{q}_1^2 + \left(\dfrac{\partial J_{11}}{\partial q_2}\right)\dot{q}_1\dot{q}_2 + \\ \qquad \left(\dfrac{\partial J_{12}}{\partial q_1}\right)\dot{q}_1\dot{q}_2 + \left(\dfrac{\partial J_{12}}{\partial q_2}\right)\dot{q}_2^2 \\ \dfrac{d}{dt}\left(\dfrac{\partial E_k}{\partial \dot{q}_2}\right) = J_{12}\ddot{q}_1 + J_{22}\ddot{q}_2 + \left(\dfrac{\partial J_{22}}{\partial q_1}\right)\dot{q}_1\dot{q}_2 + \left(\dfrac{\partial J_{22}}{\partial q_2}\right)\dot{q}_2^2 + \\ \qquad \left(\dfrac{\partial J_{12}}{\partial q_1}\right)\dot{q}_1^2 + \left(\dfrac{\partial J_{12}}{\partial q_2}\right)\dot{q}_1\dot{q}_2 \end{cases} \tag{4.14}$$

Substituting these equations into Eq. (4.1) and rearranging the terms yields the following governing equation of motion:

$$\begin{cases} J_{11}\ddot{q}_1 + J_{12}\ddot{q}_2 + \dfrac{1}{2}\dfrac{\partial J_{11}}{\partial q_1}\dot{q}_1^2 + \dfrac{\partial J_{11}}{\partial q_2}\dot{q}_1\dot{q}_2 + \left(\dfrac{\partial J_{12}}{\partial q_2} - \dfrac{1}{2}\dfrac{\partial J_{22}}{\partial q_1}\right)\dot{q}_2^2 = Q_1 \\ J_{22}\ddot{q}_2 + J_{12}\ddot{q}_1 + \left(\dfrac{\partial J_{12}}{\partial q_1} - \dfrac{1}{2}\dfrac{\partial J_{11}}{\partial q_2}\right)\dot{q}_1^2 + \dfrac{\partial J_{22}}{\partial q_1}\dot{q}_1\dot{q}_2 + \dfrac{1}{2}\dfrac{\partial J_{22}}{\partial q_2}\dot{q}_2^2 = Q_2. \end{cases} \tag{4.15}$$

Generally, it is difficult to find closed-form solutions to these second-order ordinary differential equations (ODEs). Numerical methods, thus, have to be relied on. For example, various Runge–Kutta methods are widely accepted in science and engineering. These numerical methods have been included in most commercial software packages, such as Matlab®. Interested readers are directed to relevant references (Chapra and Canale, 2011; Pratap, 2010).

Example 4.1. A differential gear drive is shown in Fig. 4.2. The gears are numbered as shown, and the tooth number of the gears are represented as z_1, z_2, z_3, and z_4, respectively. Gears 1 and 4 have moments of inertia

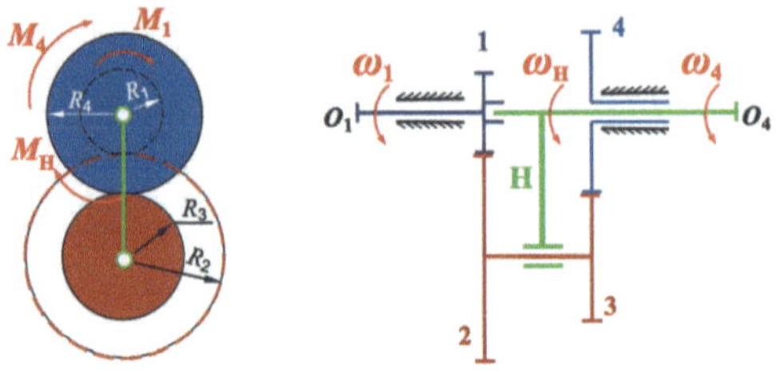

Figure 4.2 A differential.

about their axis J_1 and J_4. Gears 2 and 3 are connected and have a combined mass m_{23} and a moment of inertia J_2. The arm (or carrier) H has a moment of inertia about its axis J'_H. M_1, M_4, and M_H are the external torques on gear 1, gear 4, and the carrier H. Develop the equation of motion of the machine under these conditions.

Solution. This is a differential drive having two DOFs. The positive direction of the angular velocities and moments are defined as shown. Take $q_1 = \varphi_1$, $q_2 = \varphi_4$, the rotational displacements of gears 1 and 4, as the generalized coordinates. The process of solution can be conducted by the following procedure.

1. Kinematic Analysis

According to the kinematics of planetary gear trains, we have

$$\frac{\omega_1 - \omega_H}{\omega_4 - \omega_H} = \frac{\dot{q}_1 - \omega_H}{\dot{q}_2 - \omega_H} = (-1)^2 \frac{z_4 z_2}{z_1 z_3} = \frac{R_4 R_2}{R_1 R_3}, \tag{a}$$

where R_i ($i = 1$ to 4) is the radius of pitch circle of the gear i. Then, ω_H can be calculated by

$$\omega_H = \frac{R_1 R_3}{R_1 R_3 - R_2 R_4}\dot{q}_1 - \frac{R_2 R_4}{R_1 R_3 - R_2 R_4}\dot{q}_2. \tag{b}$$

Taking advantage of $R_1 + R_2 = R_3 + R_4$, we have

$$R_1 R_3 - R_2 R_4 = R_1 R_3 - R_2(R_1 + R_2 - R_3) = (R_1 + R_2)(R_3 - R_2). \tag{c}$$

ω_H can then be recast into

$$\omega_H = \frac{R_1 R_3}{(R_1 + R_2)(R_3 - R_2)}\dot{q}_1 - \frac{R_2 R_4}{(R_1 + R_2)(R_3 - R_2)}\dot{q}_2. \tag{d}$$

Similarly, we have

$$\frac{\omega_2 - \omega_H}{\omega_1 - \omega_H} = -\frac{R_1}{R_2} \tag{e}$$

and

$$\omega_2 = \frac{R_1}{R_3 - R_2}\dot{q}_1 - \frac{R_4}{R_3 - R_2}\dot{q}_2. \tag{f}$$

2. Equivalent Moments of Inertia
With angular velocities in hand, we can calculate the potential energy E_k as

$$\begin{aligned} E_k =& \frac{1}{2}J_1\omega_1^2 + \frac{1}{2}J_2\omega_2^2 + \frac{1}{2}m_{23}(R_1+R_2)^2\omega_H^2 + \frac{1}{2}J'_H\omega_H^2 + \frac{1}{2}J_4\omega_4^2 \\ =& \frac{1}{2}J_1\dot{q}_1^2 + \frac{1}{2}J_2\omega_2^2 + \frac{1}{2}\left[m_{23}(R_1+R_2)^2 + J'_H\right]\omega_H^2 + \frac{1}{2}J_4\dot{q}_2^2. \end{aligned} \tag{g}$$

Letting $J_H = m_{23}(R_1 + R_2)^2 + J'_H$ and substituting ω_H in Eq. (d) and ω_2 in Eq. (f) into the previous equation, the following can be obtained after rearrangement;

$$E_k = \frac{1}{2}J_{11}\dot{q}_1^2 + J_{12}\dot{q}_1\dot{q}_2 + \frac{1}{2}J_{22}\dot{q}_2^2, \tag{h}$$

where J_{11}, J_{12}, and J_{22} are the equivalent moments of inertia. They have the following forms:

$$\begin{cases} J_{11} = J_1 + J_2\left(\dfrac{R_1}{R_3 - R_2}\right)^2 + J_H\left[\dfrac{R_1R_3}{(R_1+R_2)(R_3-R_2)}\right]^2 \\ J_{12} = -J_2\dfrac{R_1R_4}{(R_3-R_2)^2} - J_H\dfrac{R_1R_2R_3R_4}{(R_1+R_2)^2(R_3-R_2)^2} \\ J_{22} = J_4 + J_2\left(\dfrac{R_4}{R_3-R_2}\right)^2 + J_H\left[\dfrac{R_2R_4}{(R_1+R_2)(R_3-R_2)}\right]^2. \end{cases} \tag{i}$$

It should be noted that J_{11}, J_{12}, and J_{22} are all constants because of the constant gear ratios.

3. Generalized Forces
The power of the external forces are

$$\begin{aligned} P =& M_1\omega_1 + M_H\omega_H + M_4\omega_4 \\ =& \left[M_1 + \frac{R_1R_3}{(R_1+R_2)(R_3-R_2)}M_H\right]\dot{q}_1 + \end{aligned}$$

$$\left[M_4 - \frac{R_2R_4}{(R_1+R_2)(R_3-R_2)}M_{\rm H}\right]\dot{q}_2. \tag{j}$$

We can obtain the Q_1 and Q_2 immediately as

$$\begin{cases} Q_1 = M_1 + \dfrac{R_1R_3}{(R_1+R_2)(R_3-R_2)}M_{\rm H} \\ Q_2 = M_4 - \dfrac{R_2R_4}{(R_1+R_2)(R_3-R_2)}M_{\rm H}. \end{cases} \tag{k}$$

4. Governing Equation of Motion
The equation of motion can be established straightforwardly by plugging the relevant terms into Eq. (4.15) as

$$\begin{cases} J_{11}\ddot{q}_1 + J_{12}\ddot{q}_2 = Q_1 \\ J_{12}\ddot{q}_1 + J_{22}\ddot{q}_2 = Q_2. \end{cases} \tag{l}$$

4.3 Dynamics of two-link manipulators

An important part in an industrial robotic manipulator is the arm, the simplest consisting of only two links connected by revolute joints as shown in Fig. 4.3. In this section, we will show the procedures for developing the dynamic equation using the Lagrange's equation. This device consists of the upper arm AB, the lower arm BC, and the gripper C holding an object of m_2. At joint A, a torque M_1 is input to drive the whole mechanism. A motor and transmission to input a torque M_2 at B are installed at the end of the upper arm to rotate the lower arm. The motor and the transmission have a total mass of m_1. In addition, the lengths of the upper and lower arms are l_1 and l_2, respectively.

The arm moves only in a plane, and the weight of the arms is neglected. Obviously, this is a system having two DOFs. To develop the equation, we establish the $A-xy$ frame as shown and choose θ_1 and θ_2 as the generalized coordinates. The coordinates of the gripper are x_C and y_C. It is straightforward to calculate these gripper coordinates as

$$\begin{cases} x_C = l_1\sin\theta_1 + l_2\sin(\theta_1+\theta_2) \\ y_C = l_1\cos\theta_1 + l_2\cos(\theta_1+\theta_2). \end{cases} \tag{4.16}$$

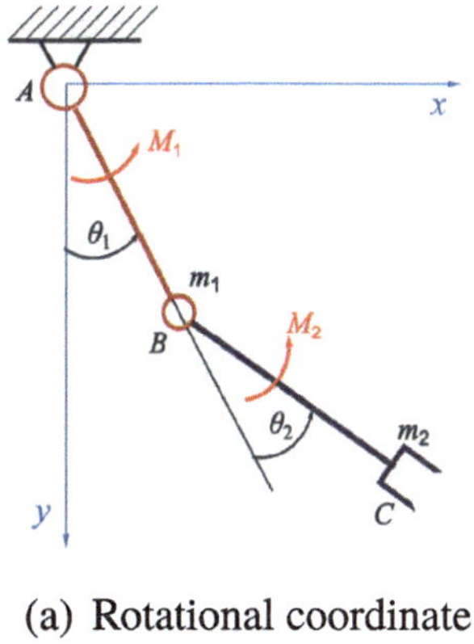

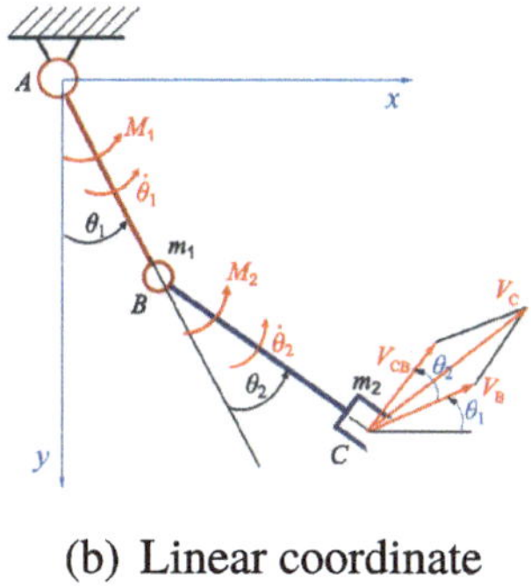

Figure 4.3 Two-bar manipulator.

Differentiating gives the velocity as

$$\begin{cases} \dot{x}_C = \dot{\theta}_1 l_1 \cos\theta_1 + (\dot{\theta}_1 + \dot{\theta}_2) l_2 \cos(\theta_1 + \theta_2) \\ \dot{y}_C = -\dot{\theta}_1 l_1 \sin\theta_1 - (\dot{\theta}_1 + \dot{\theta}_2) l_2 \sin(\theta_1 + \theta_2). \end{cases} \tag{4.17}$$

Then, the values of velocity at points B and C are obtained as

$$\begin{cases} v_B = l_1 \dot{\theta}_1 \\ v_C = \sqrt{\dot{x}_C^2 + \dot{y}_C^2} = \sqrt{l_1^2 \dot{\theta}_1^2 + l_2^2 (\dot{\theta}_1 + \dot{\theta}_2)^2 + 2 l_1 l_2 \dot{\theta}_1 (\dot{\theta}_1 + \dot{\theta}_2) \cos\theta_2}. \end{cases} \tag{4.18}$$

The kinetic energy of the mechanism is

$$\begin{aligned} E_k &= \frac{1}{2} m_1 v_B^2 + \frac{1}{2} m_2 v_c^2 \\ &= \frac{1}{2}(m_1 + m_2) l_1^2 \dot{\theta}_1^2 + m_2 l_1 l_2 \dot{\theta}_1 (\dot{\theta}_1 + \dot{\theta}_2) \cos\theta_2 + \frac{1}{2} m_2 l_2^2 (\dot{\theta}_1 + \dot{\theta}_2)^2. \end{aligned} \tag{4.19}$$

The potential energy can not be neglected in this case, but it is if point A is taken as the zero potential:

$$E_p = -m_1 g l_1 \cos\theta_1 - m_2 g \left[l_1 \cos\theta_1 + l_2 \cos(\theta_1 + \theta_2) \right]. \tag{4.20}$$

Obviously, E_k depends on θ_1, θ_2, $\dot{\theta}_1$, and $\dot{\theta}_2$, while E_p is a function only of θ_1 and θ_2.

From Eqs. (4.19) and (4.20) the following can be obtained:

$$\begin{cases} \dfrac{\partial E_k}{\partial \dot{\theta}_1} = (m_1+m_2)l_1^2\dot{\theta}_1 + m_2l_1l_2\dot{\theta}_2\cos\theta_2 + 2m_2l_1l_2\dot{\theta}_1\cos\theta_2 + m_2l_2^2\left(\dot{\theta}_1+\dot{\theta}_2\right) \\ \dfrac{\mathrm{d}}{\mathrm{d}t}\left(\dfrac{\partial E_k}{\partial \dot{\theta}_1}\right) = (m_1+m_2)l_1^2\ddot{\theta}_1 + m_2l_1l_2\ddot{\theta}_2\cos\theta_2 - m_2l_1l_2\dot{\theta}_2^2\sin\theta_2 \\ \qquad +2m_2l_1l_2\ddot{\theta}_1^2\cos\theta_2 - 2m_2l_1l_2\dot{\theta}_1\dot{\theta}_2\sin\theta_2 + m_2l_2^2\left(\ddot{\theta}_1+\ddot{\theta}_2\right) \\ \dfrac{\partial E_k}{\partial \dot{\theta}_2} = m_2l_2^2\left(\dot{\theta}_1+\dot{\theta}_2\right) + m_2l_1l_2\dot{\theta}_1\cos\theta_2 \\ \dfrac{\mathrm{d}}{\mathrm{d}t}\left(\dfrac{\partial E_k}{\partial \dot{\theta}_2}\right) = m_2l_2^2\left(\ddot{\theta}_1+\ddot{\theta}_2\right) + m_2l_1l_2\ddot{\theta}_1\cos\theta_2 - m_2l_1l_2\dot{\theta}_1\dot{\theta}_2\sin\theta_2 \\ \dfrac{\partial E_k}{\partial \theta_1} = 0 \\ \dfrac{\partial E_k}{\partial \theta_2} = -m_2l_1l_2\dot{\theta}_1\left(\dot{\theta}_1+\dot{\theta}_2\right)\sin\theta_2 \\ \dfrac{\partial E_p}{\partial \theta_1} = (m_1+m_2)gl_1\sin\theta_1 + m_2gl_2\sin(\theta_1+\theta_2) \\ \dfrac{\partial E_p}{\partial \theta_2} = m_2gl_2\sin(\theta_1+\theta_2). \end{cases} \tag{4.21}$$

M_1 and M_2 are directly applied to the two generalized coordinates θ_1 and θ_2, respectively. Thus, we have

$$\begin{cases} Q_1 = M_1 \\ Q_2 = M_2. \end{cases} \tag{4.22}$$

Substituting Eqs. (4.21) and (4.22) into Eq. (4.1) and rearranging terms in shorter form yields the following governing equation of motion of the arm:

$$\begin{cases} D_{11}\ddot{\theta}_1 + D_{12}\ddot{\theta}_2 + D_{111}\dot{\theta}_1^2 + D_{122}\dot{\theta}_2^2 + D_{112}\dot{\theta}_1\dot{\theta}_2 + D_{121}\dot{\theta}_2\dot{\theta}_1 + D_1 = M_1 \\ D_{21}\ddot{\theta}_1 + D_{22}\ddot{\theta}_2 + D_{211}\dot{\theta}_1^2 + D_{222}\dot{\theta}_2^2 + D_{212}\dot{\theta}_1\dot{\theta}_2 + D_{221}\dot{\theta}_2\dot{\theta}_1 + D_2 = M_2. \end{cases} \tag{4.23}$$

This can then be cast into more compact matrix form:

$$\begin{bmatrix} D_{11} & D_{12} \\ D_{21} & D_{22} \end{bmatrix} \begin{Bmatrix} \ddot{\theta}_1 \\ \ddot{\theta}_2 \end{Bmatrix} + \begin{bmatrix} D_{111} & D_{122} \\ D_{211} & D_{222} \end{bmatrix} \begin{Bmatrix} \dot{\theta}_1^2 \\ \dot{\theta}_2^2 \end{Bmatrix} +$$

$$\begin{bmatrix} D_{112} & D_{121} \\ D_{212} & D_{221} \end{bmatrix} \begin{Bmatrix} \dot{\theta}_1\dot{\theta}_2 \\ \dot{\theta}_2\dot{\theta}_1 \end{Bmatrix} + \begin{Bmatrix} D_1 \\ D_2 \end{Bmatrix} = \begin{Bmatrix} M_1 \\ M_2 \end{Bmatrix}. \tag{4.24}$$

Each term in this equation has a specific physical meaning, which is outlined next.

1. Inertial term: The first term represents the inertial force. The two diagonal coefficients, D_{11} and D_{22}, are the generalized moment of inertia at joint i and j, respectively. They have the following forms:

$$\begin{cases} D_{11} = (m_1 + m_2)l_1^2 + m_2 l_2^2 + 2m_2 l_1 l_2 \cos\theta_2 \\ D_{22} = m_2 l_2^2. \end{cases} \tag{4.25}$$

   The other two coefficients, D_{12} and D_{21}, are the coupling moment of inertia between joint i and j. Their forms are

$$D_{12} = D_{21} = m_2 l_2^2 + m_2 l_1 l_2 \cos\theta_2. \tag{4.26}$$

2. Centrifugal term: The second term represents the centrifugal force. The coefficients are

$$\begin{cases} D_{111} = 0 \\ D_{122} = -m_2 l_1 l_2 \sin\theta_2 \\ D_{222} = 0 \\ D_{211} = m_2 l_1 l_2 \sin\theta_2. \end{cases} \tag{4.27}$$

3. Coriolis term: The third terms represents the forces caused by the Coriolis acceleration. The coefficients are

$$\begin{cases} D_{112} = D_{121} = -m_2 l_1 l_2 \sin\theta_2 \\ D_{212} = D_{221} = 0. \end{cases} \tag{4.28}$$

4. Weight term: The fourth term is the force caused by the weight. D_i represents the weight transformed to joint i. They have the following forms

$$\begin{cases} D_1 = (m_1 + m_2)gl_1 \sin\theta_1 + m_2 gl_2 \sin(\theta_1 + \theta_2) \\ D_2 = m_2 gl_2 \sin(\theta_1 + \theta_2). \end{cases} \tag{4.29}$$

Obviously, the coefficients are functions of position (θ_1 and θ_2) and masses (m_1 and m_2), as well as arm lengths (l_1 and l_2). Once the arms are designed

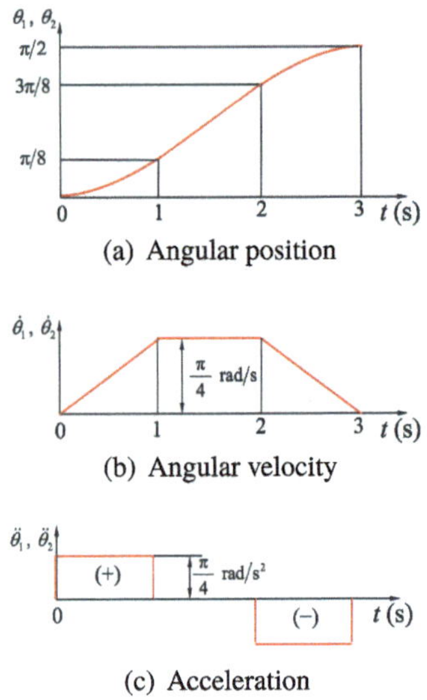

Figure 4.4 Motion of a two-bar manipulator.

(m_i and l_i determined), the desired motion (θ_1 and θ_2) can be achieved through proper control of $M_1(t)$ and $M_2(t)$.

Example 4.2. In the robotic arm in Fig. 4.3, the following parameters are given: $m_1 = 2$, $m_2 = 4$, $l_1 = l_2 = 1$ (all are dimensionless for simplicity). Weights are neglected. The required motion for the upper and lower arms is given in Fig. 4.4. The motion from position C_1 to position C_2, shown in Fig. 4.5, is required to be completed in three sec. Find the required torques M_1 and M_2.

Solution. Due to the weight term being neglected, the equation of motion can be shortened to

$$\begin{cases} D_{11}\ddot{\theta}_1 + D_{12}\ddot{\theta}_2 + D_{122}\dot{\theta}_2^2 + D_{112}\dot{\theta}_1\dot{\theta}_2 = M_1 \\ D_{21}\ddot{\theta}_1 + D_{22}\ddot{\theta}_2 + D_{211}\dot{\theta}_1^2 = M_2. \end{cases} \tag{a}$$

The coefficients are

$$\begin{cases} D_{11} = 10 + 8\cos\theta_2 \\ D_{22} = 4 \\ D_{12} = D_{21} = 4 + 4\cos\theta_2 \\ D_{122} = -D_{211} = -4\sin\theta_2 \\ D_{112} = D_{121} = -4\sin\theta_2. \end{cases} \tag{b}$$

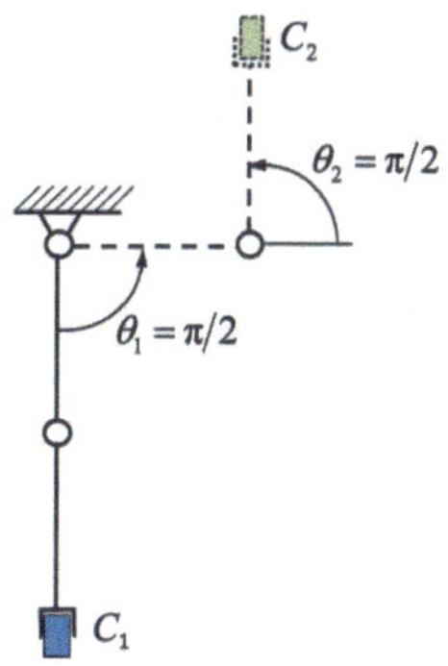

Figure 4.5 Starting and stopping position of the manipulator.

The required motion given in Fig. 4.4 can be represented mathematically as

$$\theta_1 = \theta_2 = \begin{cases} \dfrac{\pi t^2}{8} & \text{if } t = 0 \sim 1 \\ \dfrac{\pi(-1+2t)}{8} & \text{if } t = 1 \sim 2 \\ \dfrac{\pi(-5+6t-t^2)}{8} & \text{if } t = 2 \sim 3 \end{cases} \tag{c}$$

$$\dot{\theta}_1 = \dot{\theta}_2 = \begin{cases} \dfrac{\pi t}{4} & \text{if } t = 0 \sim 1 \\ \dfrac{\pi}{4} & \text{if } t = 1 \sim 2 \\ \dfrac{\pi(3-t)}{4} & \text{if } t = 2 \sim 3 \end{cases} \tag{d}$$

$$\ddot{\theta}_1 = \ddot{\theta}_2 = \begin{cases} \dfrac{\pi}{4} & \text{if } t = 0 \sim 1 \\ 0 & \text{if } t = 1 \sim 2 \\ \dfrac{-\pi}{4} & \text{if } t = 2 \sim 3. \end{cases} \tag{e}$$

Discretize the time interval [1, 3] into a series of time instants, and compute the values of M_1 and M_2 at these instants. The results are shown in Fig. 4.6.

Obviously, $M_1(t)$ and $M_2(t)$ change abruptly at some points in time. This is caused by the abrupt change in $\ddot{\theta}_1$ and $\ddot{\theta}_2$ and is impossible to be implemented in reality. To improve the controllability, we can modify the motion required as in Fig. 4.7, which is continuous everywhere.

The $M_1(t)$ and $M_1(t)$ calculated in this case are shown in Fig. 4.8. It can be seen that both $M_1(t)$ and $M_2(t)$ are continuous everywhere without abrupt jumps or drops.

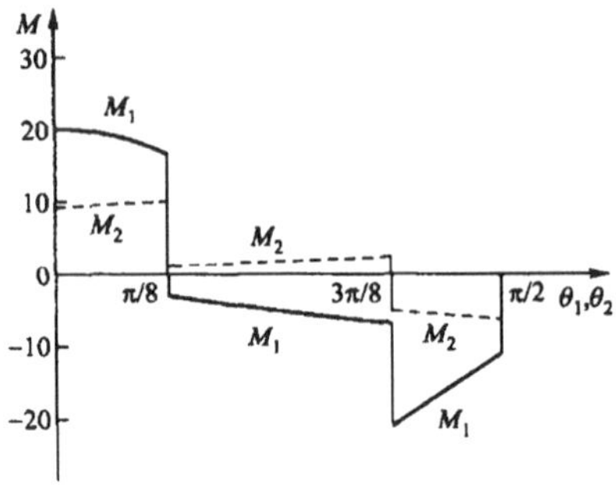

Figure 4.6 Torque vs position.

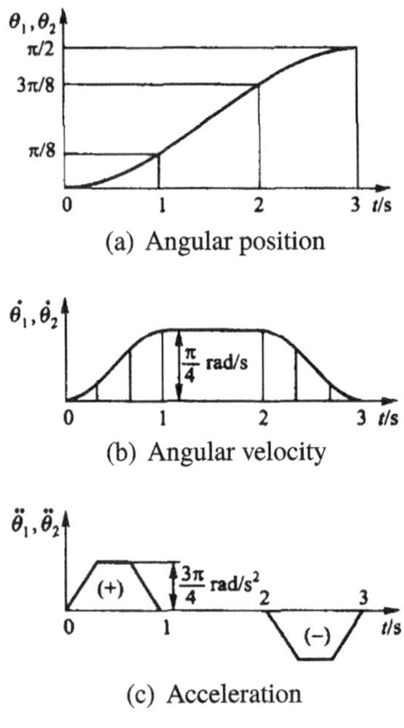

Figure 4.7 Motion program of continuous acceleration.

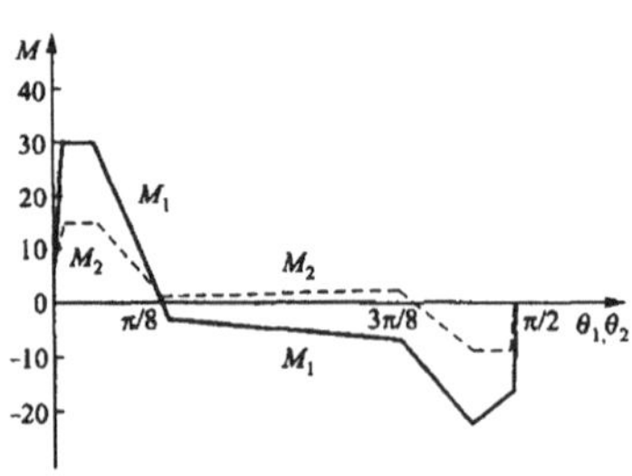

Figure 4.8 Torque vs position.

4.4 Brief introduction to the dynamics of robotic manipulators

Robots are probably one of the most representative high-tech products. Industrial robots, in particular, have been widely applied in many manufacturing and other industrial settings and will continuously expand their fields of application. In the previous section, we developed the equation of motion of a two-link planar manipulator arm, while in the section following, we present a brief introduction to the dynamics of general robotic manipulators.

4.4.1 Robots and robotic manipulators

There are many types of robots, such as industrial robots, service robots, humanoid robots, etc. Industrial robots are an important subset of modern robots, which are widely used in painting, welding, material handling, and assembly, as well as many other operations (Mihelj et al., 2019). The mechanical part in an industrial robot is called a manipulator. In the early days, manipulators were designed for repetitive operations of mass production and were not reprogrammable. Modern robots, however, can be reprogrammed for different tasks with a much higher level of flexibility, which is regarded as a fundamental difference between manipulators and robots.

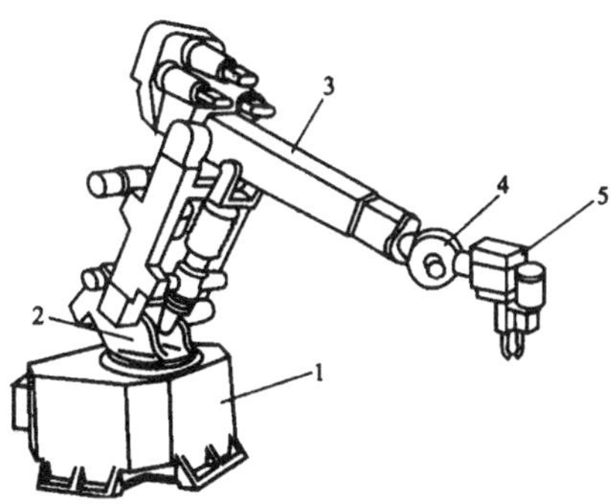

Figure 4.9 A serial manipulator.

An industrial robot can be either serial and parallel in structure. Fig. 4.9 shows a typical serial industrial robot that consists of the following subsystems (units):

1. Mechanical unit: The mechanical unit, also called a manipulator or arm, is the part that implements the required motion of the robot. It is

generally a mechanism having multiple DOFs, consisting of links connected through revolute or prismatic joints, wrists, and end-effectors.

2. Actuator unit: The actuation unit generally consists of drivers and transmissions. The driver may be electric, hydraulic, or pneumatic. The transmissions may be harmonic drives, screws, or other gear drives .
3. Controller unit: The controller unit generally consists of computers and servo controllers. The computers are for programming, and the servo controllers control the motion of the robot.
4. Sensor unit: Various sensors are used in robots to collect information, such as force and displacement, and to feed the information into the control system.

Serial manipulators are often open-loop mechanisms. Parallel manipulators, however, are closed-loop mechanisms. One example of the parallel manipulators is the Steward platform shown in Fig. 4.10. It has a moving triangular platform $A_1A_2A_3$, and a stationary one $B_1B_2B_3$. Six prismatic actuators move the platform; thus, it has six DOFs. Compared with its serial counterpart, parallel manipulators generally have higher rigidity, higher precision, and higher speed of movement, but lower mobility. For this reason, parallel robots are often used as flight simulators and machine tools.

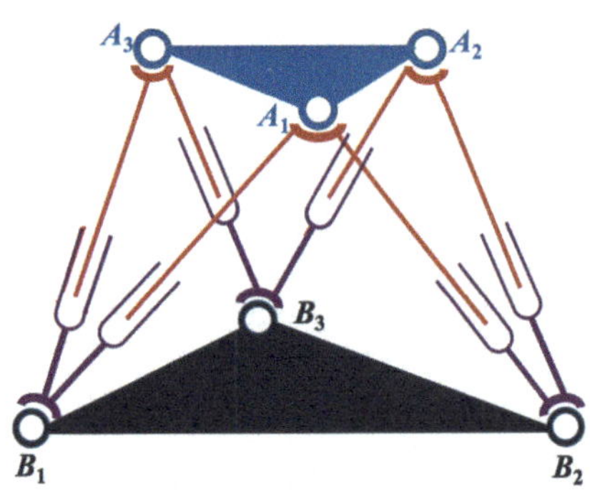

Figure 4.10 Stewart platform.

4.4.2 Introduction to robotic kinematics

Kinematics is the basis for the analysis and control of robots. In general, robotic manipulators are multiple DOF mechanisms, and their kinematics has some special features.

In the kinematics of a manipulator, the relationship between the end-effector position and orientation (pose) and the joint variables is to be established. Taking the two-link manipulator in Fig. 4.3 as an example, the end-effector C can be represented by Cartesian coordinates x and y. To achieve this position, two actuators at the revolute joints A and B are needed, their orientations are represented by θ_1 and θ_2. The relationship between the end-effector coordinates, x and y, and the joint coordinates, θ_1 and θ_2 are formulated in Eq. (4.16).

For manipulators having more DOFs, this relationship is more complicated. For example, the robot shown in Fig. 4.9 has six DOFs. If the end-effector is a welding gun, its pose, including orientation and position, needs six coordinates to be represented, including three translational x_p, y_p, and z_p, as well as three rotational θ_x, θ_y and θ_z as shown in Fig. 4.11. This group of coordinates are given as functions of time, based on the specific job requirements.

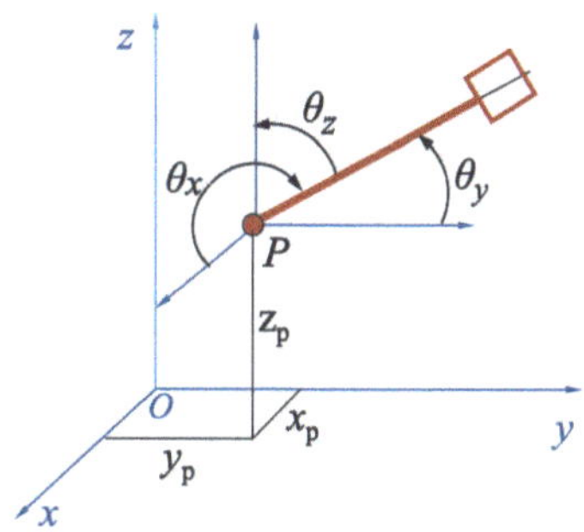

Figure 4.11 Pose of end-effector.

To achieve the required motion of the end-effector, six actuators, each having one coordinate, are needed to drive the links and make up the joint coordinates. The relationship between the end-effector's pose and the joint coordinates, however, in this case is more complicated than the two-link manipulator.

There are two types of problems in robotic kinematics, namely, *direct kinematics* and *inverse kinematics*. In the direct kinematics, the joint coordinates are given, and the end-effector's motion is calculated, whereas inverse kinematics determines the required joint coordinates with the end-effector's motion given. The distinction between the two kinematics will be further illustrated through the following examples.

Trajectory planning is critical to generate the reference inputs to the control system for the end-effector to follow the desired trajectory. In some cases, only the start point and the end point are specified, and there is no requirement about the path between the start and end points. For these cases, the evolution of the joint variables can be specified directly to achieve the required end-effector motion; thus, direct kinematics is appropriate. This type of trajectory planning is referred to as *joint trajectory planning*. In some other cases, however, the complete trajectory, not only the start and end points, needs to be specified. Welding robots are typical examples of this group. In this case, the position and orientation of the end-effector are first specified, based on which joint variables are being calculated. Inverse kinematics is needed and the planning is generally done in the Cartesian space (Jazar, 2010).

Direct kinematics of robots, serial robots, in particular, is relatively easier. Taking the two-link manipulator as an example, substituting the value of θ_1 and θ_2 into Eq. (4.16) gives the Cartesian coordinates, x and y, of the end-effector. The solution of direct kinematics always exists and is unique. However, inverse kinematics is much more complicated. All the positions that a robot can reach is defined as the *workspace*. For a robot, although a position of the end-effector, (x, y, z), in the workspace can be reached, the orientation, $(\theta_x, \theta_y, \theta_z)$ may be impossible. This means that there may be no solution of inverse kinematics for some end-effector poses. In some other cases, although the solution exists, but there may be multiple solutions. Taking the simplest two-link case in Fig. 4.12 as an example, the end-effector position corresponds to two sets of possible joint variables. In general, the equation of kinematics contains nonlinear triangular functions. Thus, it is difficult to solve in closed form. Numerical methods have to be relied on.

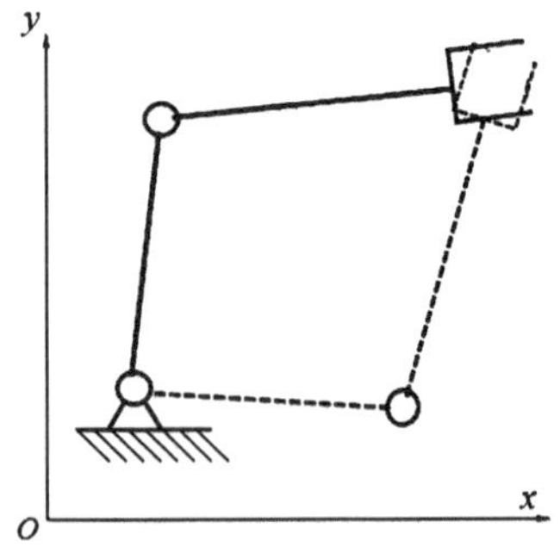

Figure 4.12 Inverse kinematics.

The inverse kinematics of parallel manipulators is much easier than that of the serial ones. If a pose of the platform $A_1A_2A_3$ in Fig. 4.10 is given, it is straightforward to calculate the actuator's position. However, its direct kinematics is more difficult to solve.

4.4.3 Introduction to robotic dynamics

Dynamics is critical to manipulator and controller design. In addition, dynamic analysis is a necessary step for any high-speed machine. In the previous section, we already derived the equation of motion for the two-link manipulator dynamics to be Eq. (4.24), using the Lagrange's equation. For manipulators with more DOFs, the dynamic equations can be derived with the Lagrange's equation as well, but the procedure is more cumbersome.

The dynamics equation of Eq. (4.24) can be used for two purposes, e.g., 1) with θ_i, $\dot{\theta}_i$ and $\ddot{\theta}_i$ given, find the required $M_i(t)$ $(i = 1, 2)$, and 2) with $M_i(t)$ given, find θ_i, $\dot{\theta}_i$ and $\ddot{\theta}_i$. The first is called *direct dynamics*, while the second is *inverse dynamics*.

In direct dynamics, the equation is generally solved numerically. Thus, it is often referred to as *dynamic simulation* of robots. The inverse dynamics constitutes the basis of controller design.

The efficiency of the numerical algorithms in robotic dynamics is critical because of the requirement of real-time computation. In feedback control, the joint variables θ_i are measured in very short time interval (sampling time). Based on the measured values, the required torque $T_i(t)$ is calculated through the inverse dynamics, which are input to the controller. The calculation time of the inverse dynamics has be shorter than the sampling time for real-time control.

Much research effort has been directed to find more efficient numerical algorithms and neater methods for the derivation of the dynamic equations. The Lagrange's equation is only one of the many methods. Some other methods include the Newton–Euler method, Kane's equation method, and Gibbs–Appell equation method, etc. These methods may lead to equations in different forms, but should yield the same calculation results because they are the same in their essential mechanics.

PART 2

Theory of mechanical vibration

The study of vibration began long before classical mechanics and the Industrial Revolution. At that time, the speed of machines in manual workshops was generally very low; thus, vibration was not a real concern for machines. In fact, the early research of vibration was related more to musical instruments and pendulum clocks (Rao, 2010; Zhang and Yang, 2020).

Musical instruments were probably the original inspiration for our ancestors to study vibration. The violin family and the drum, as well as the piano invented in the late 17th century, make sound with vibrating strings or membranes; the consonant cavities use a plate structure. In fact, the vibration of strings, beams, membranes, and plates had been a hot topic in the field of mechanics for centuries. Several prominent scientist, including L. Euler, D. Bernoulli, and J. D'alembert, got involved in research into musical vibration.

In 1602, G. Galileo, an Italian scientist, discovered the isochronicity of the simple pendulum and calculated the period of the pendulum using his free-fall formula. J. Rayleigh is among the greatest contributors to the establishment of classical vibration theory. In 1878, he published his masterpiece *The Theory of Sound*, in which he almost formularized linear vibration theory.

During the Second Industrial Revolution, various machines were invented, including motors, steam turbines, cars, and aircraft, etc. Vibration problems became prominent and many practical vibration issues were encountered in high-speed and heavily loaded machines. In particular, the bending vibration of shafts in transmission units attracted wide attention. In parallel, the speed of internal combustion engines went through a phase of dramatic increase, causing a series of dynamic problems. This led to the need to analyze the crankshaft vibration of multicylinder engines. In machine tools, the torsional vibration of shafts was also critical. Since then, the study of shaft systems has developed into a new scientific branch, rotor dynamics.

By the mid-20th century, the theory of linear discrete vibration had been fully developed. However, due to the limitation of computing power, the scale of problems which could be solved with the theory was still limited. At the same time, the theory of vibration of continuous systems developed along analytic methods since closed-form solutions were available only for very simple geometries and boundary conditions.

The advent of computers after WWII dramatically changed the landscape of vibration in both theoretical developments and applications. In particular, the following two should be highlighted:

- Without computers, the scale of vibration problems that can be feasibly handled is very limited. Thus, only very rough vibration models were used before WWII. Thereafter, computers made it possible to model and solve vibration problems with hundreds of thousands of DOFs. This greatly expanded the application of vibration theory. In addition, more refined models were possible to be developed, leading to further theoretical development.
- Combination of the finite element method (FEM) and the vibration theory of continuous systems fundamentally changed the picture of the subject. By pure analytic methods, only very simple problems with simple boundary conditions can be solved. With modern computers, very complicated structures having complex geometries can be treated using the theory of continuous vibration and FEM. Actually, a new branch of solving vibration problems with FEM has been formed.

Many nonlinear factors exist in an actual mechanical system; thus, nonlinear vibration is needed in some cases. Poincaré laid the foundation of nonlinear vibration theory around the end of the 19th century. Nonlinearity greatly complicates the problem, and this causes many unique response features that a linear vibration system does not have. In addition, there is no one-for-all solution method, such as the modal superposition method in linear systems, available for nonlinear vibration systems. The solution method for a specific problem has to be chosen according to the specific nature of the problem (Nayfeh and Mook, 2008).

The vibration theory just introduced assumes the excitation and the parameters are *deterministic*; thus, the field is referred to as *deterministic vibration theory*. However, uncertainty exists in all mechanical systems. For example, joint clearance is formed by the tolerance of the shaft and the hole, and it is nondeterministic or *random*. Some typical random excitation includes road roughness, ocean-wave load, and wind load, etc. The theory dealing with uncertainties in vibration is *random vibration theory* (Lutes and Sarkani, 2004).

In summary, the entire vibration theory can be classified into *deterministic linear vibration*, *deterministic nonlinear vibration*, *random linear vibration*, and *random nonlinear vibration*.

This part provides a brief introduction to the basic deterministic linear and nonlinear vibration theory. Random vibration is not covered here due to space considerations. In addition, the finite element method (FEM) in solving vibration problems is also presented because mechanism links are approximated by discrete models.

CHAPTER 5

Vibration of systems with a single DOF

5.1 Introduction

Many practical systems are described with continuous vibration theory, while some others can be simplified as discrete vibration systems having a finite number of DOFs. In the discrete systems, the theory of single DOF is of particular importance for two reasons. First, the theory of single DOF is simple, but capable of capturing the main features of vibrations, such as natural frequency, resonance, etc. Second, many practical problems can be modeled as single-DOF system with adequate accuracy. Thus, the theory of single-DOF vibration is the basis for more complicated multiple-DOF and continuous system vibrations.

This chapter introduces the basic theory of vibration of single-DOF systems. The chapter is organized in two parts:

- A short introduction is presented in Section 5.1.
- Free vibration and force vibration are introduced in Sections 5.2 and 5.3, respectively.

5.2 Free vibration

Free vibration means the motion of a system after an initial disturbance, while forced vibration is the motion of the system when a time-varying disturbance is applied. A single DOF spring–damper–mass system is shown in Fig. 5.1. If the mass m moves a distance x downward from the static equilibrium position, the FBD of the mass is shown in the figure to the right. The equation of motion of the mass can be obtained according to Newton's second law of motion:

$$m\ddot{x} + c\dot{x} + kx = F(t), \tag{5.1}$$

where m is the mass of the vibrating object, k is the spring stiffness, c is the damping coefficient, and $F(t)$ is the external force excitation.

Machinery Dynamics
https://doi.org/10.1016/B978-0-12-815785-5.00015-7

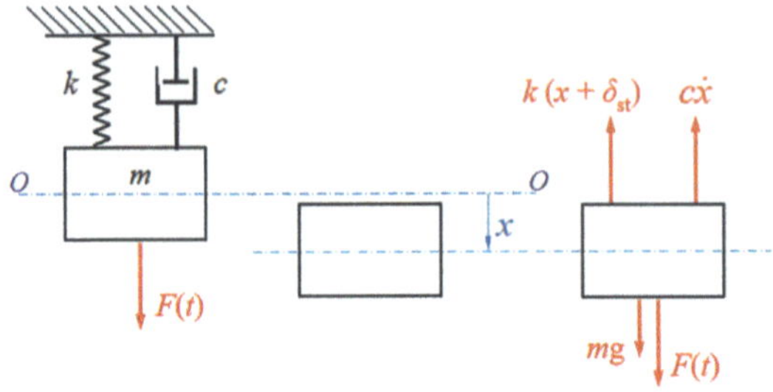

Figure 5.1 A single-DOF system.

At the static equilibrium, m actually already has a static displacement, δ_{st}, under its own weight. δ_{st} does not enter the equation because x is measured from the static equilibrium position.

Eq. (5.1) is a secondorder linear ordinary differential equation (ODE). Such a vibration system is called *linear vibration system.*

5.2.1 Undamped free vibration

The undamped free-vibration equation is obtained by dropping off the external excitation $F(t)$ and the damping term $c\dot{x}$ from Eq. (5.1):

$$m\ddot{x} + kx = 0. \tag{5.2}$$

This is the simplest case in vibration, called *undamped free vibration.* Eq. (5.2) is often written in the following form:

$$\ddot{x} + \lambda x = 0, \tag{5.3}$$

where λ is a positive constant defined as:

$$\lambda = \frac{k}{m}. \tag{5.4}$$

The solution of Eq. (5.3) can be found by assuming

$$x = A_1 \cos \omega_n t + A_2 \sin \omega_n t, \tag{5.5}$$

where A_1 and A_2 are constants, and

$$\omega_n = \sqrt{\frac{k}{m}} \tag{5.6}$$

is the *circular natural frequency*. It has the unit of rad/s, which can be easily changed to Hz by the following relationship:

$$f = \frac{1}{2\pi}\omega_n = \frac{1}{2\pi}\sqrt{\frac{k}{m}}. \tag{5.7}$$

f is often referred to as *natural frequency*. In practice, ω_n is often shortened as natural frequency as long as no confusion is caused. The *period of vibration* T is then

$$T = \frac{1}{f} = \frac{2\pi}{\omega_n}. \tag{5.8}$$

T generally has a unit of sec. Eq. (5.5) can be recast in the following form:

$$x = A\sin(\omega_n t + \varphi), \tag{5.9}$$

where A is called the *amplitude* and φ is called the *phase angle*. A and φ (or A_1 and A_2) are determined by the *initial conditions* of the following form:

$$\begin{cases} x(0) = x_0. \\ \dot{x}(0) = \dot{x}_0. \end{cases} \tag{5.10}$$

Substituting it into Eq. (5.9) gives

$$\begin{cases} A = \sqrt{x_0^2 + \left(\dfrac{\dot{x}_0}{\omega_n}\right)^2} \\ \varphi = \arctan\left(\dfrac{x_0\omega_n}{\dot{x}_0}\right). \end{cases} \tag{5.11}$$

Vibration often enlarges the deformation of and stress within the mechanical parts or structural members. In general, the dynamic stress is much larger than the stress calculated from static analysis. This can be clearly seen from the following examples.

Example 5.1. A cantilever beam is shown in Fig. 5.2(a), in which $l = 3500$ mm. The cross section of the beam is shown in Fig. 5.2(b) (units in the figure are mm). The material of the beam is aluminum that has a modulus of elasticity $E = 7 \times 10^4$ MPa and density $\rho = 2700$ kg/m^3. An object $G_1 = 2400$ N freely falls from a height of $h = 25$ mm, and hits the midpoint of the beam. Calculate the natural frequency and the maximum deflection of the beam.

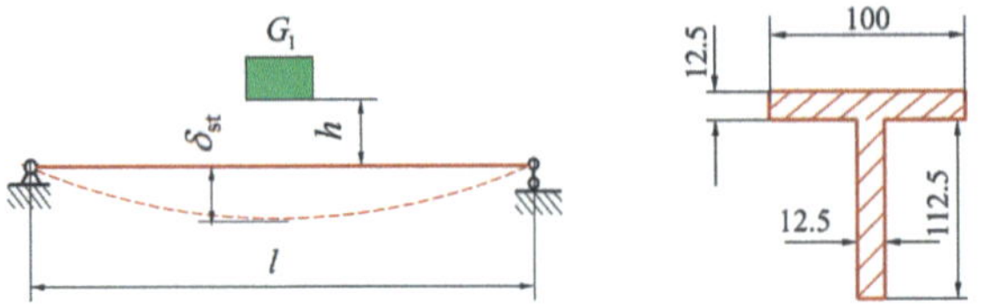

(a) A beam (b) The cross section

Figure 5.2 Vibration of a beam.

Solution. Before G_1 hits the beam, it is free falling. After hitting the beam, G_1 vibrates up and down together with the beam (free vibration).

1) The beam

From the cross section given in Fig. 5.2(b), the moment of inertia I can be calculated as $I = 4.06 \times 10^{-6}$ m^4. Then the deflection of the beam at the midpoint under the static action of G_1 is

$$\delta_{st} = \frac{G_1 l^3}{48EI} = \frac{(2400)(3.5)^3}{(48)[7(10^4)(10^6)][4.06(10^{-6})]} = 7.5(10^{-3}) \text{ m}. \qquad (a)$$

It is easy to calculate that the weight of the beam is $G_0 = 251$ N. This is very small compared with G_1. Thus, the mass of the beam can be neglected without significant deterioration in accuracy.

If the beam is modeled as a massless spring, the spring stiffness is obtained by the formula

$$k = \frac{G_1}{\delta_{st}}. \qquad (b)$$

2) The object G_1

At the instant it hits the beam, the downward velocity v_0 is

$$v_0 = \sqrt{2gh} = \sqrt{2(9.81)(2.5)(10^{-2})} = 0.7 \text{ m/s}. \qquad (c)$$

3) Free vibration

After G_1 hits the beam, it will vibrate up and down freely. The natural frequency is calculated by

$$\omega_n = \sqrt{\frac{k}{m}} = \sqrt{\frac{G_1/\delta_{st}}{G_1/g}} = \sqrt{\frac{g}{\delta_{st}}} = \sqrt{\frac{9.81}{7.5(10^{-3})}} = 36 \text{ rad/s}. \qquad (d)$$

Establish a coordinate x with its origin at the static equilibrium position of the beam under the action of G_1, and downward positive. Then, the initial conditions are the displacement and velocity at the moment of contact. They are

$$\begin{cases} x_0 = -\delta_{st} \\ \dot{x}_0 = v_0. \end{cases} \tag{e}$$

The amplitude of free vibration is

$$A = \sqrt{x_0^2 + \left(\frac{\dot{x}_0}{\omega_n}\right)^2} = \sqrt{[(-0.0075)^2 + \left(\frac{0.7}{36}\right)^2} = 0.0209 \text{ m}. \tag{f}$$

The maximum dynamic deflection of the beam is

$$\delta_{max} = A + \delta_{st} = 0.0209 + 0.0075 = 0.0284 \text{ m} \tag{g}$$

It is clear that δ_{max} is much larger than δ_{st}. The ratio of the two is often referred to as the *dynamic magnification factor*, β:

$$\beta = \frac{\delta_{max}}{\delta_{st}} \approx 3.8. \tag{h}$$

Example 5.2. A hoisting machine is shown in Fig. 5.3. When the machine lowers a object $G = 20,000$ N at a speed of $v_0 = 0.6$ m/s, braking is applied. The parameters given include the area of cross section of the rope $A = 2.51 \times 10^{-4}$ m^2, the length of the rope $l = 16$ m, and the modulus of elasticity of the rope $E = 2 \times 10^{11}$ Pa. Find the maximum force in the rope caused by the braking.

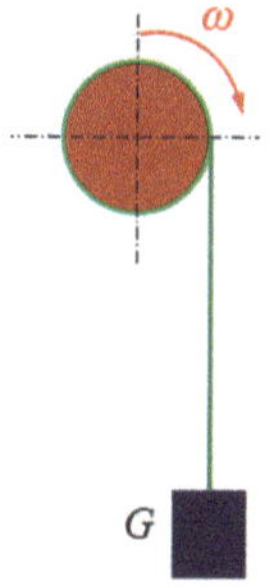

Figure 5.3 A hoisting machine.

Solution. At the moment of braking, although the pulley is stopped suddenly, the object G keeps moving downward at a velocity of $v_0 = 0.6$ m/s due to the inertial effect. The rope thus works as a spring and the object experiences free vibration. Establish the coordinate x with the origin located at the static equilibrium position, and downward positive. The initial conditions at the moment of braking ($t=0$) are $x_0=0$ and $\dot{x}_0=v_0$. Then the free vibration is represented as

$$x=\frac{v_0}{\omega_n}\sin\omega_n t. \tag{a}$$

The amplitude of vibration is

$$x_{max}=A=\frac{v_0}{\omega_n}. \tag{b}$$

The maximum elongation of the rope is

$$\delta_{max}=\delta_{st}+A=\frac{G}{k}+\frac{v_0}{\omega_n}. \tag{c}$$

The spring constant k of the rope is calculated by

$$k=\frac{EA}{l}=\frac{2(10^{11})(2.51)(10^{-4})}{16}=3.138(10^6)\ \text{N/m}. \tag{d}$$

The maximum tension force within the rope is

$$\begin{aligned} F_{max}=k\delta_{max}&=k\left(\frac{G}{k}+v_0\sqrt{\frac{m}{k}}\right)=G\left(1+v_0\sqrt{\frac{k}{Gg}}\right)\\ &=20(10^3)\left(1+0.6\sqrt{\frac{3.138(10^6)}{20(10^3)(9.81)}}\right)=6.8(10^4)\ \text{N}. \end{aligned} \tag{e}$$

The dynamic magnification factor can also be defined as the ratio of dynamic and static forces as

$$\beta=\frac{F_{max}}{G}=\frac{6.8(10^4)}{20(10^3)}=3.4. \tag{f}$$

In the spring–mass model, the spring is treated as massless. If the mass of spring is indeed necessary to be considered for some reason, for example, higher accuracy, an equivalent mass can be calculated based on the principle of equal kinetic energy as subsequently outlined.

For the spring shown in Fig. 5.4, we want to find an equivalent mass M sitting on top of the spring. For this purpose, we take a differential element

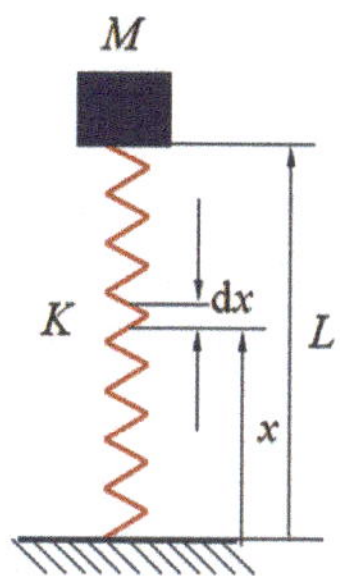

Figure 5.4 Equivalent mass of spring.

of the spring *dx*, whose kinetic energy is

$$dE = \frac{1}{2} q\dot{x}^2 dx, \tag{5.12}$$

where q is the spring mass of unit length. If the velocity of the top of the spring is $\dot{X}$, the velocity at arbitrary x is

$$\dot{x} = \frac{x}{L}\dot{X}. \tag{5.13}$$

The total kinetic energy is then the integral:

$$E = \int_0^L \frac{1}{2} q \left(\frac{x}{L} \dot{X} \right)^2 dx = \frac{1}{2} \frac{qL}{3} \dot{X}^2. \tag{5.14}$$

The kinetic energy calculated from the equivalent mass M is

$$E = \frac{1}{2} M\dot{X}^2. \tag{5.15}$$

These two should equal each other. Obviously,

$$M = \frac{qL}{3}. \tag{5.16}$$

This means that the equivalent mass of the spring is one third of the total spring mass.

5.2.2 Damped free vibration

Damping may take many forms, such as dry friction between sliding objects, resistance from surrounding medium, the material's internal friction, etc. It is very difficult to accurately account for the different damping in real calculations; common practice is to simplify it to *viscous damping,* which is proportional to the velocity. This is the way we derive Eq. (5.1). The main consideration in this treatment is the simplification in mathematical manipulation, together with acceptable accuracy in most cases.

Dropping off the external force excitation $F(t)$ in Eq. (5.1), the equation of free vibration with viscous damping is

$$m\ddot{x} + c\dot{x} + kx = 0. \tag{5.17}$$

Dividing both sides of Eq. (5.17) by m and letting $2a = c/m$ gives another form of the equation:

$$\ddot{x} + 2a\dot{x} + \omega_{\mathrm{n}}^2 x = 0. \tag{5.18}$$

The two roots of the characteristic equation are

$$r = -a \pm \sqrt{a^2 - \omega_{\mathrm{n}}^2}. \tag{5.19}$$

Introducing a dimensionless quantity, the *damping ratio*

$$\zeta = \frac{a}{\omega_{\mathrm{n}}}, \tag{5.20}$$

the solution of Eq. (5.17) can be classified into one of the following three cases:

1) Overdamped: When $\zeta > 1$ or $a > \omega_{\mathrm{n}}$, the system is overdamped. In this case, the two roots in Eq. (5.19) are distinct and real:

$$\begin{cases} r_1 = -a + \sqrt{a^2 - \omega_{\mathrm{n}}^2} \\ r_2 = -a - \sqrt{a^2 - \omega_{\mathrm{n}}^2}, \end{cases} \tag{5.21}$$

x is then represented as

$$x = e^{-at}\left[A_1 \cosh(\beta t) + A_2 \sinh(\beta t)\right], \tag{5.22}$$

where $\beta = \sqrt{a^2 - \omega_n^2}$. The motion in this case is aperiodic, not a vibration.

2) Critically damped: When $\zeta = 1$ or $a = \omega_{\mathrm{n}}$, the two characteristic roots are equal, i.e.,

$$r_1 = r_2 = -a. \tag{5.23}$$

According to the ODE theory, the solution of Eq. (5.17) is of the following form in this case:

$$x = e^{-at}(A_1 + A_2 t). \tag{5.24}$$

This is not a vibration either.

3) Underdamped: When $\zeta < 1$ or $a < \omega_\mathrm{n}$, the two characteristic roots are complex conjugates and

$$\begin{cases} r_1 = -a + i\sqrt{\omega_\mathrm{n}^2 - a^2} \\ r_2 = -a - i\sqrt{\omega_\mathrm{n}^2 - a^2}. \end{cases} \tag{5.25}$$

Letting

$$\omega_\mathrm{d} = \sqrt{\omega_\mathrm{n}^2 - a^2} = \sqrt{1 - \zeta^2}\,\omega_\mathrm{n}, \tag{5.26}$$

ω_d is referred to as *damped natural frequency*. The solution therefore can be represented as

$$x = e^{-at}(A_1 \cos\omega_\mathrm{d} t + A_2 \sin\omega_\mathrm{d} t), \tag{5.27}$$

where A_1 and A_2 are constants to be determined through the initial conditions. Substituting the initial conditions $x(t) = x_0$, $\dot{x}(0) = \dot{x}_0$ into the response, it becomes

$$\begin{aligned} x &= e^{-at}\left(x_0 \cos\omega_\mathrm{d} t + \frac{\dot{x}_0 + a x_0}{\omega_\mathrm{d}} \sin\omega_d t\right) \\ &= A e^{-at} \sin(\omega_\mathrm{d} t + \varphi), \end{aligned} \tag{5.28}$$

where

$$\begin{cases} A = \sqrt{x_0^2 + \left(\dfrac{\dot{x}_0 + a x_0}{\omega_\mathrm{d}}\right)^2} \\ \varphi = \arctan\left(\dfrac{\omega_\mathrm{d} x_0}{\dot{x}_0 + a x_0}\right). \end{cases} \tag{5.29}$$

Eq. (5.28) represents a periodic vibration. The frequency ω_d is slightly smaller than the undamped natural frequency ω_n. The amplitude decays exponentially with time. The response curve of this attenuated vibration is shown in Fig. 5.5.

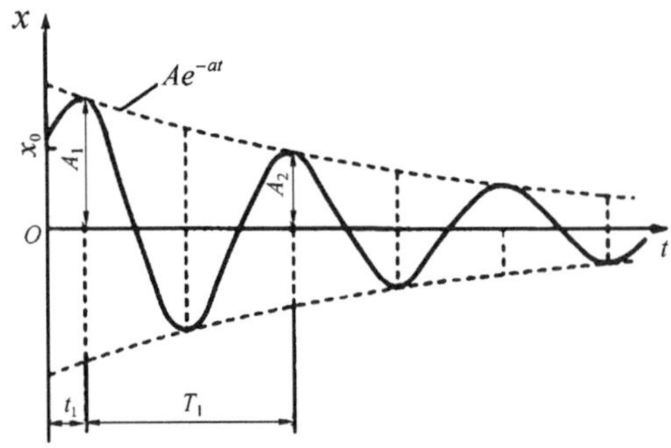

Figure 5.5 Underdamped response.

Fig. 5.5 provides a practical way to measure the damping coefficient (or damping ratio). The dashed line, or the amplitude-decay envelope, represents Ae^{-at}. Through measurement of points A_0, A_1, A_2, which are curve fit to A, Ae^{-at_1}, Ae^{-at_2} ($t_2 = t_1 + T_1$), the value of $a = \zeta\omega_n$ can be calculated. If m and k is given, ω_n is known, and ζ can be calculated then. This method is often used in vibration labs.

5.3 Forced vibration

When there is external excitation applied to the system, the response is called forced vibration. The excitation may be in the form of either a force or base motion (displacement). In the section following, the responses of a single-DOF system to various forms of excitation, including a harmonic force, a harmonic base motion, a periodic force and a general nonperiodic force, are discussed.

5.3.1 Response to harmonic force excitation

The equation of motion under the action of a *harmonic force excitation* can be expressed as

$$m\ddot{x} + c\dot{x} + kx = F_0 \cos(\omega t + \varphi), \tag{5.30}$$

where ω is the frequency of the harmonic excitation. The equation can also be transformed into the following form:

$$\ddot{x} + 2\zeta\omega_n\dot{x} + \omega_n^2 x = h\cos(\omega t + \varphi), \tag{5.31}$$

where

$$2\zeta\omega_n = 2a = \frac{c}{m}$$
$$\omega_n^2 = \frac{k}{m}$$
$$h = \frac{F_0}{m}.$$

Eq. (5.31) is nonhomogeneous; its general solution is the sum of the homogeneous solution and a particular solution. The homogeneous solution is given in Eq. (5.27) for the underdamped case:

$$x_1 = e^{-\zeta\omega_n t}\left(C_1 \cos\omega_d t + C_2 \sin\omega_d t\right), \tag{5.32}$$

where C_1 and C_2 are constants to be determined. One particular solution to Eq. (5.31) can be assumed as

$$x_2 = B_1 \cos(\omega t + \varphi) + B_2 \sin(\omega t + \varphi). \tag{5.33}$$

It can also be written in an alternative form as:

$$x_2 = B\cos(\omega t + \varphi - \theta), \tag{5.34}$$

where B and θ are the amplitude and phase difference of the particular solution. Substituting Eq. (5.34) into Eq. (5.30) and collecting the cosine and sine terms, we can determine B and θ as

$$B = \frac{\delta_{st}}{\sqrt{(1-\lambda^2)^2 + 4\zeta^2\lambda^2}} = \beta\delta_{st} \tag{5.35}$$

$$\theta = \arctan\left(\frac{2\zeta\lambda}{1-\lambda^2}\right), \tag{5.36}$$

where δ_{st} is the static deflection of the spring under F_0, λ is the frequency ratio, and β is the *dynamic magnification factor*. They are calculated so:

$$\delta_{st} = \frac{F_0}{k} = \frac{h}{\omega_n^2} \tag{5.37}$$

$$\lambda = \frac{\omega}{\omega_n} \tag{5.38}$$

$$\beta = \frac{1}{\sqrt{(1-\lambda^2)^2 + 4\zeta^2\lambda^2}}. \tag{5.39}$$

Then, the general solution of Eq. (5.31) is the sum of Eqs. (5.32) and (5.34):

$$x = x_1 + x_2 = e^{-\zeta\omega_n t}(C_1 \cos\omega_d t + C_2 \sin\omega_d t) + B\cos(\omega t + \varphi - \theta). \quad (5.40)$$

Substituting C_1, C_2 into this and rearranging the terms, the complete forced vibration response can be formulated as:

$$x = X_1 + X_2 + X_3, \quad (5.41)$$

where

$$X_1 = B\cos(\omega t + \varphi - \theta) \quad (5.42)$$

$$X_2 = e^{-\zeta\omega_n t}\left(x_0 \cos\omega_d t + \frac{\dot{x}_0 + \zeta\omega_n x_0}{\omega_d}\sin\omega_d t\right) \quad (5.43)$$

$$X_3 = -Be^{-\zeta\omega_n t}\left[\cos(\varphi - \theta)\cos\omega_d t + \frac{\zeta\omega_n \cos(\varphi - \theta) - \omega\sin(\varphi - \theta)}{\omega_d}\sin\omega_d t\right]. \quad (5.44)$$

The first term X_1 is the particular solution, which is the response when time t approaches infinity. For this reason, it is also called the *steady-state response*. The second term X_2 and the third one X_3 disappear when time t approaches infinity; thus, they are called *a transient response*. In practice, the steady-state response is generally of more interest because the transient response decays to zero eventually. Obviously, the magnification factor β (or the amplitude B) and the phase lag θ are dependent on the frequency ratio λ and the damping ratio ζ. Fig. 5.6 shows the variation of β and θ

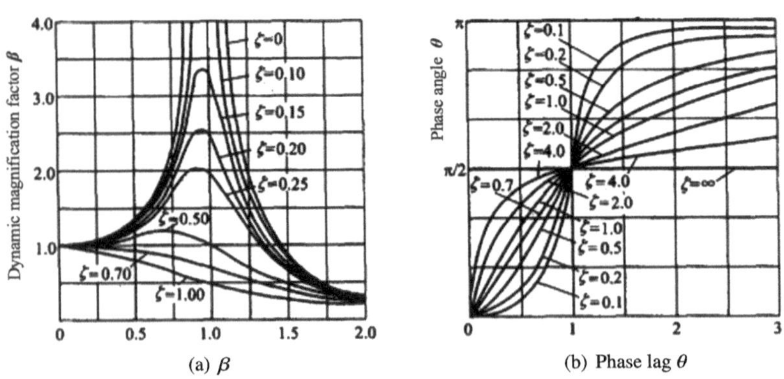

Figure 5.6 Variation of β and θ with frequency ratio λ.

with the frequency ratio λ. Fig. 5.6(a) and Fig. 5.6(b) are referred to as the *amplitude–frequency curve* and the *phase-frequency curve*, respectively. From Fig. 5.6(a), we have the following observation:

1) When $\lambda << 1$ or $\omega << \omega_n$, β is very close to unity, meaning that the response amplitude B is very close to the static deflection δ_{st}. In this case, the inertia and damping forces are insignificant, and elastic spring force is dominant. Thus, it is basically a quasistatic problem in this case.
2) When $\lambda = 1$ or $\omega = \omega_n$, β becomes infinite for $\zeta = 0$, and very large but finite for $\zeta \neq 0$. With further increase of λ, β drops from the peak. When $\lambda = 1$, it is called *resonance*.
3) When $\lambda >> 1$ or $\omega >> \omega_n$, β approaches zero. In this case, the inertial force is large, and the external force almost entirely works to overcome the large inertial force.

Due to the damping, the response always lags behind the excitation. In other words, there is always a θ, as long as $\zeta \neq 0$. The following observation is made from Fig. 5.6(b):

1) When $\lambda << 1$ or $\omega << \omega_n$, θ is almost zero.
2) With the increase of λ, θ increases as well. At $\lambda = 1$ or $\omega = \omega_n$, $\theta = \pi/2$. This fact is often used in vibration tests.
3) When λ is very large, or $\omega >> \omega_n$, θ approaches π. For the speed of θ to go from 0 to π depends on the damping ratio ζ.

In rotating machinery, such as ventilators, electric motors, pumps, centrifugal compressors, steam turbines, etc., unbalanced rotating components often excite a machine's vibration as illustrated in the following example.

Example 5.3. In a front-loading washing machine, the inevitable imbalance of the clothing causes the machine to vibrate. To isolate the vibration, one vibration isolation mount, which is combination of a spring and a damper, is installed underneath the machine at each corner. The four vibration mounts are identical; thus, the vibration can be studied with a single-DOF model as shown in Fig. 5.7. The stiffness of each spring is $k = 8.3 \times 10^4$ N/m, and the overall damping ratio is $\zeta = 0.15$. The machine has a mass of $M = 2000$ kg. The clothing imbalance is $m = 13$ kg at a distance to the center of $e = 0.5$ m. Assuming the machine spinning at $n = 300$ rpm, find the vibration amplitude.

Solution. The excitation is the centrifugal force caused by the imbalance of the clothing; thus, the vibration equation is

$$\ddot{x} + 2\zeta\omega_n\dot{x} + \omega_n^2 x = \frac{me\omega^2}{M}\sin\omega t, \tag{a}$$

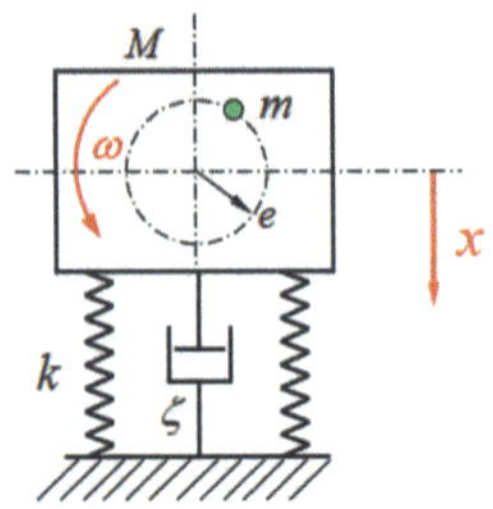

Figure 5.7 Model of a washing machine.

where ω is the excitation frequency calculated by

$$\omega = \frac{2\pi n}{60} = \frac{2\pi(300)}{60} = 31.4 \text{ rad/s}. \tag{b}$$

The total spring constant $K = 4k = 4(8.3)(10^4) = 3.32 \times 10^5$ N/m. Then the natural frequency ω_n is

$$\omega_{\rm n} = \sqrt{\frac{K}{M}} = \sqrt{\frac{3.32(10^5)}{2000}} = 12.8 \text{ rad/s}. \tag{c}$$

The frequency ratio λ is

$$\lambda = \frac{\omega}{\omega_n} = \frac{31.4}{12.8} = 2.4. \tag{d}$$

This indicates that the system is far from resonance. The amplitude of the vibration can be calculated by Eq. (5.35)

$$B = \frac{me}{M}\frac{\lambda^2}{\sqrt{(1-\lambda^2)^2 + 4\zeta^2\lambda^2}} = 3.82(10^{-3}) \text{ m}. \tag{e}$$

This example indicates that the imbalance mass m and its radius e are crucial to the machine's vibration. Thus, balancing introduced in Chapter 2 is considered the primary measure for vibration control. No matter how well the balancing is done, however, the imbalance can't be completely eliminated, and measures of vibration isolation are often needed in practice. Generally vibration isolation problems are roughly classified unto two groups (Preumont, 2018; Rivin, 2003):

1) **Vibration-producing machines**: The goal of vibration isolation in this type of problem is to reduce the dynamic force or vibration transmitted from the vibration-producing machine to the supporting struc-

ture. Typical examples include forging and stamping machines, as well as automobile engines, etc.

2) **Vibration-sensitive machines**: In this group of problems, the vibration source is the supporting structure, and the goal is to reduce and control the force and vibration transmitted from the supporting structure to the vibration-sensitive machine. Examples include precision machine tools in a workshop and electronic devices in a car, as well as telescopes in spacecrafts, etc.

Vibration-isolation measures are roughly divided into two groups as well, namely, *active vibration isolation* and *passive vibration isolation*. In this book, we will not dig into this topic further. Details can be found in specialized monographs, such as (Preumont, 2018; Rivin, 2003).

The effectiveness of vibration isolation is generally evaluated by the *force transmissibility* or *displacement transmissibility* that are defined as the ratio of forces (or displacements) with and without the isolation. In the following example, the vibration isolation of the washing machine in the previous example is further examined.

Example 5.4. In Example 5.3, springs and dampers are installed under each corner of the machine to prevent the vibration being transmitted to the surrounding. Find the force transmissibility of the vibration isolation.

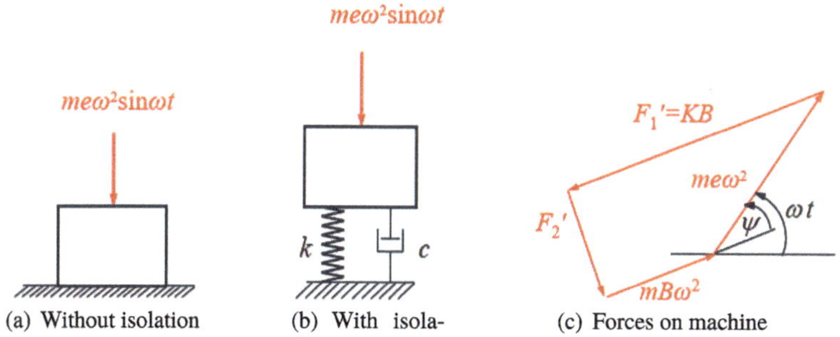

Figure 5.8 Vibration isolation of a washing machine.

Solution. The machine without and with vibration isolation measures are shown in Fig. 5.8(a) and Fig. 5.8(b), respectively. The forces on the machine are shown in Fig. 5.8(c).

If no vibration isolation exists, the centrifugal force caused by the clothing imbalance will be directly transmitted to the ground. The maximum value is

$$F = me\omega^2. \tag{a}$$

With the vibration-isolation mount installed, the forces transmitted to the floor include two components: the elastic spring force F_1' and the damping force F_2'. Keeping in mind that $c = 2\zeta\omega_n M$, we have

$$F_1' = Kx = KB\sin(\omega t - \psi) \tag{b}$$

$$F_2' = c\dot{x} = 2M\zeta\omega_n B\omega\cos(\omega t - \psi). \tag{c}$$

These two forces have the same frequency, ω, as shown in Fig. 5.8(c). The maximum value of the resultant force is

$$\begin{aligned} F' &= \sqrt{(F_{1\max}')^2 + (F_{2\max}')^2} = \sqrt{(KB)^2 + (2M\zeta\omega_n B\omega)^2} \\ &= \frac{Kme\lambda^2\sqrt{1 + 4\zeta^2\lambda^2}}{M\sqrt{(1 - \lambda^2)^2 + 4\zeta^2\lambda^2}}. \end{aligned} \tag{d}$$

The force transmissibility η is defined as

$$\eta = \frac{F'}{F} = \frac{\sqrt{1 + 4\zeta^2\lambda^2}}{\sqrt{(1 - \lambda^2)^2 + 4\zeta^2\lambda^2}} = 0.2475. \tag{e}$$

Obviously, this belongs to the first type of vibration-isolation problem. It can be seen that the force transmitted to the ground with an isolation mount is only about one quarter of that without the mount.

5.3.2 Response to harmonic base-motion excitation

Sometimes, the excitation is in the form of harmonic base motion. This is shown in Fig. 5.9(a). The forces transmitted to mass m through the spring and damper are shown in Fig. 5.9(b).

The equation of motion of the mass m is represented as

$$m\ddot{x} + c(\dot{x} - \dot{h}) + k(x - h) = 0. \tag{5.45}$$

The harmonic base motion is represented as

$$h = H\sin\omega t. \tag{5.46}$$

Substituting it into Eq. (5.45) and rearranging the terms give

$$m\ddot{x} + c\dot{x} + kx = c\omega H\cos\omega t + kH\sin\omega t. \tag{5.47}$$

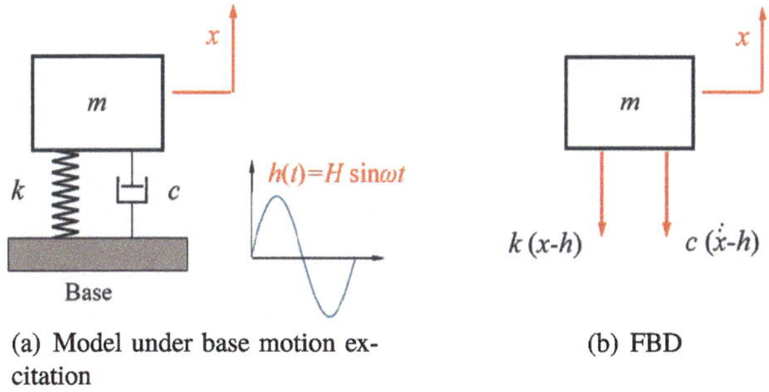

Figure 5.9 Base-motion excitation.

This can be transformed to the following form:

$$m\ddot{x} + c\dot{x} + kx = F_0 \cos(\omega t + \beta), \tag{5.48}$$

where

$$F_0 = H\sqrt{k^2 + c^2\omega^2} \tag{5.49}$$

$$\beta = \arctan\left(-\frac{k}{c\omega}\right). \tag{5.50}$$

Eq. (5.48) can be treated now with the techniques introduced in the force-excited system.

Example 5.5. A moving car is often studied with a simple quarter-car model (single DOF) as shown in Fig. 5.10. The suspension is modeled as a spring plus a damper, and the displacement y acts in the vertical direction. The road surface is simplified as a sinusoidal waveform with parameters H and L, which applies a displacement excitation to the tires. M, C, and K are the mass of the car body, the damping coefficient, and the spring stiffness of the suspension, respectively. The speed of the car is assumed as v. The parameters are given as $H = 0.025$ m, $L = 6$ m, $v = 100$ km/h, $\zeta = 0.85$, and $\omega_n = 7.5$ rad/s. Find the displacement transmissibility provided by the suspension.

Solution. According to the road waveform given, the displacement excitation applied to the tire of the car is

$$h = H \sin \omega t. \tag{a}$$

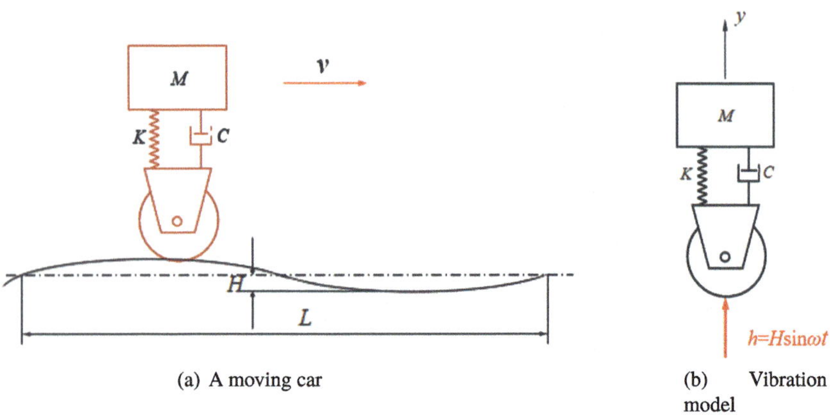

Figure 5.10 A quarter-car model.

ω is the excitation frequency that is calculated by

$$\omega = \frac{2\pi v}{L}. \tag{b}$$

Assuming the vertical displacement of the car body is y as shown in Fig. 5.10(b), the equation of motion of the car body is given by Eq. (5.45):

$$M\ddot{y} + C(\dot{y} - \dot{h}) + K(y - h) = 0. \tag{c}$$

With the road waveform, this equation can be changed to the following form according to Eq. (5.48):

$$M\ddot{y} + C\dot{y} + Ky = F_0 \cos(\omega t + \beta). \tag{d}$$

F_0 and β are calculated by Eqs. (5.49) and (5.50).

Then the solution of (d) can be represented by Eq. (5.34):

$$y = Y\cos(\omega t + \beta - \theta), \tag{e}$$

where Y and θ are

$$Y = \frac{F_0}{K}\frac{1}{\sqrt{(1-\lambda^2)^2 + 4\zeta^2\lambda^2}} \tag{f}$$

$$\theta = \arctan\frac{2\zeta\lambda}{1-\lambda^2}. \tag{g}$$

If no suspension is installed, the displacement caused by the road waveform is directly transmitted to the car body. Thus, the maximum displacement would be H. With the suspension, the car body experiences harmonic vibration whose amplitude is Y. Then the displacement transmissibility η is

$$\eta = \frac{Y}{H} = \frac{\sqrt{1 + 4\zeta^2\lambda^2}}{\sqrt{(1-\lambda^2)^2 + 4\zeta^2\lambda^2}}. \tag{5.51}$$

Substitution of the given numbers into the relevant equation gives $\lambda = 3.88$ and $\eta = 0.43$, indicating that the vibration amplitude with suspension is less than half of that without a suspension.

This quarter-car model is the simplest model used in vehicle dynamics. In a realistic study of vehicle dynamics, more complicated models, such as half-car or complete-car models, are often used.

5.3.3 Response to periodic force excitation

In many engineering problems, the excitation is a general periodic function of time. For example, the gas pressure inside the cylinder of a four-stroke internal combustion engine is shown in Fig. 5.11.

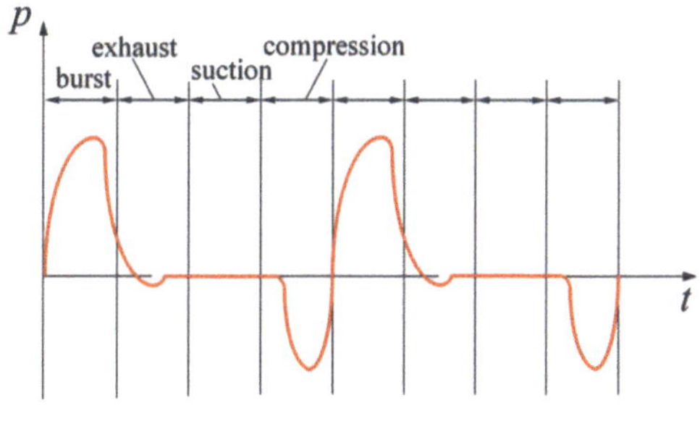

Figure 5.11 Gas pressure curve of a four-stroke internal combustion engine.

One powerful tool for solving vibration problems under periodic excitation is the *superposition principle* (Weaver et al., 1990). In this method, the periodic excitation is first expanded to a *Fourier series*, and then the overall response is obtained through the superposition principle.

5.3.3.1 Fourier series

A periodic excitation can be expanded mathematically into infinite Fourier series. Assuming the period of the force excitation $F(t)$ is T (or $\omega = 2\pi/T$), the following is true:

$$F(t+T) = F(t). \tag{5.52}$$

If the *Dirichlet conditions* are met, $F(t)$ can be represented by the following Fourier series:

$$F(t) = \frac{1}{2}a_0 + \sum_{i=1}^{\infty}\left(a_i \cos i\omega t + b_i \sin i\omega t\right), \tag{5.53}$$

where a_0, a_i, and b_i are the Fourier coefficients in these forms:

$$\begin{cases} a_0 = \dfrac{2}{T}\displaystyle\int_{-\frac{T}{2}}^{\frac{T}{2}} F(t)\,\mathrm{d}t \\ a_i = \dfrac{2}{T}\displaystyle\int_{-\frac{T}{2}}^{\frac{T}{2}} F(t)\cos i\omega t\,\mathrm{d}t \quad i = 1, 2, \cdots, \\ b_i = \dfrac{2}{T}\displaystyle\int_{-\frac{T}{2}}^{\frac{T}{2}} F(t)\sin i\omega t\,\mathrm{d}t \quad i = 1, 2, \cdots. \end{cases} \tag{5.54}$$

Eq. (5.53) can also be written in the following form:

$$F(t) = c_0 + \sum_{i=1}^{\infty} c_i \cos\left(i\omega t - \varphi_i\right), \tag{5.55}$$

where

$$\begin{cases} c_0 = \dfrac{a_0}{2} \\ c_i = \sqrt{a_i^2 + b_i^2} \\ \varphi_i = \arctan\left(\dfrac{b_i}{a_i}\right). \end{cases} \tag{5.56}$$

5.3.3.2 Superposition principle

The superposition principle simply states that the response caused by two or more excitation forces is the sum of the responses that would have been caused by each individual excitation. In other words, if $x_i(t)$ is the response to an individual excitation $F_i(t)$, the response to a combined excitation $c_1F_1(t) + c_2F_2(t) + \cdots$ is $x(t) = c_1x_1(t) + c_2x_2(t) + \cdots$ (c_i are constants). Attention should be paid to the fact that the superposition principle only applies to linear systems. Thus, if the system is nonlinear, it cannot be used.

After a periodic excitation force $F(t)$ is expanded into Fourier series, the vibration equation is

$$m\ddot{x} + c\dot{x} + kx = c_0 + \sum_{i=1}^{\infty} c_i \cos(i\omega t - \varphi_i), \tag{5.57}$$

where c_0 is a static force; it does not cause vibration. According to the superposition principle, the response can be represented as the sum of the response caused by each individual term in the series:

$$x(t) = \sum_{i=1}^{\infty} \frac{c_i}{k} \beta_i \cos(i\omega t - \varphi_i - \theta_i), \tag{5.58}$$

where β_i and θ_i are calculated by Eq. (5.39) and Eq. (5.36), respectively.

$$\begin{cases} \beta_i = \dfrac{1}{\sqrt{(1-\lambda_i^2)^2 + 4\zeta^2\lambda_i^2}} \\ \theta_i = \arctan \dfrac{2\zeta\lambda_i}{1-\lambda_i^2}, \end{cases} \tag{5.59}$$

where $\lambda_i = i\omega/\omega_n$ is the i-th-order frequency ratio.

In real problems, it is neither possible nor necessary to include the infinite Fourier series in Eq. (5.55). Common practice is to truncate the series, and keep only the several low-order terms. The number of terms retained depends on the accuracy requirement.

Example 5.6. A single-DOF undamped system is subjected to force excitation described by Fig. 5.12. Develop the response of the system.

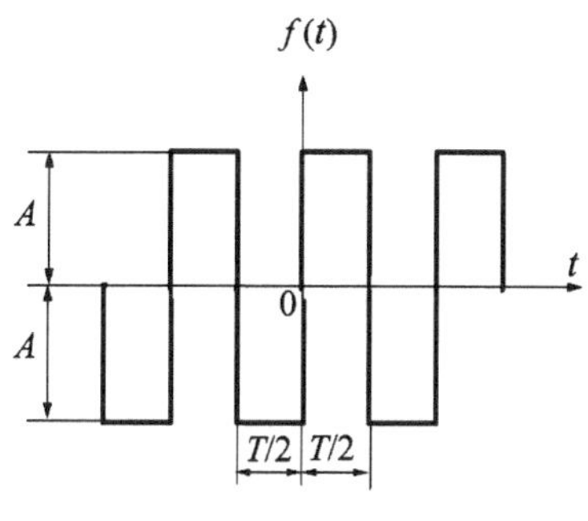

Figure 5.12 Force excitation of rectangular form.

Solution. The rectangular-form excitation is mathematically represented as

$$f(t) = \begin{cases} -A & \text{if } -\frac{T}{2} \le t \le 0 \\ A & \text{if } 0 \le t \le \frac{T}{2}. \end{cases} \tag{a}$$

It is straightforward to know that $a_0 = a_i = 0$ without conducting the integration because $f(t)$ is an odd function.

b_i is calculated by

$$b_i = \frac{2}{T}\left[\int_{-\frac{T}{2}}^{0} (-A)\sin i\omega t\,\mathrm{d}t + \int_{0}^{\frac{T}{2}} A\sin i\omega t\,\mathrm{d}t\right] = \frac{4A}{\pi i}. \tag{b}$$

The excitation $f(t)$ can be written as

$$f(t) = \frac{4A}{\pi} \sum_{i=1,3,\cdots,}^{\infty} \frac{1}{i} \sin i\omega t. \tag{c}$$

The response is

$$x(t) = \frac{4A}{\pi} \sum_{i=1,3,\cdots,}^{\infty} \frac{1}{i[1-(i\omega/\omega_n)]} \sin i\omega t. \tag{d}$$

5.3.3.3 Response in frequency domain

To this point, vibration problems have been treated in such a way that a given excitation $f(t)$ results in the response $x(t)$. The response $x(t)$ is a function of time, and it is commonly visualized by a diagram with time t as the abscissa and x as the ordinate. Such a diagram is referred to as *time history*. This type of analysis is termed *time-domain* analysis.

With Fourier series, a general periodic-force excitation is expressed as a sum of an infinite number of harmonics. Resonance will happen when any one of the harmonics is close to the natural frequency. Fig. 5.6 provides a convenient way to examine the resonance, in particular, for multiple harmonics excitation. In contrast to a time history, this figure has the frequency as the abscissa. Vibration analysis of this type is referred to as *frequency-domain* analysis. It is easy to imagine that the amplitude–frequency curve under a general periodic-force excitation has multiple peaks.

The bridge between time domain and frequency domain is the *Fourier transform*.

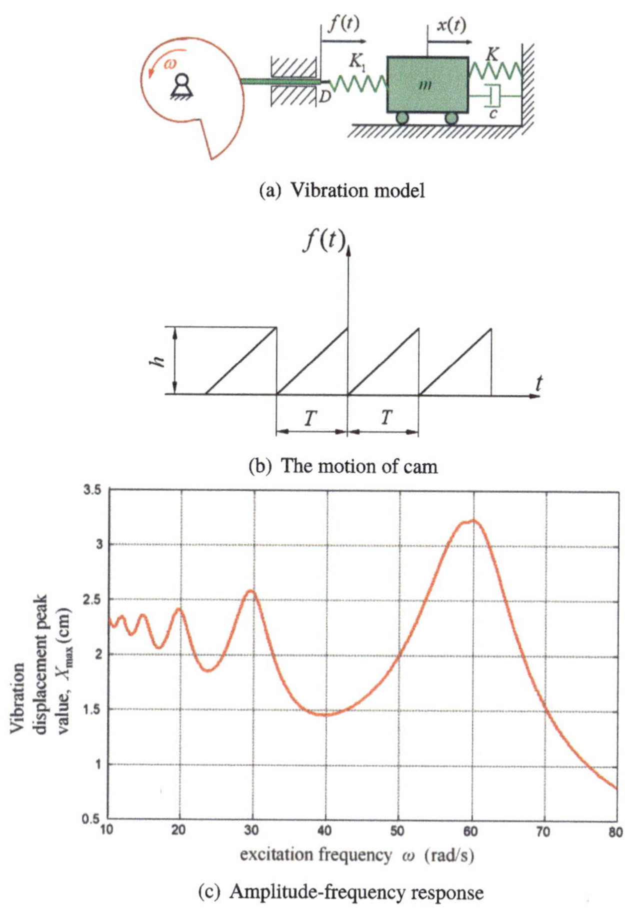

Figure 5.13 A machine-tool feeding system.

Example 5.7. Fig. 5.13(a) shows a simplified model of the cam-feeding system in a machine tool. m is the total mass of the sliding table and the tool holder. K and c are the stiffness and damping coefficient in the cutting interface. K_1 represents the stiffness of the feeding system. The cam pushes the feeding system through a rod D, horizontally. The displacement of the motion program at the end of D is shown in Fig. 5.13(b). The rotation speed of the cam is n. Find the response of the mass m.

Solution. Obviously the excitation is in the form of base motion. It is straightforward to write the equation of motion for Fig. 5.13(a) as

$$m\ddot{x} + c\dot{x} + (K + K_1)x = K_1 f(t), \tag{a}$$

where $f(t)$ is the horizontal displacement given in Fig. 5.13(b) at the end D. It can be formulated in the following form:

$$f(t)=\begin{cases} h+\dfrac{h}{T}t & \text{if } -\dfrac{T}{2}\le t\le 0 \\ \dfrac{h}{T}t & \text{if } 0\le t\le \dfrac{T}{2}. \end{cases} \tag{b}$$

No damping is included in between m and the rod D. The excitation frequency is

$$\omega=\frac{2\pi n}{60}, \tag{c}$$

and the period is

$$T=\frac{2\pi}{\omega}. \tag{d}$$

The Fourier coefficients of $f(t)$ calculated by Eq. (5.54) are

$$a_0=\frac{2}{T}\left[\int_{-\frac{T}{2}}^{0}\left(h+\frac{h}{T}t\right)\mathrm{d}t+\int_{0}^{\frac{T}{2}}\left(\frac{h}{T}t\right)\mathrm{d}t\right]=h \tag{e}$$

$$a_i=\frac{2}{T}\left[\int_{-\frac{T}{2}}^{0}\left(h+\frac{h}{T}t\right)\cos i\omega t\mathrm{d}t+\int_{0}^{\frac{T}{2}}\left(\frac{h}{T}t\right)\cos i\omega t\mathrm{d}t\right]=0 \tag{f}$$

$$b_i=\frac{2}{T}\left[\int_{-\frac{T}{2}}^{0}\left(h+\frac{h}{T}t\right)\sin i\omega t\mathrm{d}t+\int_{0}^{\frac{T}{2}}\left(\frac{h}{T}t\right)\sin i\omega t\mathrm{d}t\right]=-\frac{h}{\pi i}. \tag{g}$$

$f(t)$ then can be represented as

$$f(t)=h\left[\frac{1}{2}-\frac{1}{\pi}\sum_{i=1}^{\infty}\frac{1}{i}\sin i\omega t\right]. \tag{h}$$

Substituting (h) into (a), the solution of (a) can be represented as

$$x(t)=\frac{K_1 h}{K_1+K}\left[\frac{1}{2}-\frac{1}{\pi}\sum_{i=1}^{\infty}\frac{1}{\sqrt{\left(1-\lambda_i^2\right)^2+4\zeta^2\lambda^2}}\sin(i\omega t-\varphi_i)\right], \tag{i}$$

where

$$\omega_{\text{n}}=\sqrt{\frac{K+K_1}{m}} \tag{j}$$

$$\lambda_i = \frac{i\omega}{\omega_n} \tag{k}$$

$$\varphi_i = \arctan\frac{2\zeta\lambda_i}{1-\lambda_i^2}. \tag{l}$$

For a given set of parameters: $\omega_n = 60$ rad/s, $h = 0.02$ m, $\zeta = 0.1$, K is much smaller than K_1, and thus is neglected. Keep the first 20 harmonics in the response ($i = 1$–20). The variation of x_{max} with ω is numerically plotted in Fig. 5.13(c) with ω in the range 10–80 rad/s. Clearly the peaks appear around $\omega \approx 10$, 12, 15, 20, 30, and 60 rad/s within the range of ω. They are caused by the harmonics ranging from the sixth order to the primary one.

Example 5.8. Fig. 5.14(a) shows a single-cylinder internal combustion engine mounted on the base with springs and dampers. Under the reciprocating inertial force of the piston, the frame of the engine vibrates in the vertical direction. Assuming the following parameters: mass of the frame $M = 50$ kg, spring stiffness $K = 2 \times 10^8$ N/m, damping ratio $\zeta = 0.2$, mass of the piston $m = 3$ kg, $r = 0.05$ m, $l = 0.25$ m, analyze the forced vibration of the engine frame.

Solution. In Chapter 2, the acceleration of the engine piston is given by Eq. (2.18) (see p. 28). If the crank rotates at a constant speed, then terms with $\ddot{\theta}$ will disappear, and the acceleration becomes

$$a_C = r\omega^2(-\cos\theta - A_2\cos 2\theta + A_4\cos 4\theta - A_6\cos 6\theta + \cdots). \tag{a}$$

Truncating and remaining six lower-order harmonics, the inertial force caused by this acceleration can be generalized as

$$f(t) = -mr\omega^2\sum_{i=1}^{6} A_i\cos i\omega t, \tag{b}$$

where $A_1 = 1$, $A_3 = A_5 = 0$. A_2, A_4, and A_6 are given in Eq. (2.17) on p. 28. The vibration equation of the frame under this inertial force is

$$M\ddot{x} + C\dot{x} + Kx = -mr\omega^2\sum_{i=1}^{6} A_i\cos i\omega t. \tag{c}$$

The response is then, according to Eqs. (5.34) through (5.39),

$$x(t) = -\frac{mr\omega^2}{K}\sum_{i=1}^{6} \beta_i\cos(i\omega t - \varphi_i), \tag{d}$$

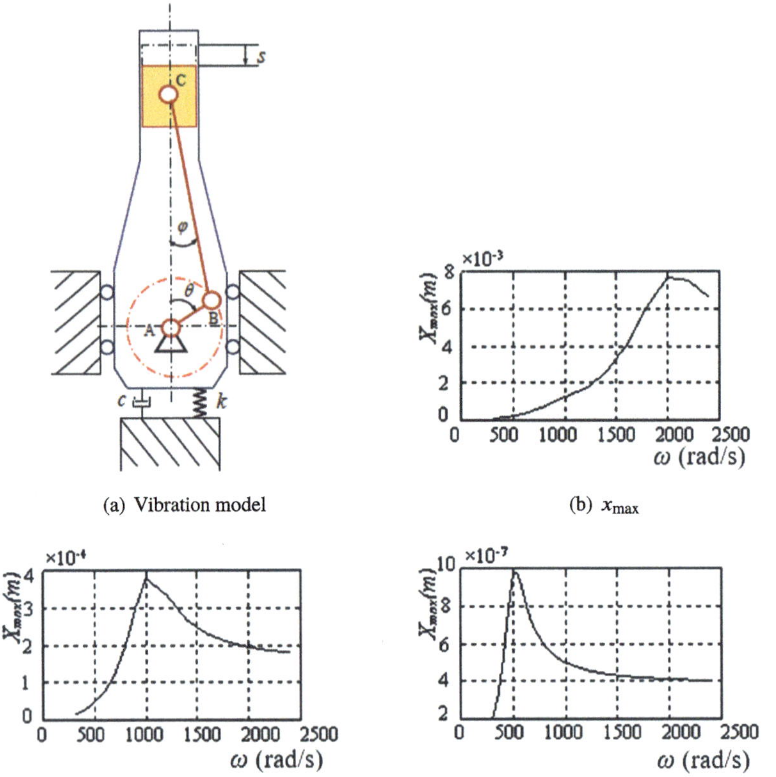

(a) Vibration model

(b) $x_{\max}$

(c) $x_{\max}$ with 1st order inertial force balanced

(d) $x_{\max}$ with 1st and 2nd order inertial forces balanced

Figure 5.14 Vibration of a single-cylinder engine.

where

$$\beta_i = \frac{1}{\sqrt{\left[1-(i\omega/\omega_{\mathrm{n}})^2\right]^2 + 4\zeta^2\,(i\omega/\omega_{\mathrm{n}})^2}} \tag{e}$$

$$\varphi_i = \arctan\left(\frac{2\zeta\omega_{\mathrm{n}} i\omega}{\omega_{\mathrm{n}}^2 - i^2\omega^2}\right). \tag{f}$$

The natural frequency ω_{n} is

$$\omega_{\mathrm{n}} = \sqrt{\frac{K}{M}} = \sqrt{\frac{2(10^8)}{50}} = 2000\ \mathrm{rad/s}. \tag{g}$$

Fig. 5.14(b) shows the $x_{\max}$ variation with ω. Clearly, resonance happens at $\omega_{\mathrm{n}} = \omega = 2000$ rad/s. Fig. 5.14(c) is the $x_{\max}$ with the primary inertial force balanced. It can be seen that x_{max} is dramatically decreased, and the resonance happens at $\omega_n = 2\omega = 2000$ rad/s. Fig. 5.14(d) is the case with both the primary and secondary inertial forces balanced. In this figure, the x_{max} decreases further, and the resonance happens at $\omega_n = 4\omega = 2000$ rad/s. It shows that the vibration caused by inertial forces of higher orders is very small and can be neglected in general.

5.3.4 Response to nonperiodic forces

The equation of motion under a nonperiodic excitation $F(t)$ can be given by

$$m\ddot{x} + c\dot{x} + kx = F(t) \tag{5.60}$$

or

$$\ddot{x} + 2\zeta\omega_n\dot{x} + \omega_n^2 x = \frac{F(t)}{m} = f(t). \tag{5.61}$$

The simplest nonperiodic force is an impulse, which has a large magnitude F, but acts over a very short time Δt. Fig. 5.15 (a) and (b) show an impulse acting at $t = 0$ and $t = \tau$, respectively. The strength of the impulse I is defined as $I = F\Delta t$.

Before dealing with a general nonperiodic force, let's first look at the mass m at rest initially, but subjected to an impulse at $t = 0$, as shown in Fig. 5.15(a). When an impulse applies on an object, it changes the momentum according to $I = F\Delta t = m\Delta v$. Thus, we have

$$\Delta v = \frac{I}{m}. \tag{5.62}$$

This means that the impulse I acting at $t = 0$ is equivalent to the free vibration under the initial conditions: $x(0) = 0$ and $\dot{x}(0) = I/m$. The free-vibration response of a underdamped system to the initial conditions is governed by Eq. (5.28). In this case, it can be represented as

$$x(t) = \frac{e^{-\zeta\omega_n t} I}{\omega_d m} \sin\omega_d t = Ih(t), \tag{5.63}$$

where $h(t)$ is called the *impulse response function*, representing the response to a unit impulse applied at $t = 0$:

$$h(t) = \frac{e^{-\zeta\omega_n t}}{\omega_d m} \sin\omega_d t. \tag{5.64}$$

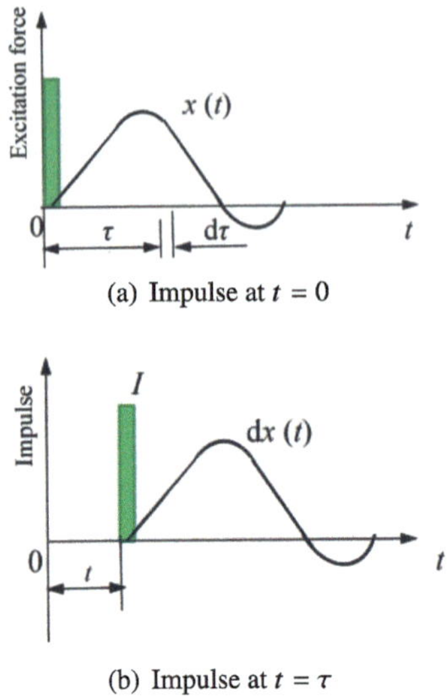

Figure 5.15 Impulse and impulse response.

Now let's move one step further to look at a more general case when an impulse acts at $t = \tau$ as shown in Fig. 5.15(b). In this case, the response is zero before $t = \tau$ and is obtained by simply replacing t with $t - \tau$ in Eq. (5.63) after $t = \tau$. Thus, we have

$$x(t) = \begin{cases} 0 & \text{if } 0 \le t \le \tau \\ Ih(t-\tau) & \text{if } \tau \le t, \end{cases} \tag{5.65}$$

where

$$h(t-\tau) = \frac{e^{-\zeta\omega_n(t-\tau)}}{\omega_d m} \sin \omega_d (t-\tau). \tag{5.66}$$

For a more general nonperiodic force excitation, we divide the force into many impulses $F(t_i)\Delta t$ as shown in Fig. 5.16. The response caused by the individual impulse $F(t_i)\Delta t$ is

$$\Delta x(t_i) = [F(t_i)\Delta t]\, h(t - t_i), \tag{5.67}$$

and the total response caused by all the impulses is the sum of

$$x(t) = \sum x(t) = \sum [F(t_i)\Delta t]\, h(t - t_i). \tag{5.68}$$

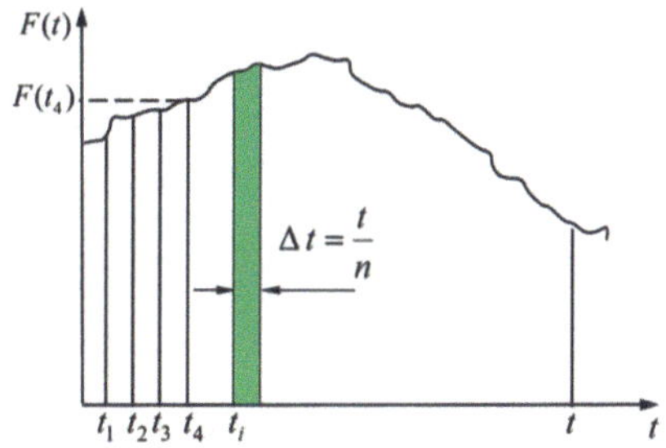

Figure 5.16 Arbitrary force excitation.

Letting $\Delta \to 0$, the summation becomes the integral

$$x(t) = \int_0^t F(\tau)h(t-\tau)d\tau = \frac{1}{\omega_d m}\int_0^t F(\tau)e^{-\zeta\omega_n(t-\tau)}\sin\omega_d(t-\tau)d\tau. \quad (5.69)$$

This relationship was first derived by J. Duhamel in 1839 and is called *Duhamel's integral.* It represents the response caused by the excitation force $F(t)$ from time 0 to t without including the response caused by the initial conditions.

Summation of the responses by the force $F(t)$ (Eq. (5.69)) and initial conditions $x_0 = x(0)$, $\dot{x}_0 = \dot{x}(0)$ (Eq. (5.28)) gives the total response:

$$\begin{aligned} x(t) =& e^{-\zeta\omega_n t}\left[x_0\cos\omega_d t + \frac{\dot{x} + \zeta\omega_n x_0}{\omega_d}\sin\omega_d t\right] + \\ & \frac{1}{\omega_d m}\int_0^t F(\tau)e^{-\zeta\omega_n(t-\tau)}\sin\omega_d(t-\tau)d\tau. \end{aligned} \quad (5.70)$$

CHAPTER 6

Vibration of systems with multiple DOFs

6.1 Introduction

In Chapter 5, the vibration theory of single DOF was introduced. Although some problems can be simplified as single-DOF systems, most real problems in engineering require more complicated multiple-DOF models to achieve satisfactory accuracy of analysis. The vibration of multiple-DOF systems has some features that single-DOF systems do not have. In addition, methods to derive the vibration equations of multiple-DOF systems are more diversified, including not only Newton's second law, but also many other methods, such as the Lagrange equation, the flexibility influence coefficient method, etc.

This chapter is dedicated to the linear vibration of multiple-DOF systems. The chapter is organized as follows:

- Two DOF systems, the simplest multiple DOF system, are first presented in Section 6.2 to bring up the basic concepts in multiple-DOF vibration systems.
- In Section 6.3, coverage transitions to general multiple-DOF systems.
- Section 6.4 introduces the solution method of multiple-DOF vibration equations. Specifically, the modal superposition method in presented.
- Section 6.5 analyzes the free and forced vibration response of multiple-DOF systems using the method introduced in Section 6.4.

To help understand the basic concepts, several examples are presented in the various sections.

6.2 Vibration of two-DOF systems

6.2.1 Equation derived through Newton's second law

Example 6.1 (A spring–mass system). Fig. 6.1(a) shows two masses, m_1 and m_2, that are connected by springs and dampers. The stiffness coefficients of the springs are k_1, k_2, and k_3, and the damping coefficients are c_1, c_2, and c_3, respectively. Two force excitations, $F_1(t)$ and $F_2(t)$, act on the two masses. Develop the equation of motion for the two masses.

Machinery Dynamics
https://doi.org/10.1016/B978-0-12-815785-5.00016-9

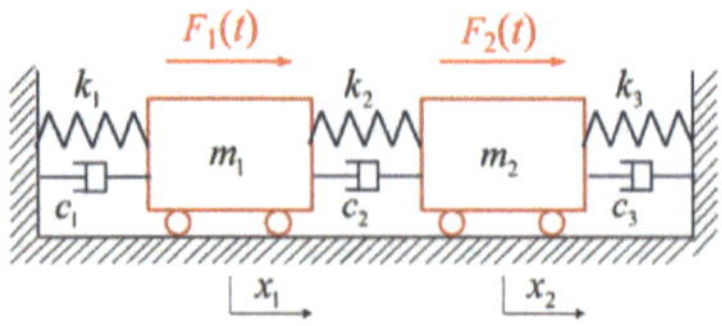

(a) Vibration model

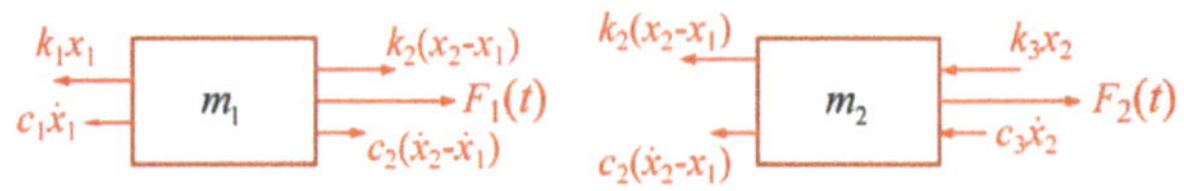

(b) FBD

Figure 6.1 Two-mass system.

Solution. To describe the motion of the two masses, two variables, x_1 and x_2, which are the linear displacements of m_1 and m_2 from their equilibrium positions, are chosen as the *generalized coordinates.*

The free-body diagrams (FBD) of the two masses are shown in Fig. 6.1(b). According to Newton's second law of motion, the equation of motion of the two masses are

$$\begin{cases} m_1\ddot{x}_1 = -k_1x_1 - k_2(x_1 - x_2) - c_1\dot{x}_1 - c_2(\dot{x}_1 - \dot{x}_2) + F_1(t) \\ m_2\ddot{x}_2 = -k_3x_2 + k_2(x_1 - x_2) - c_3\dot{x}_2 + c_2(\dot{x}_1 - \dot{x}_2) + F_2(t). \end{cases} \tag{6.1}$$

Organizing this into matrix form yields

$$\begin{bmatrix} m_1 & 0 \\ 0 & m_2 \end{bmatrix} \begin{Bmatrix} \ddot{x}_1 \\ \ddot{x}_2 \end{Bmatrix} + \begin{bmatrix} c_1 + c_2 & -c_2 \\ -c_2 & c_2 + c_3 \end{bmatrix} \begin{Bmatrix} \dot{x}_1 \\ \dot{x}_2 \end{Bmatrix} + \begin{bmatrix} k_1 + k_2 & -k_2 \\ -k_2 & k_2 + k_3 \end{bmatrix} \begin{Bmatrix} x_1 \\ x_2 \end{Bmatrix} = \begin{Bmatrix} F_1(t) \\ F_2(t) \end{Bmatrix}. \tag{6.2}$$

It can be written in more compact form as

$$\boldsymbol{M}\ddot{\boldsymbol{x}} + \boldsymbol{C}\dot{\boldsymbol{x}} + \boldsymbol{K}\boldsymbol{x} = \boldsymbol{F}(t), \tag{6.3}$$

where $\boldsymbol{M}$, $\boldsymbol{C}$, and $\boldsymbol{K}$ are the *mass matrix*, *damping matrix*, and *stiffness matrix*, respectively. $\boldsymbol{x}$ and $\boldsymbol{F}(t)$ are the *generalized coordinate vector* and the *generalized force vector*. Their components are

$$\boldsymbol{M}=\begin{bmatrix} m_1 & 0 \\ 0 & m_2 \end{bmatrix}, \quad \boldsymbol{C}=\begin{bmatrix} c_1+c_2 & -c_2 \\ -c_2 & c_2+c_3 \end{bmatrix}, \quad \boldsymbol{K}=\begin{bmatrix} k_1+k_2 & -k_2 \\ -k_2 & k_2+k_3 \end{bmatrix}$$

$$\boldsymbol{x}=\begin{Bmatrix} x_1 & x_2 \end{Bmatrix}^{\mathrm{T}}, \quad \boldsymbol{F}(t)=\begin{Bmatrix} F_1(t) & F_2(t) \end{Bmatrix}^{\mathrm{T}}.$$

Eq. (6.3) formulates the forced vibration. Letting $F_1(t)=F_2(t)=0$, it reduces to the damped free vibration as

$$\boldsymbol{M}\ddot{\boldsymbol{x}}+\boldsymbol{C}\dot{\boldsymbol{x}}+\boldsymbol{K}\boldsymbol{x}=\boldsymbol{0}. \tag{6.4}$$

Example 6.2 (A torsional shaft–disk system). A torsional shaft–disk system is shown Fig. 6.2. The moments of inertia of the two disks are J_1 and J_2, respectively. The three shaft segments are modeled as torsional springs, k_1, k_2, and k_3. Two torques $T_1(t)$ and $T_2(t)$ act on the two disks. The damping of the shaft is neglected. Derive the vibration equation.

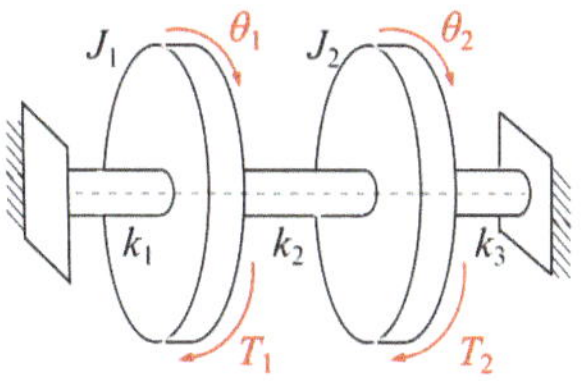

Figure 6.2 Two-DOF torsional vibration.

Solution. Choosing the angular displacements of the disks from the equilibrium positions, θ_1 and θ_2, as the generalized coordinates, the equation of motion of the two disks can be obtained according to Newton's second law of motion:

$$\begin{cases} J_1\ddot{\theta}_1=-k_1\theta_1-k_2(\theta_1-\theta_2)+T_1(t) \\ J_2\ddot{\theta}_2=-k_3\theta_2+k_2(\theta_1-\theta_2)+T_2(t). \end{cases} \tag{6.5}$$

It can be written in matrix form as

$$\boldsymbol{J}\ddot{\boldsymbol{\theta}}+\boldsymbol{K}\boldsymbol{\theta}=\boldsymbol{T}(t), \tag{6.6}$$

where

$$\boldsymbol{J}=\begin{bmatrix} J_1 & 0 \\ 0 & J_2 \end{bmatrix}, \quad \boldsymbol{K}=\begin{bmatrix} k_1+k_2 & -k_2 \\ -k_2 & k_2+k_3 \end{bmatrix}$$

$$\boldsymbol{\theta} = \begin{Bmatrix} \theta_1 & \theta_2 \end{Bmatrix}^{\mathrm{T}}, \quad \boldsymbol{T}(t) = \begin{Bmatrix} T_1(t) & T_2(t) \end{Bmatrix}^{\mathrm{T}}.$$

Letting $T_1(t) = T_2(t) = 0$, Eq. (6.6) becomes this undamped free-vibration equation:

$$\boldsymbol{J}\ddot{\boldsymbol{\theta}} + \boldsymbol{K}\boldsymbol{\theta} = \boldsymbol{0}. \tag{6.7}$$

6.2.2 Equation derived through Lagrange's equation

Lagrange's equation without damping was already introduced in Chapter 4 (Eq. (4.1) on p. 87). If damping is included, it will take the following form. Given the need for brevity, we directly give the equation without presenting the derivation:

$$\frac{\mathrm{d}}{\mathrm{d}t}\left(\frac{\partial E_{\mathrm{k}}}{\partial \dot{q}_i}\right) - \frac{\partial E_{\mathrm{k}}}{\partial q_i} + \frac{\partial E_{\mathrm{p}}}{\partial q_i} + \frac{\partial E_{\mathrm{d}}}{\partial \dot{q}_i} = Q_i \quad (i = 1, 2, \cdots, n), \tag{6.8}$$

where q_i $(i = 1, 2, \cdots, n)$ are the generalized coordinates. Q_i is the generalized force corresponding to q_i. E_{k}, E_{p}, and E_{d} are the kinetic energy, the potential energy, and the dissipation function of the system, respectively.

E_{d} is calculated by

$$E_{\mathrm{d}} = \frac{1}{2}\sum_{i=1}^{n}\sum_{j=1}^{n} c_{ij}\dot{q}_i\dot{q}_j \tag{6.9}$$

where c_{ij} is the damping coefficient. For undamped system, E_{d} is dropped off.

Lagrange's equation formulates the motion of mechanical systems from the point of view of energy. The procedures to derive the vibration equation with Lagrange's equation is outlined as: 1) choosing the generalized coordinates of the system; 2) representing the kinetic energy, the potential energy, the dissipation function, and the generalized force in terms of the generalized coordinates and generalized velocities; and 3) substituting into Eq. (6.8) to obtain the vibration equation of the system. We will illustrate the procedures through the following examples.

Example 6.3 (Vibration of a car body). Fig. 6.3(a) shows a simplified vibration model of a car body. In this model, the car body is treated as rigid. The front and rear wheels, including tires and suspension springs, are modeled as springs k_1 and k_2. If only the pitch and heaving are considered, the car body can be represented by a mass M and a moment of inertia J

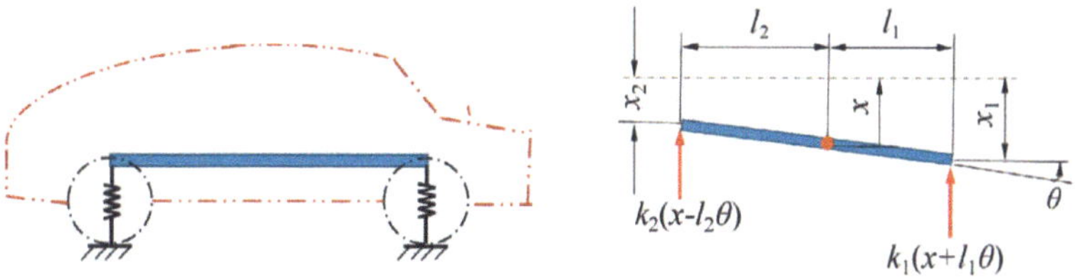

(a) The vibration of a car (b) Motion of the car body

Figure 6.3 Car-body vibration.

around its center of mass, whose location is shown in Fig. 6.3(b). Develop the vibration equation using Lagrange's equation; do not consider damping.

Solution. The first step is to choose the generalized coordinates. In this case, we choose the vertical displacement of the mass center x (heaving) (downward positive) and the rotation angle θ (pitch) as the generalized coordinates (Fig. 6.3(b)) to represent the motion of the car body. That is, $q_1 = x$ and $q_2 = \theta$. As expected, x is measured from the static equilibrium position so that the weight of the car body and the static deformation of the springs do not appear in the equation of motion.

Then, the kinetic energy and potential energy need to be represented as functions of the generalized coordinates and velocities. The kinetic energy is

$$E_k = \frac{1}{2}M\dot{x}^2 + \frac{1}{2}J\dot{\theta}^2, \tag{6.10}$$

where $\dot{x}$ and $\dot{\theta}$ are the vertical velocity and the angular velocity of rotation, respectively.

The potential energy contains only the elastic energy of the two springs because the static deformation is not included:

$$E_p = \frac{1}{2}k_1x_1^2 + \frac{1}{2}k_2x_2^2, \tag{6.11}$$

where x_1 and x_2 are the vertical displacements of the body at the front and rear wheel, respectively. Because the car body is treated as rigid, the following relationships exist:

$$\begin{cases} x_1 = x + l_1\theta \\ x_2 = x - l_2\theta. \end{cases} \tag{6.12}$$

Substitution of x_1 and x_2 into Eq. (6.11) yields

$$E_p = \frac{1}{2}k_1(x + l_1\theta)^2 + \frac{1}{2}k_2(x - l_2\theta)^2. \quad (6.13)$$

From Eqs. (6.10) and (6.13), we can obtain the following:

$$\frac{\partial E_k}{\partial \dot{q}_1} = \frac{\partial E_k}{\partial \dot{x}} = M\dot{x}, \quad \frac{d}{dt}\left(\frac{\partial E_k}{\partial \dot{q}_1}\right) = M\ddot{x}, \quad \frac{\partial E_k}{\partial q_1} = 0$$

$$\frac{\partial E_k}{\partial \dot{q}_2} = \frac{\partial E_k}{\partial \dot{\theta}} = J\dot{\theta}, \quad \frac{d}{dt}\left(\frac{\partial E_k}{\partial \dot{q}_2}\right) = J\ddot{\theta}, \quad \frac{\partial E_k}{\partial q_2} = 0$$

$$\frac{\partial E_p}{\partial q_1} = \frac{\partial E_p}{\partial x} = k_1\left(x + l_1\theta\right) + k_2\left(x - l_2\theta\right)$$

$$\frac{\partial E_p}{\partial q_2} = \frac{\partial E_p}{\partial \theta} = l_1k_1\left(x + l_1\theta\right) - l_2k_2\left(x - l_2\theta\right).$$

There are no external forces applied to the system; thus, the generalized forces are zero:

$$Q_1 = Q_2 = 0. \quad (6.14)$$

Substituting these into Eq. (6.8) gives

$$\begin{cases} M\ddot{x} + k_1\left(x + l_1\theta\right) + k_2\left(x - l_2\theta\right) = 0 \\ J\ddot{\theta} + k_1l_1\left(x + l_1\theta\right) - k_2l_2\left(x - l_2\theta\right) = 0. \end{cases} \quad (6.15)$$

Writing it in matrix form gives

$$\boldsymbol{M}\ddot{\boldsymbol{U}} + \boldsymbol{K}\boldsymbol{U} = \boldsymbol{0}, \quad (6.16)$$

where $\boldsymbol{M}$, $\boldsymbol{K}$, and $\boldsymbol{U}$ are the mass matrix, stiffness matrix, and the generalized coordinate vector, respectively:

$$\boldsymbol{M} = \begin{bmatrix} M & 0 \\ 0 & J \end{bmatrix}, \quad \boldsymbol{K} = \begin{bmatrix} k_1 + k_2 & k_1l_1 - k_2l_2 \\ k_1l_1 - k_2l_2 & k_1l_1^2 + k_2l_2^2 \end{bmatrix}, \quad \boldsymbol{U} = \begin{Bmatrix} x \\ \theta \end{Bmatrix}.$$

Lagrange's equation method does not show much advantage over Newton's method in this example. For a more complicated system with a higher number of DOFs, however, the Lagrange's equation method is generally more organized than Newton's second law. Thus, it is preferable in developing the equation of multiple-DOF systems.

Many other engineering systems, such as industrial vibratory screen equipment and lathe beds, can be studied with similar two-DOF vibration models.

6.2.3 Vibration absorbers

An important application of two-DOF vibration models is vibration absorbers. In this section, we briefly introduce this industrial device without involving the theory of the solution method for multiple-DOF vibration equations, which will be introduced in Section 6.4.

Mass imbalance causes harmonic vibration in many rotating machines. When the frequency of rotation is close to the natural frequency of the system, resonance occurs. Altering the mass or stiffness of the system can change the natural frequency, thus reducing the vibration; however, it this is at times hard to implement for existing machines.

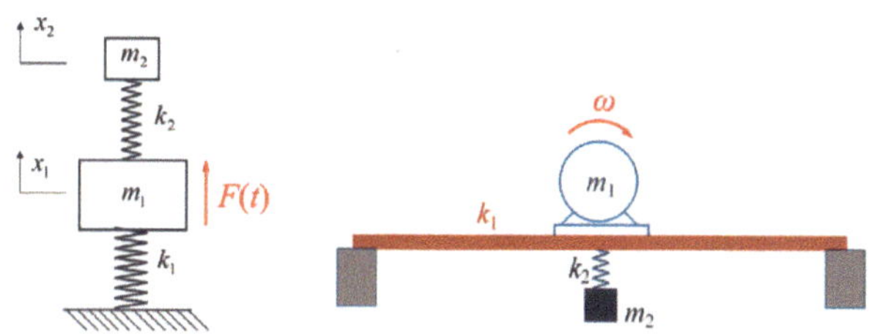

(a) Vibration model (b) Application

Figure 6.4 Vibration absorber.

A *vibration absorber* is able to control the vibration of a mass by adding an extra spring–mass element to the system. As shown in Fig. 6.4(a), a machine having a mass m_1 is supported by a spring of stiffness k_1. A harmonic excitation force $F(t) = F_1 \sin \omega t$ acts on the machine. The vibration absorber is a small mass m_2 that is connected to the machine through a spring k_2. After the addition of the vibration absorber, the original single-DOF machine becomes a two-DOF vibration system. The motion equations are

$$\begin{cases} m_1\ddot{x}_1 + (k_1 + k_2)x_1 - k_2x_2 = F_1 \sin \omega t \\ m_2\ddot{x}_2 - k_2x_1 + k_2x_2 = 0. \end{cases} \tag{6.17}$$

Given the excitation is harmonic, and no damping is included, we assume the response is also harmonic. So we can assume

$$\begin{cases} x_1(t) = X_1 \sin \omega t \\ x_2(t) = X_2 \sin \omega t. \end{cases} \tag{6.18}$$

Substitution of Eq. (6.18) into Eq. (6.17) yields

$$\begin{bmatrix} k_1 + k_2 - \omega^2 m_1 & -k_2 \\ -k_2 & k_2 - \omega^2 m_2 \end{bmatrix} \begin{Bmatrix} X_1 \\ X_2 \end{Bmatrix} = \begin{Bmatrix} F_1 \\ 0 \end{Bmatrix}. \tag{6.19}$$

From this we can solve for X_1 and X_2:

$$\begin{cases} X_1 = \dfrac{(k_2 - \omega^2 m_2) F_1}{(k_1 + k_2 - \omega^2 m_1)(k_2 - \omega^2 m_2) - k_2^2} \\ X_2 = \dfrac{k_2 F_1}{(k_1 + k_2 - \omega^2 m_1)(k_2 - \omega^2 m_2) - k_2^2}. \end{cases} \tag{6.20}$$

Let

$$k_2 - \omega^2 m_2 = 0 \quad \text{or} \quad \omega = \sqrt{\frac{k_2}{m_2}} = \omega_2, \tag{6.21}$$

where ω and ω_2 are the excitation frequency and the natural frequency of the vibration absorber, respectively. From Eq. (6.20), we have $X_1 = 0$ and $X_2 = -F_1/k_2$. Thus, the responses of the machine and the absorber are

$$\begin{cases} x_1(t) = 0 \\ x_2(t) = -\dfrac{F_1}{k_2} \sin \omega t. \end{cases} \tag{6.22}$$

This means that, if choosing the absorber's natural frequency ω_2 equals the excitation frequency ω, the vibration will be completely transferred to the absorber, and the machine does not vibrate at all.

A closer look at the force applied to m_1 through the spring k_2 reveals that $k_2 x_2(t) = -F_1 \sin \omega t$. This means that this force $k_2 x_2(t)$ and the excitation force $F(t)$ are equal to each other in magnitude, but are opposite in direction at all times. Hence the two forces cancel out each other, and no vibration happens to the machine. An application of the vibration absorber is shown in Fig. 6.4(b), in which the machine is supported from a beam.

The vibration absorber introduced above does not include a damping element; thus, it is called *undamped vibration absorber*. If damping is included, the absorber is referred to as *damped vibration absorber*, which also find wide application in industry. The analysis of damped absorbers is very similar to the undamped ones except a little more complicated because of the involvement of damping (Rao, 2010).

6.3 Vibration of multiple-DOF systems

6.3.1 Discretization of continuous systems

Literally, any actual mechanical component is a continuous elastic body. The form of a continuous-system vibration equation is a partial differential equation (PDF), and the theory is much more complicated than the discrete vibration theory (Leissa and Qatu, 2011). In many cases, however, the continuous systems can be studied with discrete vibration models. Thus, the continuous system needs to be translated into a discrete system, generally in two ways. One is to use the lumped-parameter model introduced in most vibration texts; the other is to use the finite element method, which will be introduced in Chapter 7. In the following, only the lumped-parameter model is introduced.

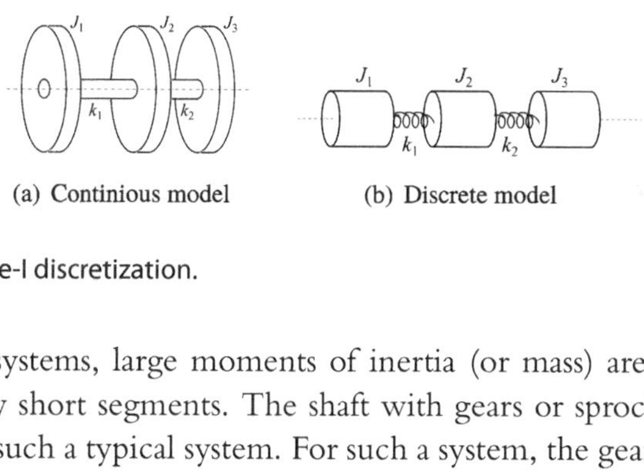

Figure 6.5 Type-I discretization.

In some systems, large moments of inertia (or mass) are concentrated on some very short segments. The shaft with gears or sprockets shown in Fig. 6.5(a) is such a typical system. For such a system, the gears or sprockets are treated as concentrated moments of inertia with the elasticity neglected because of the very small width of the installed parts. The shaft segments without parts attached are simplified as massless springs since their moments of inertia are relatively small compared to the installed gears or sprockets. With this treatment, the continuous-shaft system is transformed into a discrete model as shown in Fig. 6.5(b). Discretization of such systems is referred to as *type-I discretization* in this book.

In some other cases, the mass and stiffness distribution along the axis is more even. Beams or pure shafts without attached gears or sprockets are

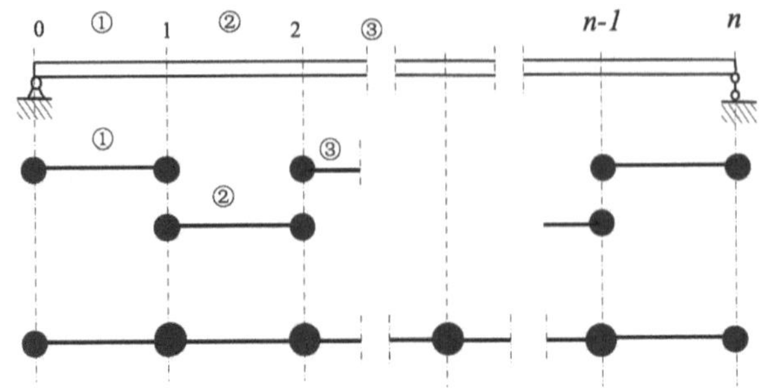

Figure 6.6 Type-II discretization.

typical examples. The discretization of such systems is termed *type-II discretization* in this book. The process is illustrated with a simply supported beam shown in Fig. 6.6. The beam is first divided into segments numbered sequentially from ① through ⓝ. Each segment is replaced by two masses at its two ends, which are called *nodes* and numbered from 0 through n. The principle for the replacement is to keep the mass and the center of mass of each segment unchanged. For each intermediate node, the total mass is obtained by summing up the mass calculated from the two adjacent segments. At the two ends of the beam, the mass comes from only one segment and has zero vertical displacement due the simple-support condition. Each segment is also a spring that connects the two lumped masses at both ends. The calculation of the spring stiffness will be introduced in Section 6.3.3. After this discretization, the original continuous beam is transformed into a discrete system having n lumped masses connected by springs.

6.3.2 Dynamic equation of multiple-DOF systems

Similar to the two-DOF system, the general forced vibration equation for systems with more DOFs has the following form:

$$\boldsymbol{M}\ddot{\boldsymbol{x}} + \boldsymbol{C}\dot{\boldsymbol{x}} + \boldsymbol{K}\boldsymbol{x} = \boldsymbol{F}(t). \tag{6.23}$$

The undamped free-vibration equation is obtained by dropping the excitation and damping terms to obtain

$$\boldsymbol{M}\ddot{\boldsymbol{x}} + \boldsymbol{K}\boldsymbol{x} = \boldsymbol{0}, \tag{6.24}$$

where $\boldsymbol{x}$ and $\boldsymbol{F}(t)$ are the $n \times 1$ displacement vector and the $n \times 1$ excitation force vector, respectively. $\boldsymbol{M}$, $\boldsymbol{C}$, and $\boldsymbol{K}$ are the mass matrix, damping matrix, and stiffness matrix. They are $n \times n$ in dimension, and, in general, the equations are coupled. The method and procedures to derive the equations are similar to that of two-DOF systems and are illustrated with the following example.

Example 6.4 (Vibration model of a car having four degrees-of-freedom). In Example 6.3, a two-DOF model has been developed for the vibration of a moving car. In that model the tire and the suspension combined are modeled as a massless spring. One way to refine the model is to separate the suspension and the tire as individual springs, k_2 and k_1, along with viscous damping c_1 and c_2 as shown in Fig. 6.7(b). In addition, the masses of the front and rear suspensions are accounted for by two equivalent masses m_1 and m_2 located in between the tire spring and the suspension. Develop the equation of motion for this vibration system.

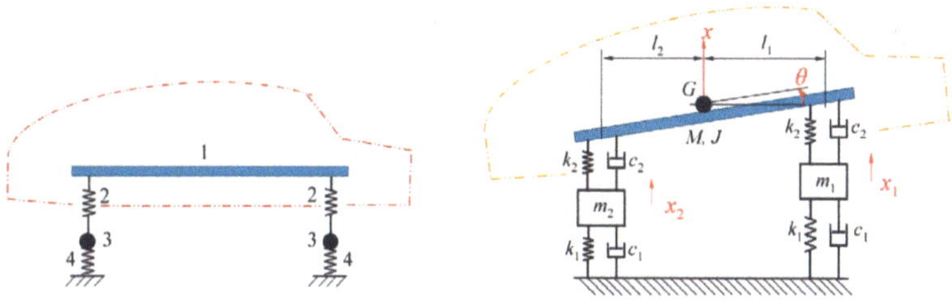

(a) The car (b) The model

Figure 6.7 Car-vibration model.

Solution. The equation will be derived with Lagrange's equation. Choose the generalized coordinate vector

$$q = \begin{Bmatrix} x & \theta & x_1 & x_2 \end{Bmatrix}^{\mathrm{T}}, \tag{a}$$

where x is the vertical displacement of the mass center G of the car body, θ is the pitch angle, and x_1 and x_2 are the vertical displacements at the front and rear suspension, respectively.

The total kinetic energy of the system can be expressed as

$$E_{\mathrm{k}} = \frac{1}{2}m_1\dot{x}_1^2 + \frac{1}{2}m_2\dot{x}_2^2 + \frac{1}{2}M\dot{x}^2 + \frac{1}{2}J\dot{\theta}^2. \tag{b}$$

Neglecting the gravitational potential, the system's potential energy consists of only the elastic potential energy in suspension E_{p1} and the tire

E_{p2} and can be formulated:

$$E_p = E_{p1} + E_{p2}. \tag{c}$$

E_{p1} and E_{p2} are given as

$$E_{p1} = \frac{1}{2}k_1\left(x_1^2 + x_2^2\right) \tag{d}$$

$$E_{p2} = \frac{1}{2}k_2\left[\left(x - \theta l_2 - x_2\right)^2 + \left(x + \theta l_1 - x_1\right)^2\right]. \tag{e}$$

Due to the damping, the dissipation function of E_d should be considered so

$$E_d = E_{d1} + E_{d2}. \tag{f}$$

E_{d1} and E_{d2} denote the dissipation function corresponding to c_1 and c_2:

$$E_{d1} = \frac{1}{2}c_1\left(\dot{x}_1^2 + \dot{x}_2^2\right) \tag{g}$$

$$E_{d2} = \frac{1}{2}c_2\left[\left(\dot{x} - \dot{\theta} l_2 - \dot{x}_2\right)^2 + \left(\dot{x} + \dot{\theta} l_1 - \dot{x}_1\right)^2\right]. \tag{h}$$

The terms in Lagrange's equation are developed as

$$\begin{cases} \frac{\mathrm{d}}{\mathrm{d}t}\left(\frac{\partial E_k}{\partial \dot{x}}\right) = M\ddot{x} \\ \frac{\mathrm{d}}{\mathrm{d}t}\left(\frac{\partial E_k}{\partial \dot{\theta}}\right) = J\ddot{\theta} \\ \frac{\mathrm{d}}{\mathrm{d}t}\left(\frac{\partial E_k}{\partial \dot{x}_1}\right) = m_1\ddot{x}_1 \\ \frac{\mathrm{d}}{\mathrm{d}t}\left(\frac{\partial E_k}{\partial \dot{x}_2}\right) = m_2\ddot{x}_2 \end{cases} \tag{i}$$

$$\begin{cases} \frac{\partial E_p}{\partial q_1} = \frac{\partial E_p}{\partial x} = k_2\left[2x - x_1 - x_2 + \theta\left(l_1 - l_2\right)\right] \\ \frac{\partial E_p}{\partial q_2} = \frac{\partial E_p}{\partial \theta} = k_2\left[l_1(x + \theta l_1 - x_1) - l_2(x - \theta l_2 - x_2)\right] \\ \frac{\partial E_p}{\partial q_3} = \frac{\partial E_p}{\partial x_1} = k_1x_1 - k_2(x + \theta l_1 - x_1) \\ \frac{\partial E_p}{\partial q_4} = \frac{\partial E_p}{\partial x_2} = k_1x_2 - k_2(x - \theta l_2 - x_2) \end{cases} \tag{j}$$

$$\begin{cases} \dfrac{\partial E_d}{\partial \dot{q}_1} = \dfrac{\partial E_d}{\partial \dot{x}} = c_2\left[2\dot{x} - \dot{x}_1 - \dot{x}_2 + \dot{\theta}\left(l_1 - l_2\right)\right] \\ \dfrac{\partial E_d}{\partial \dot{q}_2} = \dfrac{\partial E_d}{\partial \dot{\theta}} = c_2\left[l_1\left(\dot{x} + \dot{\theta}l_1 - \dot{x}_1\right) - l_2\left(\dot{x} - \dot{\theta}l_2 - \dot{x}_2\right)\right] \\ \dfrac{\partial E_d}{\partial \dot{q}_3} = \dfrac{\partial E_d}{\partial \dot{x}_1} = c_1\dot{x}_1 - c_2\left(\dot{x} + \dot{\theta}l_1 - \dot{x}_1\right) \\ \dfrac{\partial E_d}{\partial \dot{q}_4} = \dfrac{\partial E_d}{\partial \dot{x}_2} = c_1\dot{x}_2 - c_2\left(\dot{x} - \dot{\theta}l_2 - \dot{x}_2\right). \end{cases} \tag{k}$$

There is no external force applied; thus, the generalized forces are all zero:

$$Q_i = 0 \quad (i = 1 \text{ to } 4). \tag{l}$$

Substituting (i), (j), (k), and (l) into Eq. (6.8) yields

$$\begin{cases} M\ddot{x} + c_2\left[2\dot{x} - \dot{x}_1 - \dot{x}_2 + \dot{\theta}(l_1 - l_2)\right] \\ +k_2\left[2x - x_1 - x_2 + \theta(l_1 - l_2)\right] = 0 \\ J\ddot{\theta} + c_2\left[l_1(\dot{x} + \dot{\theta}l_1 - \dot{x}_1) - l_2(\dot{x} - \dot{\theta}l_2 - \dot{x}_2)\right] \\ +k_2\left[l_1(x + \theta l_1 - x_1) - l_2(x - \theta l_2 - x_2)\right] = 0 \\ m_1\ddot{x}_1 + c_1\dot{x}_1 - c_2(\dot{x} + \dot{\theta}l_1 - \dot{x}_1) + k_1x_1 - k_2(x + \theta l_1 - x_1) = 0 \\ m_2\ddot{x}_2 + c_1\dot{x}_2 - c_2(\dot{x} - \dot{\theta}l_2 - \dot{x}_2) + k_1x_2 - k_2(x - \theta l_2 - x_2) = 0. \end{cases} \tag{m}$$

It can be rewritten in matrix form as

$$\boldsymbol{M}\ddot{\boldsymbol{q}} + \boldsymbol{C}\dot{\boldsymbol{q}} + \boldsymbol{K}\boldsymbol{q} = \boldsymbol{0}, \tag{n}$$

where $\boldsymbol{M}$, $\boldsymbol{C}$ and $\boldsymbol{K}$ are in the following form:

$$\begin{cases} \boldsymbol{M} = \begin{bmatrix} M & & & \\ & J & & \\ & & m_1 & \\ & & & m_2 \end{bmatrix} \\ \boldsymbol{C} = \begin{bmatrix} 2c_2 & c_2(l_1 - l_2) & -c_2 & -c_2 \\ c_2(l_1 - l_2) & c_2(l_1^2 + l_2^2) & -c_2l_1 & c_2l_2 \\ -c_2 & -c_2l_1 & c_1 + c_2 & 0 \\ -c_2 & c_2l_2 & 0 & c_1 + c_2 \end{bmatrix} \\ \boldsymbol{K} = \begin{bmatrix} 2k_2 & k_2(l_1 - l_2) & -k_2 & -k_2 \\ k_2(l_1 - l_2) & k_2(l_1^2 + l_2^2) & -k_2l_1 & k_2l_2 \\ -k_2 & -k_2l_1 & k_1 + k_2 & 0 \\ -k_2 & k_2l_2 & 0 & k_1 + k_2 \end{bmatrix}. \end{cases} \tag{o}$$

It can be seen that, for a linear viscously damped system, the structure of $\boldsymbol{C}$ is similar to that of $\boldsymbol{K}$, except that k_i is replaced by c_i.

6.3.3 Flexibility matrix

In Section 6.3.1, the discretization of a type-II continuous system is discussed. The calculation of mass has been presented in detail. In this section, the calculation of the stiffness will be presented with the concept of *flexibility influence coefficients*. Fig. 6.8 shows the discretization of a beam. Following

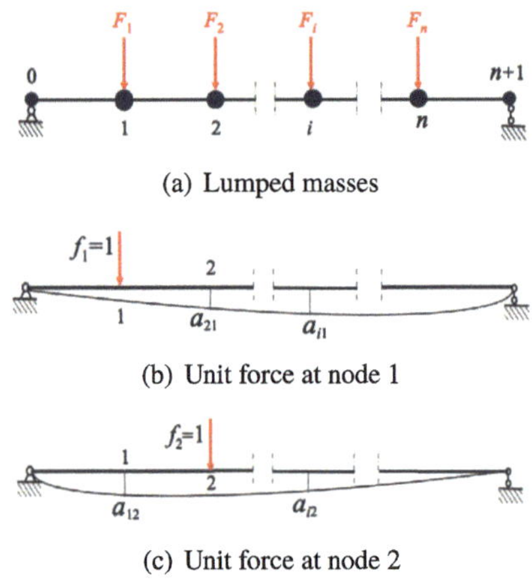

Figure 6.8 Flexibility-influence coefficient.

the procedures in Section 6.3.1, the continuous beam is replaced by the lumped-mass model as Fig. 6.8(a). Assuming that there is a unit force $f_1 = 1$ applied at node 1, and no force at other nodes as shown in Fig. 6.8(b), the nodal deformations under the unit force $f_1 = 1$ are $a_{11}, a_{21}, \cdots, a_{n1}$. The double subscripts a_{ij} are used here, with i indicating the node number of the deformation and j representing the node number where the unit force is applied. a_{ij}, meaning the deformation at node i caused by a unit force at node j, is called the *flexibility-influence coefficient*.

If an arbitrary force F_j is applied at the node j, it will cause the node i to move vertically by an amount of $a_{ij}F_j$.

If a force F_j is applied at every node, the deformation at node i will be the summation as follows according to the superposition principle:

$$u_i = \sum_{j=1}^{n} a_{ij} F_j. \tag{6.25}$$

Now, let's revisit the physical meaning of stiffness. For the same beam just treated, the relationship between the deformation and the force applied at the nodes is

$$\boldsymbol{F} = \boldsymbol{K}\boldsymbol{u}, \tag{6.26}$$

where

$$\boldsymbol{F} = \begin{Bmatrix} F_1 & F_2 & \cdots & F_i & \cdots & F_n \end{Bmatrix}^{\mathrm{T}} \tag{6.27}$$

$$\boldsymbol{u} = \begin{Bmatrix} u_1 & u_2 & \cdots & u_i & \cdots & u_n \end{Bmatrix}^{\mathrm{T}} \tag{6.28}$$

and

$$F_i = \sum_{j=1}^{n} k_{ij} u_j. \tag{6.29}$$

Comparing Eqs. (6.29) and (6.25), k_{ij} can be better referred to as *stiffness influence coefficients*, meaning the force required to be applied at node i to have a unit displacement at node j and zero displacement at all other nodes.

Eq. (6.25) can be written in vector form

$$\boldsymbol{u} = \boldsymbol{A}\boldsymbol{F}, \tag{6.30}$$

where $\boldsymbol{A}$ is referred to as the *flexibility matrix* and its entries a_{ij} are the flexibility influence coefficients.

The stiffness $\boldsymbol{K}$ and the flexibility $\boldsymbol{A}$ are reciprocal to each other. That is,

$$\boldsymbol{A} = \boldsymbol{K}^{-1}, \quad \boldsymbol{K} = \boldsymbol{A}^{-1}. \tag{6.31}$$

Vibration equations in terms of stiffness can be easily transformed into one with flexibility. For example, multiplying Eq. (6.24) by $\boldsymbol{A}$ gives

$$\boldsymbol{A}\boldsymbol{M}\ddot{\boldsymbol{x}} + \boldsymbol{x} = \boldsymbol{0}. \tag{6.32}$$

Eq. (6.32) and Eq. (6.24) are interchangeable. $\boldsymbol{A}$ can be easily calculated with the knowledge of the mechanics of materials.

Example 6.5 (Flexibility matrix of a simply supported beam). A simply supported beam having uniform equal cross section is shown in Fig. 6.9. The beam needs to be discretized into a lumped-parameter system having

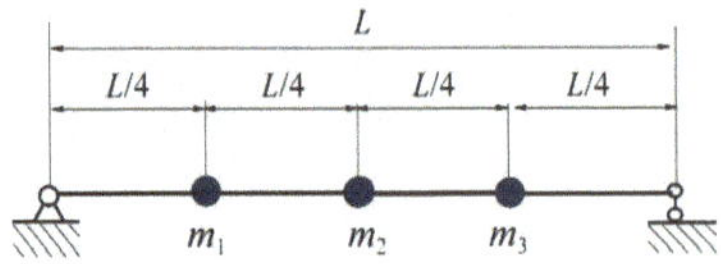

Figure 6.9 Simply supported beam.

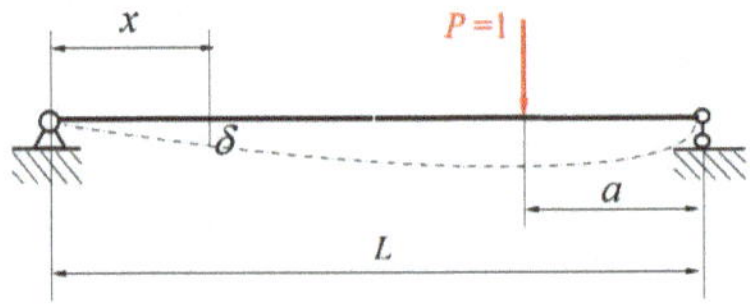

Figure 6.10 Deflection of beam.

three intermediate masses m_1, m_2, and m_3. The modulus of elasticity of the material is E, and moment of inertia of cross-section is I. Develop the flexibility matrix of the system.

Solution. The deflection of a simply supported beam can be calculated as follows with reference to Fig. 6.10:

$$\delta = \frac{Pax}{6LEI}\left(L^2 - x^2 - a^2\right). \tag{a}$$

The flexibility-influence coefficients can be calculated from (a). For example, letting $P = 1$, $a = L/2$, and $x = L/4$ gives

$$a_{12} = \frac{11L^3}{768EI}. \tag{b}$$

All the coefficients a_{ij} can be calculated in the same way. Then, the flexibility matrix $\boldsymbol{A}$ is obtained as

$$\boldsymbol{A} = \frac{L^3}{768EI}\begin{bmatrix} 9 & 11 & 7 \\ 11 & 16 & 11 \\ 7 & 11 & 9 \end{bmatrix}. \tag{c}$$

6.4 Solution of multiple-DOF vibration

6.4.1 Coordinate coupling

Comparing Eq. (6.23) and Eq. (5.60) reveals that the vibration equation of multiple DOFs is similar to that of a single DOF. In the single-DOF equation, the coordinate x and the parameters (m, c and k) are single dimensional. In the multiple-DOF equation, however, they become matrices of dimensions either $n \times 1$ or $n \times n$. This causes a big difference between the single- and multiple-DOF systems. In the multiple-DOF equation, the coordinates are generally coupled, which is referred to as *coordinate coupling*. Taking the two-DOF case, Eq. (6.1), as an example, both x_1 and x_2 appear in each equation. This is a sign of the coordinate coupling. Coordinate coupling hinders the direct application of the method developed for the single-DOF system for solving multiple-DOF problems.

Coordinate coupling highly depends on the selection of generalized coordinates. To illustrate this statement, let's revisit the car-vibration problem in Example 6.3, where x and θ were chosen as the generalized coordinates and the vibration equation is Eq. (6.15). It can be seen that x and θ appear in each of the two equations, meaning that the elastic restoring forces are coupled. This type of coupling is referred to as *stiffness coupling* or *static coupling*. A sign of stiffness coupling is that the stiffness matrix $\boldsymbol{K}$ is nondiagonal. For the same problem, if we choose the vertical displacements of the car body at the front and the rear suspensions, x_1 and x_2 as shown in Fig. 6.3(b), as the generalized coordinates, and the coordinate coupling will be different. The relationship between x_1, x_2 and x, θ is already given in Eq. (6.12). The equations in terms of x_1 and x_2 can be obtained through similar procedures in Example 6.3. They are

$$\begin{cases} \dfrac{(ml_2^2+J)}{l^2}\ddot{x}_1 + \dfrac{(ml_1l_2-J)}{l^2}\ddot{x}_2 + k_1x_1 = 0 \\[2ex] \dfrac{(ml_1l_2-J)}{l^2}\ddot{x}_1 + \dfrac{(ml_1^2+J)}{l^2}\ddot{x}_2 + k_2x_2 = 0, \end{cases} \tag{6.33}$$

where $l = l_1 + l_2$. These two equations can be put into matrix form as

$$\bar{\boldsymbol{M}}\ddot{\bar{\boldsymbol{U}}} + \bar{\boldsymbol{K}}\bar{\boldsymbol{U}} = \boldsymbol{0}, \tag{6.34}$$

where $\bar{\boldsymbol{M}}$, $\bar{\boldsymbol{K}}$ and $\bar{\boldsymbol{U}}$ are

$$\bar{\boldsymbol{U}} = \begin{Bmatrix} x_1 \\ x_2 \end{Bmatrix}, \quad \bar{\boldsymbol{M}} = \begin{bmatrix} J+ml_2^2 & ml_1l_2-J \\ ml_1l_2-J & J+ml_1^2 \end{bmatrix}, \quad \bar{\boldsymbol{K}} = \begin{bmatrix} k_1 & 0 \\ 0 & k_2 \end{bmatrix}.$$

Obviously, in Eq. (6.33) the two coordinates are still coupled but differently. In this equation, $\bar{\boldsymbol{K}}$ is diagonal, but $\bar{\boldsymbol{M}}$ is not, meaning that the coupling is between the inertial forces involving $\ddot{x}_1$ and $\ddot{x}_2$. This type of coordinate coupling is referred to as *inertial coupling* or *dynamic coupling*.

This clearly indicates that choosing different displacements lead to different couplings. In fact, the coupling can be completely eliminated if the generalized coordinates are properly selected. The procedures for this are presented next.

6.4.2 Natural frequency and principal mode

Let's start with the undamped free-vibration equation

$$\boldsymbol{M}\ddot{\boldsymbol{X}} + \boldsymbol{K}\boldsymbol{X} = \boldsymbol{0}, \tag{6.35}$$

where the displacement vector is

$$\boldsymbol{X} = \begin{Bmatrix} x_1 & x_2, & \cdots, & x_n \end{Bmatrix}^{\mathrm{T}}.$$

$\boldsymbol{M}$ and $\boldsymbol{K}$ are symmetric and positive. We assume the solution of Eq. (6.35) has the form

$$\boldsymbol{X} = \boldsymbol{A}\sin(\omega t + \phi), \tag{6.36}$$

where $\boldsymbol{A}$ is the amplitude vector. Differentiating Eq. (6.36) twice gives

$$\ddot{\boldsymbol{X}} = -\omega^2 \boldsymbol{A}\sin(\omega t + \phi). \tag{6.37}$$

Substituting $\boldsymbol{X}$ and $\ddot{\boldsymbol{X}}$ into Eq. (6.35) yields

$$\left(\boldsymbol{K} - \omega^2 \boldsymbol{M}\right)\boldsymbol{A} = \boldsymbol{0}. \tag{6.38}$$

This is a set of homogeneous linear equations in which $\boldsymbol{A}$ is the unknown. To have a nontrivial solution, the determinant of the coefficients must equal zero:

$$\det\left(\boldsymbol{K} - \omega^2 \boldsymbol{M}\right) = \left|\boldsymbol{K} - \omega^2 \boldsymbol{M}\right| = 0. \tag{6.39}$$

This is called the *characteristic equation* or *frequency equation*. Expansion of Eq. (6.39) gives a polynomial of n-th order in ω^2. The roots of the characteristic equation are the *eigenvalues*. Substituting ω^2 into Eq. (6.38) gives the nontrivial solution of $\boldsymbol{A}$, which is referred to as the *eigenvector* corresponding to ω^2. In the theory of algebra, this is called *eigenvalue problem*. In the Matlab® package, there is an embedded solver for the eigenvalue problem. Both the eigenvalues and eigenvectors can be obtained easily.

The eigenvalues are nonnegative real numbers when $\boldsymbol{M}$ is positive definite and $\boldsymbol{K}$ is positive definite or positive semidefinite. The square root of the eigenvalue gives the natural frequency of the system. The natural frequencies are generally distinct and arranged in an increasing order so:

$$\omega_1 < \omega_2 < \cdots < \omega_n,$$

where ω_i is referred to as the i-th-order natural frequency, and the corresponding eigenvector is designated as $\boldsymbol{A}^{(i)}$. From Eq. (6.36) we know that $\boldsymbol{A}^{(i)}$ gives the relative amplitudes of each mass at the corresponding natural frequency. It is referred to as the i-th *principal mode* or simply as the *mode.*

6.4.3 Orthogonality and normalization of the principal mode

The concept of principal mode plays a critical role in the vibration theory of multiple-DOF systems. It is the theoretical base of solving multiple-DOF vibration problems with the *modal superposition method.*

6.4.3.1 Orthogonality of the principal mode

For a system having n DOFs, the n natural frequencies $\omega_1, \omega_2, \cdots, \omega_n$ correspond to n principal modes $\boldsymbol{A}^{(1)}, \boldsymbol{A}^{(2)}, \cdots, \boldsymbol{A}^{(n)}$. The principal modes have a very important property, namely, *orthogonality*, which is illustrated next.

For two different frequencies ω_i and ω_j, we have the following relationship according to Eq. (6.38):

$$\boldsymbol{K}\boldsymbol{A}^{(i)} = \omega_i^2 \boldsymbol{M}\boldsymbol{A}^{(i)} \tag{6.40}$$

$$\boldsymbol{K}\boldsymbol{A}^{(j)} = \omega_j^2 \boldsymbol{M}\boldsymbol{A}^{(j)}. \tag{6.41}$$

Multiplying the left-hand term of Eqs. (6.40) by $\boldsymbol{A}^{(j)\mathrm{T}}$ and the right-hand terms, the transpose of (6.41), by $\boldsymbol{A}^{(i)}$ gives the following relationship:

$$\boldsymbol{A}^{(j)\mathrm{T}}\boldsymbol{K}\boldsymbol{A}^{(i)} = \omega_i^2 \boldsymbol{A}^{(j)\mathrm{T}}\boldsymbol{M}\boldsymbol{A}^{(i)} \tag{6.42}$$

$$\boldsymbol{A}^{(j)\mathrm{T}}\boldsymbol{K}\boldsymbol{A}^{(i)} = \omega_j^2 \boldsymbol{A}^{(j)\mathrm{T}}\boldsymbol{M}\boldsymbol{A}^{(i)}. \tag{6.43}$$

Note that the property of symmetry of $\boldsymbol{M}$ and $\boldsymbol{K}$ has been used in arriving at these relationships. Subtracting Eq. (6.43) from Eq. (6.42) yields

$$\left(\omega_i^2 - \omega_j^2\right)\boldsymbol{A}^{(j)\mathrm{T}}\boldsymbol{M}\boldsymbol{A}^{(i)} = \boldsymbol{0}. \tag{6.44}$$

Then, from Eqs. (6.43) and (6.44), we have

$$\boldsymbol{A}^{(j)\mathrm{T}}\boldsymbol{M}\boldsymbol{A}^{(i)} = \boldsymbol{0} \quad (i \neq j) \tag{6.45}$$

$$A^{(j)\mathrm{T}}KA^{(i)} = 0 \quad (i \neq j). \tag{6.46}$$

These indicate that the principal modes $A^{(i)}$ and $A^{(j)}$ are orthogonal with respect to M and K.

6.4.3.2 Normalization of principal mode

Letting $i = j$ in Eq. (6.42), we have

$$A^{(i)\mathrm{T}}KA^{(i)} = \omega_i^2 A^{(i)\mathrm{T}}MA^{(i)}. \tag{6.47}$$

Given that the mass matrix M is positive definite, $A^{(i)T}MA^{(i)}$ is of positive-definite quadratic form. That is,

$$A^{(i)\mathrm{T}}MA^{(i)} = M_i \quad (i = 1, 2, \cdots, n). \tag{6.48}$$

Similarly, if K is positive definite, then we have

$$A^{(i)\mathrm{T}}KA^{(i)} = K_i \quad (i = 1, 2, \cdots, n), \tag{6.49}$$

where K_i and M_i are referred to as the *normal stiffness* and the *normal mass*, respectively. From Eq. (6.47), we have

$$\omega_i^2 = \frac{A^{(i)\mathrm{T}}KA^{(i)}}{A^{(i)\mathrm{T}}MA^{(i)}} = \frac{K_i}{M_i}. \tag{6.50}$$

This indicates that the square of the i-th-order natural frequency equals the ratio of the i-th-order normal stiffness to the i-th-order normal mass.

If we arrange each principal mode as a column, we can obtain the *modal matrix* as

$$A = \begin{bmatrix} A^{(1)} & A^{(2)} & \cdots & A^{(n)} \end{bmatrix} = \begin{bmatrix} A_1^{(1)} & A_1^{(2)} & \cdots & A_1^{(n)} \\ A_2^{(1)} & A_2^{(2)} & \cdots & A_2^{(n)} \\ \vdots & \vdots & \cdots & \vdots \\ A_n^{(1)} & A_n^{(2)} & \cdots & A_n^{(n)} \end{bmatrix}. \tag{6.51}$$

We can easily have the following by a simple extension of Eqs. (6.48) and (6.49):

$$A^{\mathrm{T}}MA = M_{\mathrm{p}} = \mathrm{diag}\begin{bmatrix} M_1 & M_2 & \cdots & M_n \end{bmatrix} \tag{6.52}$$

$$A^{\mathrm{T}}KA = K_{\mathrm{p}} = \mathrm{diag}\begin{bmatrix} K_1 & K_2 & \cdots & K_n \end{bmatrix}. \tag{6.53}$$

As stated before, the principal mode gives the relative vibration amplitude of m_i. Thus, a principal mode multiplied or divided by a constant is still a principal mod:

$$\boldsymbol{\varphi}^{(i)} = \frac{1}{C_i}\boldsymbol{A}^{(i)}. \tag{6.54}$$

$\boldsymbol{\varphi}^{(i)}$ is also a principal mode. If C_i is taken as

$$C_i = \pm\sqrt{M_i} \tag{6.55}$$

to make

$$\boldsymbol{\varphi}^{(i)\mathrm{T}}\boldsymbol{M}\boldsymbol{\varphi}^{(i)} = 1, \tag{6.56}$$

C_i is called the *normalization factor* and $\boldsymbol{\varphi}^{(i)}$ is called the i-th *normalized principal mode*. The modal matrix designated by $\boldsymbol{\Phi}$ consisting of the normalized principal modes is referred as to *normalized modal matrix:*.

$$\boldsymbol{\Phi} = \left[\boldsymbol{\varphi}^{(1)} \quad \boldsymbol{\varphi}^{(2)} \quad \cdots \quad \boldsymbol{\varphi}^{(n)}\right] = \begin{bmatrix} \varphi_1^{(1)} & \varphi_1^{(2)} & \cdots & \varphi_1^{(n)} \\ \varphi_2^{(1)} & \varphi_2^{(2)} & \cdots & \varphi_2^{(n)} \\ \vdots & \vdots & \cdots & \vdots \\ \varphi_n^{(1)} & \varphi_n^{(2)} & \cdots & \varphi_n^{(n)} \end{bmatrix}. \tag{6.57}$$

It has the feature

$$\boldsymbol{\Phi}^{\mathrm{T}}\boldsymbol{M}\boldsymbol{\Phi} = \boldsymbol{I} \tag{6.58}$$

and

$$\boldsymbol{\Phi}^{\mathrm{T}}\boldsymbol{K}\boldsymbol{\Phi} = \boldsymbol{\Omega} = \mathrm{diag}\left[\omega_1^2 \quad \omega_2^2 \quad \cdots \quad \omega_n^2\right], \tag{6.59}$$

where $\boldsymbol{\Omega}$ is called the *eigenvalue matrix*. By using the normalized modal matrix, the mass matrix is normalized to a unit matrix, and the stiffness matrix becomes the eigenvalue matrix. This is the theoretical base of the decoupling method to solve multiple-DOF vibration problems.

6.4.3.3 Decoupling of vibration equation of multiple-DOF systems

The decoupling of the equations is the key to the solution of the vibration problem of the multiple-DOF linear discrete system. D'alembert first used this method for the solution of a simple vibration problem, but he did not formulate the method into a theory. The eigenvalue problem was finally solved by A. Cauchy, a French mathematician, from 1820s to 1850s, and so laying the foundation of multiple-DOF vibration theory.

The n eigenvectors corresponding to the n eigenvalues are linearly independent. They form a n-dimensional linear space. Any vector in that space can be represented by a linear combination of the eigenvectors

$$\boldsymbol{X} = \eta_1 \boldsymbol{\varphi}^{(1)} + \eta_2 \boldsymbol{\varphi}^{(2)} + \cdots + \eta_n \boldsymbol{\varphi}^{(n)} = \boldsymbol{\Phi}\boldsymbol{\eta}, \tag{6.60}$$

where

$$\boldsymbol{\eta} = \begin{Bmatrix} \eta_1 & \eta_2 & \cdots & \eta_n \end{Bmatrix}^{\mathrm{T}} \tag{6.61}$$

is called *a modal coordinate vector* and its entries η_i are called the *modal coordinates*. In contrast, the displacement vector $\boldsymbol{X}$ and its entries x_i are called the *physical coordinate vector* (Eq. (6.35)) and the *physical coordinate*, respectively.

From Eq. (6.60) we have

$$\ddot{\boldsymbol{X}} = \boldsymbol{\Phi}\ddot{\boldsymbol{\eta}}. \tag{6.62}$$

Substituting Eqs. (6.60) and (6.62) into Eq. (6.35) gives

$$\boldsymbol{\Phi}^{\mathrm{T}}\boldsymbol{M}\boldsymbol{\Phi}\ddot{\boldsymbol{\eta}} + \boldsymbol{\Phi}^{\mathrm{T}}\boldsymbol{K}\boldsymbol{\Phi}\boldsymbol{\eta} = \boldsymbol{0}. \tag{6.63}$$

Taking advantage of Eqs. (6.58) and (6.59), this reduces to

$$\ddot{\boldsymbol{\eta}} + \boldsymbol{\Omega}\boldsymbol{\eta} = \boldsymbol{0}. \tag{6.64}$$

Because $\boldsymbol{\Omega}$ is diagonal, Eq. (6.64) consists of n decoupled equations which are independent from each other:

$$\begin{cases} \ddot{\eta}_1 + \omega_1^2 \eta_1 = 0 \\ \ddot{\eta}_2 + \omega_2^2 \eta_2 = 0 \\ \cdots \quad \cdots \\ \ddot{\eta}_n + \omega_n^2 \eta_n = 0. \end{cases} \tag{6.65}$$

Each of these equations can be solved as a single-OF vibration problem. Once $\boldsymbol{\eta}$ is obtained, the physical response $\boldsymbol{X}$ is calculated by Eq. (6.60). Through this method, the multiple-DOF vibration problem is transformed into n independent single-DOF problems. This method is called the *modal superposition method*. In the following, an example is presented to illustrate the procedure and get more insight into the concept of the *modal superposition*.

Example 6.6 (Visualization of the principal mode). For the torsional vibration system given in Fig. 6.2, the parameters are $k_1 = k_2 = k_3 = k$ and $J_1 = J_2 = J$. Find the natural frequencies and the principal modes.

Solution. According to Eq. (6.7), the free-vibration equation of this system is

$$\begin{bmatrix} J & 0 \\ 0 & J \end{bmatrix} \begin{Bmatrix} \ddot{\theta}_1 \\ \ddot{\theta}_2 \end{Bmatrix} + \begin{bmatrix} 2k & -k \\ -k & 2k \end{bmatrix} \begin{Bmatrix} \theta_1 \\ \theta_2 \end{Bmatrix} = \begin{Bmatrix} 0 \\ 0 \end{Bmatrix}. \tag{a}$$

From Eq. (6.38), we have

$$\begin{bmatrix} 2k - J\omega^2 & -k \\ -k & 2k - J\omega^2 \end{bmatrix} \begin{Bmatrix} A_1 \\ A_2 \end{Bmatrix} = 0. \tag{b}$$

The characteristic equation is

$$\begin{vmatrix} 2k - J\omega^2 & -k \\ -k & 2k - J\omega^2 \end{vmatrix} = 0. \tag{c}$$

Solving (c) yields the two natural frequencies so:

$$\omega_1 = \sqrt{k/J}, \quad \omega_2 = \sqrt{3k/J}. \tag{d}$$

Substituting ω_1 into (b) yields

$$\begin{bmatrix} k & -k \\ -k & k \end{bmatrix} \begin{Bmatrix} A_1^{(1)} \\ A_2^{(1)} \end{Bmatrix} = \begin{Bmatrix} 0 \\ 0 \end{Bmatrix}. \tag{e}$$

We obtain

$$A_1^{(1)} = A_2^{(1)} \tag{f}$$

or

$$\boldsymbol{A}^{(1)} = \begin{Bmatrix} 1 & 1 \end{Bmatrix}^{\mathrm{T}}. \tag{g}$$

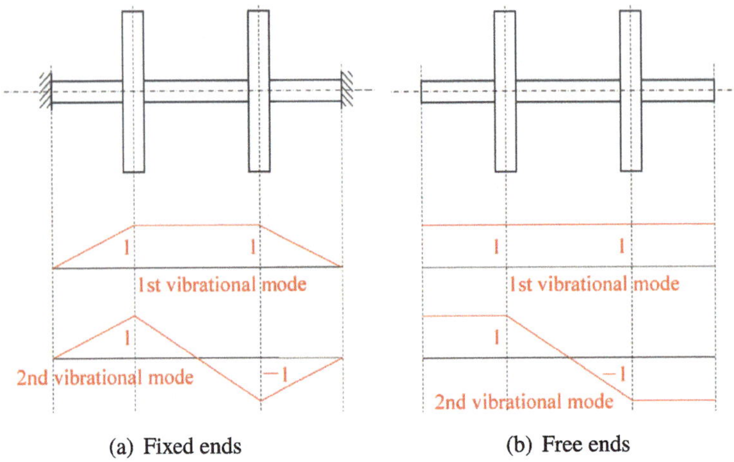

Figure 6.11 Two-vibration modes of torsional systems.

Substituting ω_2 into (b) gives

$$\boldsymbol{A}^{(2)} = \begin{Bmatrix} 1 & -1 \end{Bmatrix}^{\mathrm{T}}. \tag{h}$$

The principal modes $\boldsymbol{A}^{(1)}$ and $\boldsymbol{A}^{(2)}$ are schematically shown in Fig. 6.11(a). In the first mode, the two disks always move in the same direction and have the same angular displacement. The shaft segment in between the two disks move in pace with the two disks as a rigid system. The shaft segments between each of the fixed ends and the disk, however, experience some torsional elastic deformation. The angular displacement changes linearly with the distance from the fixed end. In the second mode, the two disks vibrate at the same amplitude, but opposite in direction. In this case, all three shaft segments have linearly changing torsional deformations, as shown. From this example, we can see that the mode of vibration reflects the shape of the system in the harmonic motion with the natural frequency (mode shape).

If the two ends are freed, $k_1 = k_3 = 0$, the vibration equation will become

$$\begin{bmatrix} J & 0 \\ 0 & J \end{bmatrix} \begin{Bmatrix} \ddot{\theta}_1 \\ \ddot{\theta}_2 \end{Bmatrix} + \begin{bmatrix} k & -k \\ -k & k \end{bmatrix} \begin{Bmatrix} \theta_1 \\ \theta_2 \end{Bmatrix} = \begin{Bmatrix} 0 \\ 0 \end{Bmatrix}. \tag{j}$$

Following the same procedure, we obtain the two natural frequencies as

$$\omega_1 = 0, \quad \omega_2 = \sqrt{2k/J}. \tag{k}$$

The two principal modes of vibration are

$$\boldsymbol{A}^{(1)} = \begin{Bmatrix} 1 & 1 \end{Bmatrix}^{\mathrm{T}} \quad \boldsymbol{A}^{(2)} = \begin{Bmatrix} 1 & -1 \end{Bmatrix}^{\mathrm{T}}. \tag{g}$$

The mode shapes are given in Fig. 6.11(b). In the first mode, the two disks move in the same direction with the same amount of displacement. The shaft will move with the disks as rigid bodies because there are no restrictions at the ends. In fact, the whole system moves as a rigid body in this case ($\omega_1 = 0$). This is a special case in vibration. A careful look at the stiffness matrix $\boldsymbol{K}$ reveals that it is *singular* and *positive semi-definite*.

6.5 Vibration response of systems with multiple DOFs

6.5.1 Damping assumption

There are different damping mechanisms, such as coulomb-friction damping, viscous damping, material internal damping, etc. In addition, the effect of damping on vibration response depends on not only the damping itself, but also the vibration status and the duration of the vibration. For example, in impact processes of very short duration, the effect of damping can be neglected. If the vibration response is far away from the resonance, the damping effect is insignificant. However, if the response is close to the resonance, the damping has significant effect on the response amplitude. Given the complicated mechanisms and its complex effect on the vibration response, damping is hard to model accurately, and is often simplified in analysis. In the single-DOF vibration problem, it is often simplified as a viscous damping coefficient C proportional to the vibration velocity. This treatment is mainly to simplify the mathematical manipulation. If we keep the viscous-damping assumption, the general vibration equation for a multiple-DOF system is in the form of Eq. (6.3). For convenience, we repeat it here as

$$\boldsymbol{M}\ddot{\boldsymbol{X}} + \boldsymbol{C}\dot{\boldsymbol{X}} + \boldsymbol{K}\boldsymbol{X} = \boldsymbol{F}. \tag{6.66}$$

To decouple the equation, the following modal transformation introduced in the previous section is performed

$$\boldsymbol{X} = \boldsymbol{\Phi}\eta. \tag{6.67}$$

Substituting into Eq. (6.66), and left-multiplying $\boldsymbol{\Phi}^{\mathrm{T}}$ on both sides yield

$$\ddot{\eta} + \bar{\boldsymbol{C}}\dot{\eta} + \boldsymbol{\Omega}\eta = \boldsymbol{\Phi}^{\mathrm{T}}\boldsymbol{F}. \tag{6.68}$$

By this linear transformation of the coordinates, the mass matrix and the stiffness matrix are diagonalized; however, the damping matrix may not be:

$$\bar{\boldsymbol{C}} = \boldsymbol{\Phi}^{\mathrm{T}}\boldsymbol{C}\boldsymbol{\Phi}. \tag{6.69}$$

If $\bar{\boldsymbol{C}}$ is not diagnosed, the equations are not decoupled, and the method of modal superposition cannot be applied to solve the multiple-DOF vibration problems.

As stated before, damping mechanisms are complicated, and accurate description is difficult. In addition, damping has insignificant influence on the vibration response as long as it is away from resonance. For these two reasons, damping is often simplified to use the modal superposition method to multiple-DOF problems. There are two simplification methods that are commonly used. One is to assume the damping is a linear combination of the mass and the stiffness so:

$$\boldsymbol{C} = \alpha\boldsymbol{M} + \beta\boldsymbol{K}, \tag{6.70}$$

where α and β are positive constants. This type of damping is referred to as *proportional damping*. With this simplification, the damping matrix $\bar{\boldsymbol{C}}$ becomes, after applying the modal decomposition,

$$\bar{\boldsymbol{C}} = \alpha\boldsymbol{I} + \beta\boldsymbol{\Omega}. \tag{6.71}$$

It is obviously diagonalized.

The other simplification method is to follow the viscous-damping assumption in the single-DOF system. After the decoupling process, the damping $\bar{\boldsymbol{C}}$ may not be diagonalized. This method simply forces all the nondiagonal elements to zero to decouple the equations:

$$\bar{\boldsymbol{C}} = \boldsymbol{\Phi}^{\mathrm{T}}\boldsymbol{C}\boldsymbol{\Phi} \approx \mathrm{diag}\left[2\xi_1\omega_1 \quad 2\xi_2\omega_2 \quad \cdots \quad 2\xi_n\omega_n\right]. \tag{6.72}$$

This is referred to as *normalized modal damping*. This treatment is based on the following considerations.

1) A diagonal $\bar{\boldsymbol{C}}$ matrix makes the decoupling of Eq. (6.66) possible. Thus, the modal superposition method can be used to solve damped multiple-DOF vibration problems.

2) Although the damping coefficient C is difficult to accurately measure, the damping ratio ξ is relatively easier to be determined through lab tests.
3) As long as the damping ratio is low ($0 \le \xi \le 0.2$), the accuracy of modal damping is acceptable.

6.5.2 Modal truncation method

Theoretically, all vibration modes can be included in the modal superposition method, but, when the number of DOFs is large, the computation becomes very intensive. In practice, however, this is not necessary because the principal modes contribute to the physical response differently. These natural frequencies close to the excitation frequency contribute significantly, while those far away from the excitation frequency contribute little. This can be easily seen from Eq. (5.35) (see p. 121). In this case, it is possible to approximate the total response by including only the several principal modes that contribute to the response significantly, neglecting all other insignificant modes. In practice, the several lower frequencies are often used, and experience indicates that using a few of these frequencies generally yields an adequate result. In real calculations, how many modes need to be retained depends on the required accuracy and can be determined by trial and error. Higher accuracy requires that more modes be retained. This method is referred to as *modal truncation*.

If the frequencies are retained up to the rth order ($r < n$), Eq. (6.60) becomes

$$\boldsymbol{X} \approx \eta_1 \boldsymbol{\varphi}^{(1)} + \eta_2 \boldsymbol{\varphi}^{(2)} + \cdots + \eta_r \boldsymbol{\varphi}^{(r)} = \sum_{i=1}^{r} \boldsymbol{\Phi}^{(i)} \eta_i. \tag{6.73}$$

The truncated modal matrix is

$$\tilde{\boldsymbol{\Phi}} = \begin{bmatrix} \phi_1^{(1)} & \phi_1^{(2)} & \cdots & \phi_1^{(r)} \\ \phi_2^{(1)} & \phi_2^{(2)} & \cdots & \phi_2^{(r)} \\ \vdots & \vdots & \cdots & \vdots \\ \phi_n^{(1)} & \phi_n^{(2)} & \cdots & \phi_n^{(r)} \end{bmatrix}. \tag{6.74}$$

Using the truncated modal matrix to diagonalize the mass and the stiffness matrices as

$$\tilde{\boldsymbol{\Phi}}^{\mathrm{T}} \boldsymbol{M} \tilde{\boldsymbol{\Phi}} = \boldsymbol{I}_{r \times r} \tag{6.75}$$

$$\tilde{\boldsymbol{\Phi}}^{\mathrm{T}}\boldsymbol{K}\tilde{\boldsymbol{\Phi}} = \mathrm{diag}\begin{bmatrix}\omega_1^2 & \omega_2^2 & \cdots & \omega_r^2\end{bmatrix} \tag{6.76}$$

and letting

$$\tilde{\eta} = \begin{Bmatrix}\eta_1 & \eta_2 & \cdots & \eta_r\end{Bmatrix}, \tag{6.77}$$

Eq. (6.73) can be shortened to

$$\boldsymbol{X} \approx \tilde{\boldsymbol{\Phi}}\tilde{\eta}. \tag{6.78}$$

Substituting Eq. (6.78) into Eq. (6.35) and left-multiplying $\tilde{\boldsymbol{\Phi}}^{\mathrm{T}}$ gives r uncoupled vibration equations:

$$\begin{cases}\ddot{\eta}_1 + \omega_1^2\eta_1 = 0\\ \ddot{\eta}_2 + \omega_1^2\eta_2 = 0\\ \cdots \quad \cdots\\ \ddot{\eta}_r + \omega_1^2\eta_r = 0.\end{cases} \tag{6.79}$$

Again, the number of modes retained depends on the required accuracy. Sometimes, it is determined by trial and error. Experience indicates that using a few modes is generally satisfactory, even for systems having hundreds of DOFs. The modal truncation method is able to cut the computational time significantly.

6.5.3 Free-vibration response of systems with multiple DOFs

If the modal damping method is applied, the n uncoupled single-DOF free-vibration equations (Eq. (6.79)) will become

$$\begin{cases}\ddot{\eta}_1 + 2\xi_1\omega_1\dot{\eta}_1 + \omega_1^2\eta_1 = 0\\ \ddot{\eta}_2 + 2\xi_2\omega_2\dot{\eta}_2 + \omega_2^2\eta_2 = 0\\ \cdots \quad \cdots\\ \ddot{\eta}_n + 2\xi_n\omega_n\dot{\eta}_n + \omega_n^2\eta_= 0.\end{cases} \tag{6.80}$$

Each equation can be solved independently because the equations are decoupled. The solution for the arbitrary i-th mode is

$$\eta_i(t) = e^{-\xi_i\omega_i t}\left[\eta_i(0)\cos\omega_{\mathrm{d}i}t + \frac{\dot{\eta}_i(0) + \xi_i\omega_i\eta_i(0)}{\omega_{\mathrm{d}i}}\sin\omega_{\mathrm{d}i}t\right] \quad (i = 1 \text{ to } n), \tag{6.81}$$

where $\eta_i(0)$ and $\dot{\eta}_i(0)$ are the initial conditions, which are not given directly, but can be obtained from the initial conditions in the physical coordinates $\boldsymbol{X}(0)$ and $\dot{\boldsymbol{X}}(0)$:

$$\boldsymbol{\eta}(0) = \boldsymbol{\Phi}^{-1}\boldsymbol{X}(0) \tag{6.82}$$

$$\dot{\boldsymbol{\eta}}(0) = \boldsymbol{\Phi}^{-1}\dot{\boldsymbol{X}}(0). \tag{6.83}$$

6.5.4 Forced-vibration response of systems with multiple DOFs

For a damped multiple DOF system, the equations are as shown in Eq. (6.66). After the modal transformation, they become Eq. (6.68). If $\bar{\boldsymbol{C}}$ is diagonalized with either the proportional damping or modal damping, the equations can be decoupled in the modal coordinate as:

$$\begin{cases} \ddot{\eta}_1 + 2\xi_1\omega_1\dot{\eta}_1 + \omega_1^2\eta_1 = N_1 \\ \ddot{\eta}_2 + 2\xi_2\omega_2\dot{\eta}_2 + \omega_2^2\eta_2 = N_2 \\ \cdots \quad \cdots \\ \ddot{\eta}_n + 2\xi_n\omega_n\dot{\eta}_n + \omega_n^2\eta_n = N_n. \end{cases} \tag{6.84}$$

In the previous equations, we have used the *modal excitation* $\boldsymbol{N}$, which is

$$\boldsymbol{N} = \boldsymbol{\Phi}^{\mathrm{T}}\boldsymbol{F}, \tag{6.85}$$

where N_i is the i-th element of $\boldsymbol{N}$ and called the i-th-order *modal excitation force*.

Now, we can solve Eq. (6.84) for each individual η_i according to Eq. (5.70).

$$\begin{aligned} \eta_i(t) =& e^{-\xi_i\omega_i t}\left[\eta_i(0)\cos\omega_{\mathrm{d}i}t + \frac{\dot{\eta}_i(0) + \xi_i\omega_i\eta_i(0)}{\omega_{\mathrm{d}i}}\sin\omega_{\mathrm{d}i}t\right] + \\ & \frac{1}{\omega_{\mathrm{d}i}}\int_0^t N_i(\tau)e^{-\xi_i\omega_i(t-\tau)}\sin\omega_{\mathrm{d}i}(t-\tau)\mathrm{d}t \quad (i = 1 \text{ to } n). \end{aligned} \tag{6.86}$$

The normalized mass $m_i = 1$ has been used in the previous equation. $\omega_{\mathrm{d}i}$ is

$$\omega_{\mathrm{d}i} = \sqrt{1 - \xi_i^2}\,\omega_i. \tag{6.87}$$

At this point, the response in the modal coordinate has been obtained. The physical coordinate can be obtained by Eq. (6.67). To save computational time, modal truncation can be applied by keeping only the vibration modes up to *r*th. In this case, the physical coordinate response will be calculated by Eq. (6.78). Due to space limitations, no examples are given here. Readers interested can refer to Example 6.17 in the book by Rao (2010).

CHAPTER 7

Finite element method for vibration problems

7.1 Introduction

Elasticity is the classical and general theory of mechanics for elastic materials. The governing equations in this theory are a set of partial differential equations (PDEs). Closed-form solutions are only available for very simple cases with simple shapes and boundary conditions; thus, their application is greatly limited. The appearance of the finite element method (FEM), however, fundamentally changed the situation. Since it was developed, FEM has become one of the most widely used engineering analysis tools in many fields, including vibration and structural dynamics. It is hard to tell who individually invented the FEM because many researchers made significant contributions. M. Turner is undoubtedly one of the earliest pioneers (Turner, 1956). Later, the in 1960s, another pioneer R. Clough (1960) made further developments and used the term finite element for the first time. In the 1980s, FEM approached its maturity with a well-developed theory and practical approaches for applications.

FEM discretizes continua into lumped parameters. However, it has much higher accuracy than the lumped-parameter method introduced in Chapter 6 for vibration and dynamic problems. It is particularly suitable for dynamic analysis of structures containing members of complicated geometries, such as gearbox housings, wind turbine blades, etc.

In FEM, the structure under investigation is first discretized into small *elements*, and each element has a certain number of *nodes*. The displacement of an arbitrary point within the element is approximated with that of the nodes through predetermined functions, called *shape functions*.

This chapter is not intended to cover FEM thoroughly. Instead, the focus is placed on the procedures and basics for the analysis of parts with one-dimensional elements. The chapter is organized as follows:

- A brief introduction of FEM is given in Section 7.1.
- Section 7.2 presents the most basic one-dimensional elements, including the bar element and beam element.

Machinery Dynamics
https://doi.org/10.1016/B978-0-12-815785-5.00017-0

- A brief introduction to two-dimensional elements, triangular elements, and rectangular element, is given in Section 7.3. Also in this section, the basic concept of isoparametric elements is introduced.

7.2 One-dimensional elements

Bars and beams are two typical forms of structural components with wide application in civil and mechanical systems, such as trusses, shafts, linkages, etc. In this section, we will show the procedures to discretize a part into these two types of elements.

7.2.1 Bar elements

From solid mechanics, we know that a bar is a part subjected to loads only in the axial direction. A bar of uniform cross section subjected to a distributed axial load $f(x)$ is shown in Fig. 7.1(a). Concentrated loads can be treated as a special case of distributed ones, thus, they are not shown here.

7.2.1.1 Shape function

First, the bar is discretized into a series of bar elements as shown in Fig. 7.1(b). The two ends of the element are called nodes, which are numbered as node 1 and 2, and have displacements $u_1(t)$ and $u_2(t)$, respectively.

The displacement at arbitrary position within the element is represented by $u(x, t)$, which needs to be approximated by the displacements at the two nodes $u_1(t)$ and $u_2(t)$. The simplest option is to use a linear relation:

$$u(x, t) = a_0 + a_1 x, \tag{7.1}$$

where x is the location coordinate. a_0 and a_1 are two constants determined by applying the *boundary conditions*:

$$\begin{aligned} u(0, t) &= u_1(t) \\ u(l, t) &= u_2(t), \end{aligned} \tag{7.2}$$

where l is the length of the element as shown. Eq. (7.1) can be rearranged to

$$u(x, t) = \phi_1(x)u_1(t) + \phi_2(x)u_2(t), \tag{7.3}$$

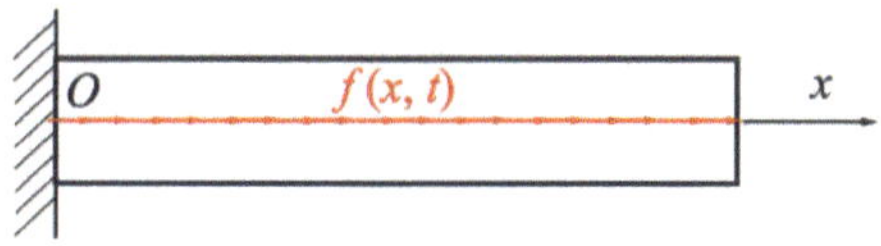

(a) A bar

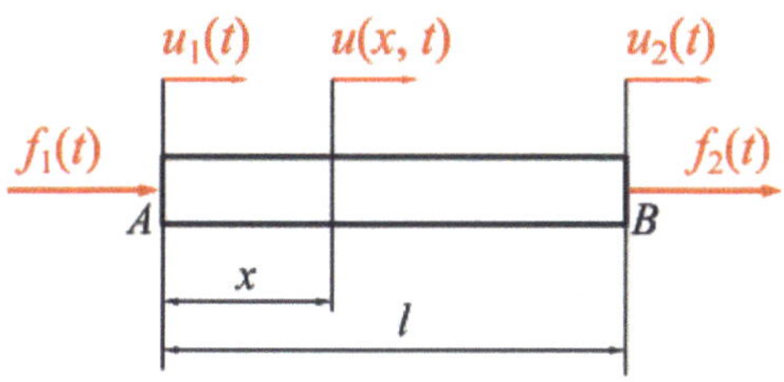

(b) A bar element

Figure 7.1 A bar and bar elements.

where $\phi_1(x)$ and $\phi_2(x)$ are the *shape functions*. They have the following forms:

$$\begin{aligned} \phi_1(x) &= 1 - e \\ \phi_2(x) &= e, \end{aligned} \tag{7.4}$$

where $e = x/l$ is the relative coordinate.

Then, Eq. (7.3) can be represented in a more compact matrix form as

$$u(x, t) = \begin{Bmatrix} \phi_1(x) & \phi_2(x) \end{Bmatrix} \begin{Bmatrix} u_1(t) \\ u_2(t) \end{Bmatrix} = \mathbf{\Phi} \boldsymbol{u}. \tag{7.5}$$

7.2.1.2 Element equations

The vibration equations for an element can be derived through Lagrange's equation.

The potential energy stored in a bar element is calculated by

$$E_{\mathrm{p}}(t) = \frac{1}{2} \int_0^l EA \left[\frac{\partial u(x, t)}{\partial x} \right]^2 \mathrm{d}x, \tag{7.6}$$

where E and A are the material's *Young modulus* and the area of cross section of the bar.

Taking advantage of Eqs. (7.4) and (7.5) and completing the integration give

$$E_{\mathrm{p}}(t) = \frac{1}{2l} EA \int_0^l [-u_1(t) + u_2(t)]^2 \, \mathrm{d}x = \frac{1}{2} \boldsymbol{u}^{\mathrm{T}} \boldsymbol{k} \boldsymbol{u}, \tag{7.7}$$

where $\boldsymbol{u}$ and $\boldsymbol{k}$ are the displacement and the stiffness of the element in vector form as

$$\boldsymbol{k} = \frac{EA}{l} \begin{bmatrix} 1 & -1 \\ -1 & 1 \end{bmatrix}. \tag{7.8}$$

The kinetic energy of the element is calculated by the equation:

$$E_{\mathrm{k}} = \frac{1}{2} \int_0^l \rho A \left[\frac{\partial u(x,t)}{\partial t} \right]^2 \mathrm{d}x. \tag{7.9}$$

Completing the integration and taking advantage of Eqs. (7.4) and (7.5) give

$$\begin{aligned} E_{\mathrm{k}} &= \frac{1}{2} \rho A \int_0^l [\phi_1(x) \dot{u}_1(t) + \phi_2(x) \dot{u}_2(t)]^2 \, \mathrm{d}x \\ &= \frac{1}{2} \dot{\boldsymbol{u}}^{\mathrm{T}} \boldsymbol{m} \dot{\boldsymbol{u}}, \end{aligned} \tag{7.10}$$

where ρ is the material density. $\dot{\boldsymbol{u}}$ is the velocity vector of the nodes. $\boldsymbol{m}$ is the mass matrix of the element represented by

$$\boldsymbol{m} = \frac{\rho A l}{6} \begin{bmatrix} 2 & 1 \\ 1 & 2 \end{bmatrix}. \tag{7.11}$$

The virtual work done by the distributed axial force $f(x, t)$ is

$$\delta W = \int_0^l f(x,t) \delta u \mathrm{d}x = \int_0^l f(x,t) \boldsymbol{\Phi} \delta \boldsymbol{u} \, \mathrm{d}x. \tag{7.12}$$

This can be expressed in the form

$$\delta W = \boldsymbol{f}^{\mathrm{T}} \delta \boldsymbol{u}, \tag{7.13}$$

where $\boldsymbol{f}$ is the generalized nodal force in the form

$$\boldsymbol{f} = \begin{bmatrix} f_1(t) & f_2(t) \end{bmatrix}^{\mathrm{T}} = \int_0^l f(x,t) \boldsymbol{\Phi}^{\mathrm{T}} \, \mathrm{d}x. \tag{7.14}$$

Then, the following equation of motion for the element can be obtained by using Lagrange's equation:

$$\boldsymbol{m}\ddot{\boldsymbol{u}} + \boldsymbol{k}\boldsymbol{u} = \boldsymbol{f}. \tag{7.15}$$

In addition to the distributed forces, concentrated forces at the nodes are directly added to the generalized force.

7.2.1.3 Global equations

After the previous procedures, the next step is to get the global equation by assembling the elemental ones properly. This is illustrated through the stepped bar shown in Fig. 7.2(a). The bar is divided into three elements

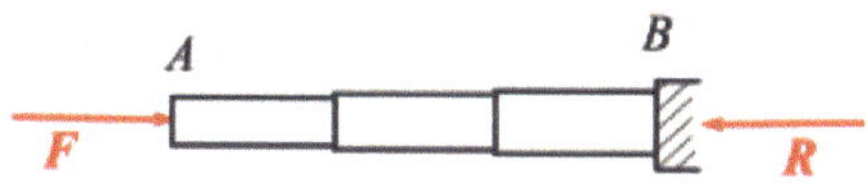

(a) A stepped bar

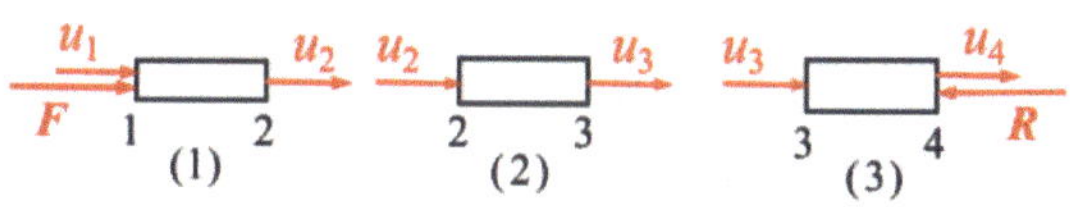

(b) Elements

Figure 7.2 Bar and bar elements.

shown in Fig. 7.2(b). The nodes and elements are numbered in sequence as shown. To differentiate from the nodes, the element number is placed in parenthesis. As can be seen, this bar has three elements and four nodes. Taking the axial displacement at each node as the generalized coordinate, the global generalized coordinate, represented by $\boldsymbol{U}$, is in the following form:

$$\boldsymbol{U} = \begin{bmatrix} u_1 & u_2 & u_3 & u_4 \end{bmatrix}^{\mathrm{T}}. \tag{7.16}$$

Each element has two nodes, which make up a part of the global generalized coordinates $\boldsymbol{U}$. Using the global $\boldsymbol{U}$ to represent the element equation, Eq. (7.15), we have

$$\boldsymbol{M}^{(1)}\ddot{\boldsymbol{U}} + \boldsymbol{K}^{(1)}\boldsymbol{U} = \boldsymbol{F}^{(1)}. \tag{7.17}$$

As usual, the convention that capital and lowercase letters represent the global and elemental, respectively. The superscript indicates the number of the element. The force $\boldsymbol{F}^{(1)}$ is of the following form:

$$\boldsymbol{F}^{(1)} = \begin{bmatrix} F & f_{21} & 0 & 0 \end{bmatrix}^{\mathrm{T}}. \tag{7.18}$$

The subscripts i and j in f_{ij} mean the nodal and elemental number, respectively. $\boldsymbol{M}^{(1)}$ and $\boldsymbol{K}^{(1)}$ are of 4 × 4 dimensions because they are global. For example, $\boldsymbol{M}^{(1)}$ has the following form:

$$\boldsymbol{M}^{(1)} = \begin{bmatrix} m_{11}^{(1)} & m_{12}^{(1)} & 0 & 0 \\ m_{21}^{(1)} & m_{22}^{(1)} & 0 & 0 \\ 0 & 0 & 0 & 0 \\ 0 & 0 & 0 & 0 \end{bmatrix}. \tag{7.19}$$

Similarly, we have the matrices for elements 2 and 3 as follows:

$$\boldsymbol{F}^{(2)} = \begin{bmatrix} 0 & f_{12} & f_{32} & 0 \end{bmatrix}^{\mathrm{T}} \tag{7.20}$$

$$\boldsymbol{M}^{(2)} = \begin{bmatrix} 0 & 0 & 0 & 0 \\ 0 & m_{11}^{(2)} & m_{12}^{(2)} & 0 \\ 0 & m_{21}^{(2)} & m_{22}^{(2)} & 0 \\ 0 & 0 & 0 & 0 \end{bmatrix} \tag{7.21}$$

$$\boldsymbol{F}^{(3)} = \begin{bmatrix} 0 & 0 & f_{23} & -R \end{bmatrix}^{\mathrm{T}} \tag{7.22}$$

$$\boldsymbol{M}^{(3)} = \begin{bmatrix} 0 & 0 & 0 & 0 \\ 0 & 0 & 0 & 0 \\ 0 & 0 & m_{11}^{(3)} & m_{12}^{(3)} \\ 0 & 0 & m_{21}^{(3)} & m_{22}^{(3)} \end{bmatrix}. \tag{7.23}$$

Combining the equations of the three elements gives

$$\boldsymbol{M}\ddot{\boldsymbol{U}} + \boldsymbol{K}\boldsymbol{U} = \boldsymbol{F}, \tag{7.24}$$

where $\boldsymbol{M}$, $\boldsymbol{K}$, and $\boldsymbol{F}$ are the global mass, global stiffness, and global load matrices. They are obtained by summing the elements as follows:

$$\begin{aligned} \boldsymbol{M} &= \sum_{i=1}^{3} \boldsymbol{M}^{(i)} \\ \boldsymbol{K} &= \sum_{i=1}^{3} \boldsymbol{K}^{(i)} \\ \boldsymbol{F} &= \sum_{i=1}^{3} \boldsymbol{F}^{(i)}. \end{aligned} \tag{7.25}$$

When assembling the global generalized force $\boldsymbol{F}$, the internal forces at the nodes cancel out each other; thus, only the external force remains. The global force vector $\boldsymbol{F}$ is greatly simplified as follows, involving only external forces:

$$\boldsymbol{F} = \begin{bmatrix} F & 0 & 0 & -R \end{bmatrix}. \tag{7.26}$$

R is reaction force at the fixed end.

Applying the *boundary condition*, Eq. (7.24) is split into two blocks.

$$\begin{bmatrix} \boldsymbol{M}_{\mathrm{FF}} & \boldsymbol{M}_{\mathrm{FC}} \\ \boldsymbol{M}_{\mathrm{CF}} & \boldsymbol{M}_{\mathrm{CC}} \end{bmatrix} \begin{Bmatrix} \ddot{\boldsymbol{U}}_{\mathrm{F}} \\ \ddot{\boldsymbol{U}}_{\mathrm{C}} \end{Bmatrix} + \begin{bmatrix} \boldsymbol{K}_{\mathrm{FF}} & \boldsymbol{K}_{\mathrm{FC}} \\ \boldsymbol{K}_{\mathrm{CF}} & \boldsymbol{K}_{\mathrm{CC}} \end{bmatrix} \begin{Bmatrix} \boldsymbol{U}_{\mathrm{F}} \\ \boldsymbol{U}_{\mathrm{C}} \end{Bmatrix} = \begin{Bmatrix} \boldsymbol{F}_{\mathrm{F}} \\ \boldsymbol{F}_{\mathrm{C}} \end{Bmatrix}, \tag{7.27}$$

where the two subscripts F and C represent the external force and boundary, respectively. For this specific case, the right end of the bar is fixed; thus, $u_4 = 0$. The submatrices in Eq. (7.27) are as follows after considering the boundary condition:

$$\boldsymbol{U}_{\mathrm{F}} = \begin{Bmatrix} u_1 & u_2 & u_3 \end{Bmatrix}^{\mathrm{T}} \tag{7.28}$$

$$\boldsymbol{U}_{\mathrm{C}} = u_4 = 0 \tag{7.29}$$

$$\boldsymbol{F}_{\mathrm{F}} = \begin{Bmatrix} F & 0 & 0 \end{Bmatrix}^{\mathrm{T}} \tag{7.30}$$

$$\boldsymbol{F}_{\mathrm{C}} = -R. \tag{7.31}$$

Eq. (7.27) can then be expanded to

$$\boldsymbol{M}_{\mathrm{FF}} \ddot{\boldsymbol{U}}_{\mathrm{F}} + \boldsymbol{K}_{\mathrm{FF}} \boldsymbol{U}_{\mathrm{F}} = \boldsymbol{F}_{\mathrm{F}} \tag{7.32}$$

$$\boldsymbol{M}_{\mathrm{CF}} \ddot{\boldsymbol{U}}_{\mathrm{F}} + \boldsymbol{K}_{\mathrm{CF}} \boldsymbol{U}_{\mathrm{F}} = \boldsymbol{F}_{\mathrm{C}}. \tag{7.33}$$

$\boldsymbol{U}_{\mathrm{F}}$ can be solved for, based on Eq. (7.32), which is substituted into the left-hand side of Eq. (7.33) to obtain the reaction force $\boldsymbol{F}_{\mathrm{C}}$.

Eq. (7.32) is in the form of vibration equation of multiple DOFs. However, M_{FF} may not be diagonal here.

Free vibration can be studied with the following eigenvalue problem:

$$\left(\boldsymbol{K}_{FF} - \omega^2 \boldsymbol{M}_{FF}\right) \boldsymbol{\Psi} = \boldsymbol{0}. \tag{7.34}$$

ω and $\boldsymbol{\Psi}$ are the *natural frequency* and the *eigen vector*, respectively.

If the inertial terms are neglected, Eqs. (7.32) and (7.33) become

$$\boldsymbol{K}_{FF} \boldsymbol{U}_F = \boldsymbol{F}_F \tag{7.35}$$

$$\boldsymbol{K}_{CF} \boldsymbol{U}_F = \boldsymbol{F}_C. \tag{7.36}$$

These are the equations for static analysis using FEM.

7.2.2 Beam element

Fig. 7.3(a) shows a more complicated element, a beam element. The concentrated and distributed forces applied to the element are shown in Fig. 7.3(b).

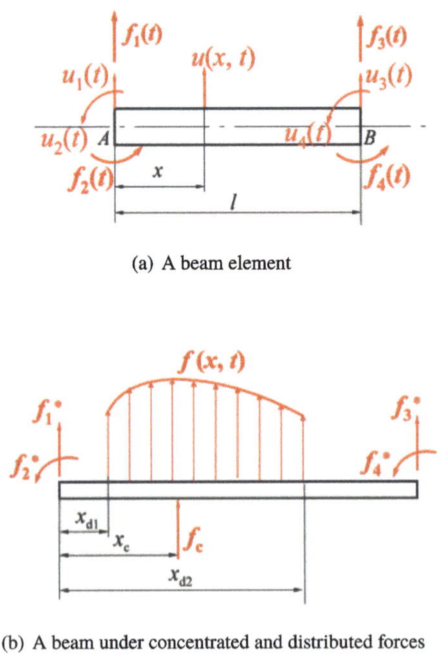

Figure 7.3 Beam and beam element.

7.2.2.1 Shape function

The element in Fig. 7.3(a) has two nodes. For each node, both the lateral deflection and the slope (rotation angle) are needed to describe the displacement. Each node has two generalized coordinates; thus, the element generalized coordinates are

$$\boldsymbol{u} = \begin{Bmatrix} u_1 & u_2 & u_3 & u_4 \end{Bmatrix}^{\mathrm{T}}. \tag{7.37}$$

The displacement at an arbitrary point within the beam is approximated by a polynomial of at least for terms:

$$u(x, t) = c_0 + c_1 x + c_2 x^2 + c_3 x^3. \tag{7.38}$$

c_0 through c_3 are constants, which can be determined through the boundary conditions

$$\begin{aligned} u(0, t) &= u_1(t); \qquad \partial u(0, t)/\partial x = u_2(t) \\ u(l, t) &= u_3(t); \qquad \partial u(l, t)/\partial x = u_4(t). \end{aligned} \tag{7.39}$$

Then, rewriting Eq. (7.38) into a form with the shape function gives

$$u(x, t) = \boldsymbol{\Phi} \boldsymbol{u}, \tag{7.40}$$

where $\boldsymbol{\Phi} = \begin{Bmatrix} \phi_1(x) & \phi_2(x) & \phi_3(x) & \phi_4(x) \end{Bmatrix}$. $\boldsymbol{u}$ is given in Eq. (7.37). $\phi_i(x)$ are as

$$\begin{aligned} \phi_1(x) &= 1 - 3e^2 + 2e^3 \\ \phi_2(x) &= l(e - 2e^2 + e^3) \\ \phi_3(x) &= 3e^2 - 2e^3 \\ \phi_4(x) &= l(-e^2 + e^3). \end{aligned} \tag{7.41}$$

As in the case of bar element, $e = x/l$ is the relative coordinate.

7.2.2.2 Element equations

The potential energy stored in a beam element due to the elastic deflection is

$$E_{\mathrm{p}} = \frac{1}{2} \int_0^l EI \left[\frac{\partial^2 u(x, t)}{\partial x^2} \right]^2 \mathrm{d}x$$

$$
\begin{aligned}
&= \frac{1}{2} EI \int_0^l \left[\sum_{i=1}^{4} \phi_i''(x) u_i(t) \right]^2 \mathrm{d}x \\
&= \frac{1}{2} EI \int_0^l \sum_{i=1}^{4} \sum_{j=1}^{4} u_i(t) \phi_i''(x) \phi_j''(x) u_j(t) \, \mathrm{d}x \\
&= \frac{1}{2} \boldsymbol{u}^{\mathrm{T}} \boldsymbol{k} \boldsymbol{u},
\end{aligned}
\tag{7.42}
$$

where E and I are the Young's modulus of the material and the moment of inertia of the cross-sectional area of the beam. The $''$ means the second-order partial derivative with respect to x. $\boldsymbol{k}$ is the element stiffness matrix. It's entries are computed by

$$
k_{ij} = EI \int_0^l \phi_i''(x) \phi_j''(x) \, \mathrm{d}x.
$$

Substituting the shape function into the previous equation and integrating results in $\boldsymbol{k}$ as

$$
\boldsymbol{k} = \frac{EI}{l^3} \begin{bmatrix} 12 & 6l & -12 & 6l \\ 6l & 4l^2 & -6l & 2l^2 \\ -12 & -6l & 12 & -6l \\ 6l & 2l^2 & -6l & 4l^2 \end{bmatrix}.
\tag{7.43}
$$

The kinetic energy of a beam element is calculated by

$$
\begin{aligned}
E_{\mathrm{k}} &= \frac{1}{2} \int_0^l \rho A \left[\frac{\partial u(x,t)}{\partial t} \right]^2 \mathrm{d}x \\
&= \frac{1}{2} \rho A \int_0^l \left[\sum_{i=1}^{4} \phi_i(x) \dot{u}_i(t) \right]^2 \mathrm{d}x \\
&= \frac{1}{2} \rho A \int_0^l \sum_{i=1}^{4} \sum_{j=1}^{4} \dot{u}_i(t) \phi_i(x) \phi_j(x) \dot{u}_j(t) \, \mathrm{d}x \\
&= \frac{1}{2} \dot{\boldsymbol{u}}^{\mathrm{T}} \boldsymbol{m} \dot{\boldsymbol{u}}.
\end{aligned}
\tag{7.44}
$$

ρ and A are the material density and the cross-sectional area. $\dot{\boldsymbol{u}}$ is the element velocity vector in the following form:

$$
\dot{\boldsymbol{u}} = \left\{ \dot{u}_1 \quad \dot{u}_2 \quad \dot{u}_3 \quad \dot{u}_4 \right\}^{\mathrm{T}}.
\tag{7.45}
$$

$\boldsymbol{m}$ is the element mass matrix; its entries are computed by

$$m_{ij} = \rho A \int_0^l \phi_i(x)\phi_j(x)\,\mathrm{d}x.$$

Substituting the shape functions into the previous equation and completing the integration give

$$\boldsymbol{m} = \frac{\rho A l}{420}\begin{bmatrix} 156 & 22l & 54 & -13l \\ 22l & 4l^2 & 13l & -3l^2 \\ 54 & 13l & 156 & -22l \\ -13l & -3l^2 & -22l & 4l^2 \end{bmatrix}. \tag{7.46}$$

The generalized forces are calculated with the virtual work principle as we did for the bar element. For the case shown in Fig. 7.3(b), the virtual work done by all the external forces is calculated by

$$\delta W = \int_{x_{\mathrm{d1}}}^{x_{\mathrm{d2}}} f(x,t)\delta u(x,t)\,\mathrm{d}x + f_c(t)\delta u(x_c,t) + \sum_{i=1}^{4} f_i^*(t)\delta u_i(t), \tag{7.47}$$

where x_{d1} and x_{d2} define the range of the force $f(x,t)$ as shown in Fig. 7.3(b).

This equals the following:

$$\delta W = \sum_{i=1}^{4} f_i(t)\delta u_i(t) = \boldsymbol{f}^{\mathrm{T}}\delta\boldsymbol{u}. \tag{7.48}$$

Then, the generalized force is obtained as

$$f_i(t) = \int_{x_{\mathrm{d1}}}^{x_{\mathrm{d2}}} f(x,t)\phi_i(x)\,\mathrm{d}x + f_c(t)\phi_i(x_c) + f_i^*(t). \tag{7.49}$$

7.2.2.3 Global equations

We use the example shown in Fig. 7.4 to illustrate the procedure of assembling the global equations.

If the beam is divided into three elements as shown in Fig. 7.5, four nodes will be created. As usual, we number the nodes and elements in sequence, but put the element number in parenthesis.

Each node has two DOFs; the global generalized coordinates are thus

$$\boldsymbol{U} = \begin{Bmatrix} u_1 & u_2 & \cdots & u_8 \end{Bmatrix}^{\mathrm{T}}. \tag{7.50}$$

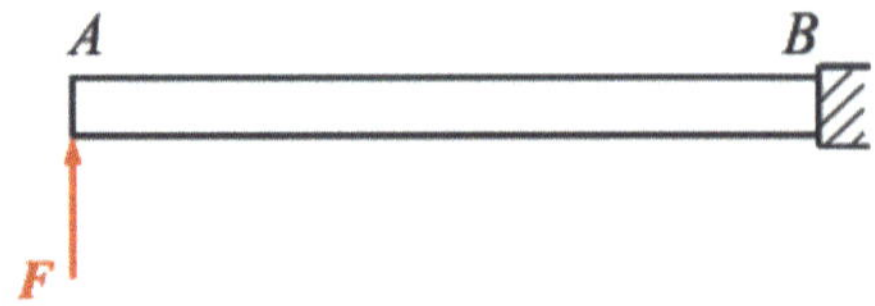

Figure 7.4 A beam.

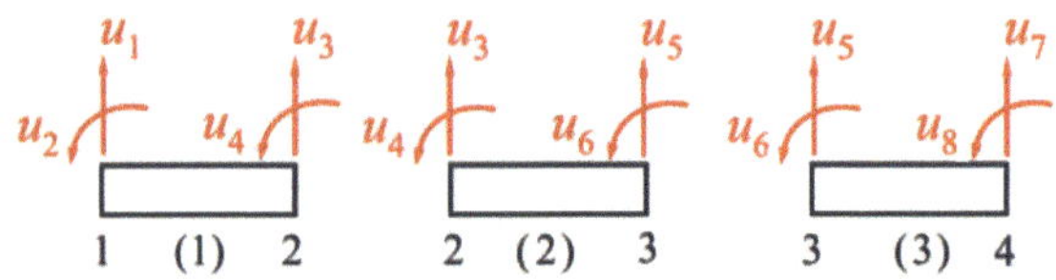

Figure 7.5 Elements.

Mapping the first element equation into the global, we have

$$\boldsymbol{M}^{(1)}\ddot{\boldsymbol{U}}+\boldsymbol{K}^{(1)}\boldsymbol{U}=\boldsymbol{F}^{(1)}. \tag{7.51}$$

$\boldsymbol{F}^{(1)}$ is the generalized force for element 1 as shown here:

$$\boldsymbol{F}^{(1)}=\begin{bmatrix} f_1^{(1)} & f_2^{(1)} & f_3^{(1)} & f_4^{(1)} & 0 & 0 & 0 & 0 \end{bmatrix}^{\mathrm{T}}. \tag{7.52}$$

$\boldsymbol{M}^{(1)}$ and $\boldsymbol{K}^{(1)}$ the are the mass and stiffness matrices mapped from element 1; both are 8 × 8 in dimension. They are constructed by placing the element matrix in the proper place and leaving other entries zero. In the following, the assembling process is illustrated with the mass matrix.

First, $\boldsymbol{M}^{(1)}$ is

$$\boldsymbol{M}^{(1)}=\begin{bmatrix} \boldsymbol{m}^{(1)} & \boldsymbol{0}_{4\times4} \\ \boldsymbol{0}_{4\times4} & \boldsymbol{0}_{4\times4} \end{bmatrix}, \tag{7.53}$$

where $\boldsymbol{0}_{4\times4}$ is a matrix of dimension 4 × 4 with all entries zero, and $\boldsymbol{m}^{(1)}$ is the mass matrix of element 1 computed by Eq. (7.46).

For element 2, we have

$$\boldsymbol{M}^{(2)}\ddot{\boldsymbol{U}}+\boldsymbol{K}^{(2)}\boldsymbol{U}=\boldsymbol{F}^{(2)}, \tag{7.54}$$

where $\boldsymbol{F}^{(2)}$ and $\boldsymbol{M}^{(2)}$ are

$$\boldsymbol{F}^{(2)}=\begin{bmatrix} 0 & 0 & f_1^{(2)} & f_2^{(2)} & f_3^{(2)} & f_4^{(2)} & 0 & 0 \end{bmatrix}^{\mathrm{T}} \tag{7.55}$$

$$\boldsymbol{M}^{(2)} = \begin{bmatrix} \mathbf{0}_{2\times2} & \mathbf{0}_{2\times4} & \mathbf{0}_{2\times2} \\ \mathbf{0}_{4\times2} & \boldsymbol{m}^{(2)} & \mathbf{0}_{4\times2} \\ \mathbf{0}_{2\times2} & \mathbf{0}_{2\times4} & \mathbf{0}_{2\times2} \end{bmatrix}. \tag{7.56}$$

For element 3, the element equation is

$$\boldsymbol{M}^{(3)}\ddot{\boldsymbol{U}} + \boldsymbol{K}^{(3)}\boldsymbol{U} = \boldsymbol{F}^{(3)}, \tag{7.57}$$

and the element force vector and mass matrix are

$$\boldsymbol{F}^{(3)} = \begin{bmatrix} 0 & 0 & & 0 & 0 & f_1^{(3)} & f_2^{(3)} & f_3^{(3)} & f_4^{(3)} \end{bmatrix}^{\mathrm{T}} \tag{7.58}$$

$$\boldsymbol{M}^{(3)} = \begin{bmatrix} \mathbf{0}_{4\times4} & \mathbf{0}_{4\times4} \\ \mathbf{0}_{4\times4} & \boldsymbol{m}^{(3)} \end{bmatrix}. \tag{7.59}$$

The global governing equation of motion is

$$\boldsymbol{M}\ddot{\boldsymbol{U}} + \boldsymbol{K}\boldsymbol{U} = \boldsymbol{F}. \tag{7.60}$$

The global matrices are assembled from the elemental ones so:

$$\begin{aligned} \boldsymbol{M} &= \sum_{i=1}^{3} \boldsymbol{M}^{(i)} \\ \boldsymbol{K} &= \sum_{i=1}^{3} \boldsymbol{K}^{(i)} \\ \boldsymbol{F} &= \sum_{i=1}^{3} \boldsymbol{F}^{(i)}. \end{aligned} \tag{7.61}$$

As in the case of bar element, $\boldsymbol{F}$ contains only the external forces because the internal forces at the nodes cancel out each other:

$$\boldsymbol{F} = \begin{Bmatrix} F & 0 & 0 & 0 & 0 & 0 & F_{\mathrm{B}} & M_{\mathrm{B}} \end{Bmatrix}^{\mathrm{T}}. \tag{7.62}$$

The boundary conditions are fixed at the left end, meaning zero deflection and zero rotation of node 4 with:

$$u_7 = 0; \qquad u_8 = 0. \tag{7.63}$$

Dividing Eq. (7.60) into these blocks

$$\begin{bmatrix} \boldsymbol{M}_{\mathrm{FF}} & \boldsymbol{M}_{\mathrm{FC}} \\ \boldsymbol{M}_{\mathrm{CF}} & \boldsymbol{M}_{\mathrm{CC}} \end{bmatrix} \begin{Bmatrix} \ddot{\boldsymbol{U}}_{\mathrm{F}} \\ \ddot{\boldsymbol{U}}_{\mathrm{C}} \end{Bmatrix} + \begin{bmatrix} \boldsymbol{K}_{\mathrm{FF}} & \boldsymbol{K}_{\mathrm{FC}} \\ \boldsymbol{K}_{\mathrm{CF}} & \boldsymbol{K}_{\mathrm{CC}} \end{bmatrix} \begin{Bmatrix} \boldsymbol{U}_{\mathrm{F}} \\ \boldsymbol{U}_{\mathrm{C}} \end{Bmatrix} = \begin{Bmatrix} \boldsymbol{F}_{\mathrm{F}} \\ \boldsymbol{F}_{\mathrm{C}} \end{Bmatrix}, \tag{7.64}$$

the submatrices in the equation are

$$\boldsymbol{U}_{\mathrm{F}}=\left\{u_1 \quad u_2 \quad u_3 \quad u_4 \quad u_5 \quad u_6\right\}^{\mathrm{T}} \tag{7.65}$$

$$\boldsymbol{U}_{\mathrm{C}}=\left\{u_7 \quad u_8\right\}^{\mathrm{T}} \tag{7.66}$$

$$\boldsymbol{F}_{\mathrm{F}}=\left\{F \quad 0 \quad 0 \quad 0 \quad 0 \quad 0\right\}^{\mathrm{T}} \tag{7.67}$$

$$\boldsymbol{F}_{\mathrm{C}}=\left\{F_{\mathrm{B}} \quad M_{\mathrm{B}}\right\}^{\mathrm{T}}. \tag{7.68}$$

Eq. (7.64) can be expanded to the following two equations in view of $\boldsymbol{U}_{\mathrm{C}}=\mathbf{0}$:

$$\boldsymbol{M}_{\mathrm{FF}}\ddot{\boldsymbol{U}}_{\mathrm{F}}+\boldsymbol{K}_{\mathrm{FF}}\boldsymbol{U}_{\mathrm{F}}=\boldsymbol{F}_{\mathrm{F}} \tag{7.69}$$

$$\boldsymbol{M}_{\mathrm{CF}}\ddot{\boldsymbol{U}}_{\mathrm{F}}+\boldsymbol{K}_{\mathrm{CF}}\boldsymbol{U}_{\mathrm{F}}=\boldsymbol{F}_{\mathrm{C}}. \tag{7.70}$$

After solving for the displacement vector $\boldsymbol{U}_{\mathrm{F}}$, the reaction force and moment at the fixed end, $\boldsymbol{F}_{\mathrm{C}}$, can be obtained by substituting $\boldsymbol{U}_{\mathrm{F}}$ into the left-hand side of Eq. (7.70).

The free-vibration characteristics can be obtained through the eigenvalue problem

$$\left(\boldsymbol{K}_{\mathrm{FF}}-\omega^2\boldsymbol{M}_{\mathrm{FF}}\right)\boldsymbol{\Psi}=\mathbf{0}. \tag{7.71}$$

$\boldsymbol{K}_{\mathrm{FF}}$ and $\boldsymbol{M}_{\mathrm{FF}}$ in these equations can also be obtained by crossing out two rows and two columns corresponding to the two restricted DOFs. Thus, they shrink to 6×6 in dimension.

7.3 Two-dimensional element

Bar and beam elements are widely used in the vibration analysis of shafts, bars, and trusses. For more complicated parts, such as plates and shells, two-dimensional elements have to be used.

Many types of two-dimensional elements are available; this book introduces only triangular elements, rectangular elements, and isoparametric elements. The focus is placed on the procedure to derive the elemental stiffness and mass matrices. Readers interested with more detailed knowledge are directed to comprehensive FEM books (Hughes, 2012; Petyt, 2010).

7.3.1 Triangular elements

The simplest two-dimensional element is the *linear triangular element* as shown in Fig. 7.6. This element has three vertex nodes. Each node has two degrees of freedom, u_i and v_i, along the x and y directions, respectively.

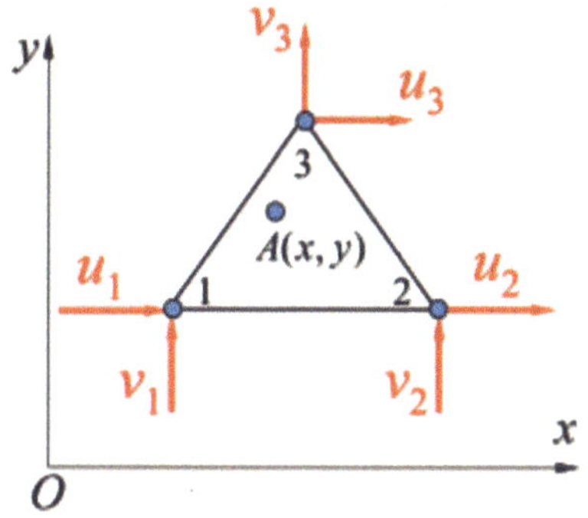

Figure 7.6 Triangular elements.

For an arbitrary point A within the element, its displacements are represented as a linear function of its coordinates as:

$$\begin{aligned} u(x,y) &= a_1 + a_2 x + a_3 y \\ v(x,y) &= a_4 + a_5 x + a_6 y. \end{aligned} \tag{7.72}$$

Letting $\boldsymbol{d} = \{u(x,y) \quad v(x,y)\}^{\mathrm{T}}$, Eq. (7.72) can be put into this more compact matrix form

$$\boldsymbol{d} = \begin{bmatrix} 1 & x & y & 0 & 0 & 0 \\ 0 & 0 & 0 & 1 & x & y \end{bmatrix} \left\{ a_1 \quad a_2 \quad a_3 \quad a_4 \quad a_5 \quad a_6 \right\}^{\mathrm{T}}. \tag{7.73}$$

The displacements of three nodal points are also in the same form. If we have the elemental displacement vector as

$$\boldsymbol{u} = \left\{ u_1 \quad v_1 \quad u_2 \quad v_2 \quad u_3 \quad v_3 \right\}^{\mathrm{T}}, \tag{7.74}$$

according to Eq. (7.73), $\boldsymbol{u}$ can be represented as

$$\boldsymbol{u}=\begin{Bmatrix} u_1\\ v_1\\ u_2\\ v_2\\ u_3\\ v_3 \end{Bmatrix}=\begin{bmatrix} 1 & x_1 & y_1 & 0 & 0 & 0\\ 0 & 0 & 0 & 1 & x_1 & y_1\\ 1 & x_2 & y_2 & 0 & 0 & 0\\ 0 & 0 & 0 & 1 & x_2 & y_2\\ 1 & x_3 & y_3 & 0 & 0 & 0\\ 0 & 0 & 0 & 1 & x_3 & y_3 \end{bmatrix}\begin{Bmatrix} a_1\\ a_2\\ a_3\\ a_4\\ a_5\\ a_6 \end{Bmatrix}=\boldsymbol{Ca}. \tag{7.75}$$

Then, one can easily obtain

$$\boldsymbol{a}=\mathbf{C}^{-1}\boldsymbol{u}. \tag{7.76}$$

Substituting Eq. (7.76) into (7.73) gives

$$\boldsymbol{d}=\boldsymbol{\Phi}\boldsymbol{u}, \tag{7.77}$$

where $\boldsymbol{\Phi}$ is the shape-function matrix in the following form:

$$\boldsymbol{\Phi}=\begin{bmatrix} \phi_1 & 0 & \phi_2 & 0 & \phi_3 & 0\\ 0 & \phi_1 & 0 & \phi_2 & 0 & \phi_3 \end{bmatrix}, \tag{7.78}$$

where

$$\begin{aligned} \phi_1 &= (\alpha_1+\beta_1 x+\gamma_1 y)/(2A)\\ \phi_2 &= (\alpha_2+\beta_2 x+\gamma_2 y)/(2A)\\ \phi_3 &= (\alpha_3+\beta_3 x+\gamma_3 y)/(2A). \end{aligned} \tag{7.79}$$

α_i, β_i, and γ_i are functions of the nodal coordinates so

$$\begin{aligned} \alpha_1 &= x_2y_3-x_3y_2; & \beta_1 &= y_2-y_3; & \gamma_1 &= x_3-x_2\\ \alpha_2 &= x_3y_1-x_1y_3; & \beta_2 &= y_3-y_1; & \gamma_2 &= x_1-x_3\\ \alpha_3 &= x_1y_2-x_2y_1; & \beta_3 &= y_1-y_2; & \gamma_3 &= x_2-x_1. \end{aligned} \tag{7.80}$$

A is the area of the triangle, which is calculated by

$$2A=(x_2-x_1)(y_3-y_1)-(x_3-x_1)(y_2-y_1). \tag{7.81}$$

It can be seen that the shape-function matrix is a function of the coordinates of the point and the nodes.

From the theory of elasticity, strain is defined as

$$\epsilon = \begin{Bmatrix} \epsilon_x \\ \epsilon_y \\ \gamma_{xy} \end{Bmatrix} = \begin{Bmatrix} \dfrac{\partial u}{\partial x} \\ \dfrac{\partial v}{\partial y} \\ \dfrac{\partial u}{\partial y} + \dfrac{\partial v}{\partial x} \end{Bmatrix}. \tag{7.82}$$

With Eq. (7.77), the strain can be represented in the following matrix form:

$$\epsilon = \begin{bmatrix} \dfrac{\partial}{\partial x} & 0 \\ 0 & \dfrac{\partial}{\partial y} \\ \dfrac{\partial}{\partial y} & \dfrac{\partial}{\partial x} \end{bmatrix} \boldsymbol{\Phi u} = \boldsymbol{Bu}. \tag{7.83}$$

$\boldsymbol{B}$ is a matrix of dimension 3×6. Stress and strain are linked by Hook's law:

$$\boldsymbol{\sigma} = \boldsymbol{E\epsilon}, \tag{7.84}$$

where $\boldsymbol{\sigma}$ is the stress vector and $\boldsymbol{E}$ is the material property. They are defined as

$$\boldsymbol{\sigma} = \begin{Bmatrix} \sigma_x & \sigma_y & \tau_{xy} \end{Bmatrix}^{\mathrm{T}} \tag{7.85}$$

$$\boldsymbol{E} = \frac{E}{1-\mu^2} \begin{bmatrix} 1 & \mu & 0 \\ \mu & 1 & 0 \\ 0 & 0 & \dfrac{(1-\mu)}{2} \end{bmatrix} \tag{7.86}$$

E and μ are the Young's modulus and the Poisson's ratio of the material.

Substituting Eq. (7.83) into Eq. (7.84) gives

$$\boldsymbol{\sigma} = \boldsymbol{EBu}. \tag{7.87}$$

Then the potential energy caused by the elastic deformation is calculated by

$$E_{\mathrm{p}} = \frac{1}{2}\int_V \boldsymbol{\sigma}^{\mathrm{T}} \boldsymbol{\epsilon}\, \mathrm{d}v = \frac{1}{2}\int_V \boldsymbol{u}^{\mathrm{T}} \boldsymbol{B}^{\mathrm{T}} \boldsymbol{EBu}\, \mathrm{d}v = \frac{1}{2}\boldsymbol{u}^{\mathrm{T}} \boldsymbol{ku}, \tag{7.88}$$

where

$$\boldsymbol{k}=\int_V \boldsymbol{B}^{\mathrm{T}}\boldsymbol{E}\boldsymbol{B}\,\mathrm{d}v \tag{7.89}$$

is the stiffness matrix of the element; it is a 6 × 6 symmetrical matrix.

The kinetic energy of the element is

$$E_{\mathrm{k}}=\frac{1}{2}\rho\int_V \dot{\boldsymbol{d}}^{\mathrm{T}}\dot{\boldsymbol{d}}\,\mathrm{d}v=\frac{1}{2}\rho\int_V \dot{\boldsymbol{u}}^{\mathrm{T}}\boldsymbol{\Phi}^{\mathrm{T}}\boldsymbol{\Phi}\dot{\boldsymbol{u}}\,\mathrm{d}v=\frac{1}{2}\dot{\boldsymbol{u}}^{\mathrm{T}}\boldsymbol{m}\dot{\boldsymbol{u}}, \tag{7.90}$$

where

$$\boldsymbol{m}=\rho\int_V \boldsymbol{\Phi}^{\mathrm{T}}\boldsymbol{\Phi}\,\mathrm{d}v \tag{7.91}$$

is the mass matrix of the element; it is also a 6 × 6 diagonal matrix.

Through this analysis, the procedures to obtain the stiffness and mass matrices of element can be summarized thus:

- Assume the form of displacement of an arbitrary point within an element, which can be interpolated by shape functions, and determine the generalized coordinates (nodal displacements). The number of DOFs should equal the number of constants contained in the assumed displacement function.
- Derive the shape functions that link the displacements of an arbitrary point within the element to the nodal displacements.
- Represent the strain in terms of the nodal displacements.
- Use Hook's law to represent the stress in terms of the nodal displacements.
- Derive the stiffness matrix of the element from the potential energy.
- Derive the mass matrix of the element from the kinetic energy.

In the previous, the displacement of a point within the element is assumed to be a linear function of the point coordinates (Eq. (7.72)), which has very limited accuracy. To improve the accuracy, higher-order polynomials can be used. However, this leads to the increase of the number of DOFs; thus, more nodes are needed. For example, the triangular element shown in Fig. 7.7 has six nodes.

The displacement function can be assumed to be:

$$\begin{aligned} u(x,y)&=a_1+a_2x+a_3y+a_4x^2+a_5xy+a_6y^2\\ v(x,y)&=a_7+a_8x+a_9y+a_{10}x^2+a_{11}xy+a_{12}y^2. \end{aligned} \tag{7.92}$$

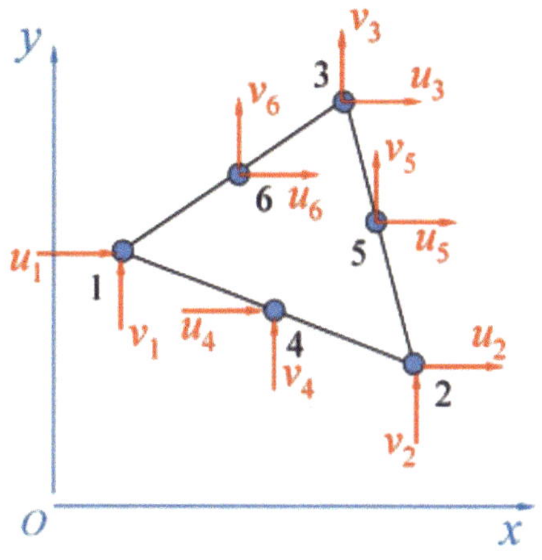

Figure 7.7 Six-node triangular element.

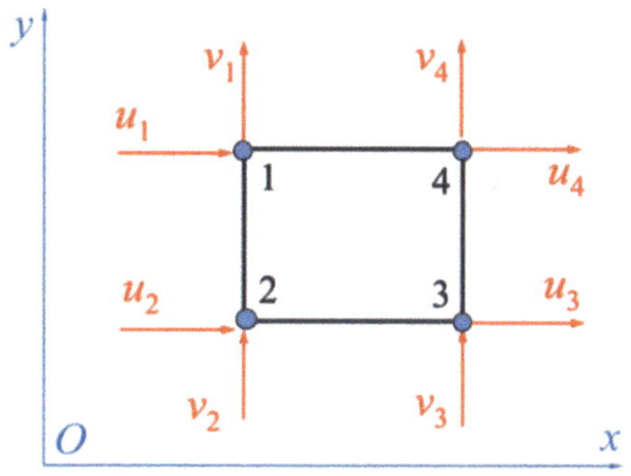

Figure 7.8 Rectangular element.

This type of element has higher accuracy than the linear triangular element.

7.3.2 Rectangular elements

Another type of commonly used elements is the rectangular element as shown in Fig. 7.8. The displacement functions of this type of elements can be assumed to be

$$\begin{aligned} u(x, y) &= a_1 + a_2x + a_3y + a_4xy \\ v(x, y) &= a_5 + a_6x + a_7y + a_8xy. \end{aligned} \tag{7.93}$$

Obviously, they contain a quadratic term xy. Thus, rectangular elements have better accuracy than the linear triangular elements. However, triangular elements have the advantage of fitting irregular boundaries better. In some cases, these two types of elements are used jointly to achieve the best effect. In the regular area, more accurate rectangular elements are used; while in the irregular border area, triangular elements are used.

Similar to the triangular elements, nodes can be added to the midpoints of the edges to increase accuracy.

7.3.3 Isoparametric element

As mentioned in the previous section, the rectangular elements have better accuracy but lower capacity to mesh complicated curve borders than triangular elements. *Isoparametric elements*, however, have good performance in both meshing irregular borders and accuracy. In the following, a brief introduction is given to the isoparametric element.

The element in Fig. 7.9(a) is a normal rectangular element with each side having a length of two. A local coordinate frame $\xi o \eta$ is attached to the element with the origin located at the center of the element as shown (also called natural coordinates). The basic idea of the isoparametric element is to transform the rectangular element into a distorted element in the global coordinate system xoy, as shown in Fig. 7.9(b); thus, better suitability will be achieved in meshing curved regions. The original rectangular element is called the *parent element*.

In the parent element, the nodal displacements are u_i and v_i ($i = 1$–4), and the displacement of a point within the element is represented by the

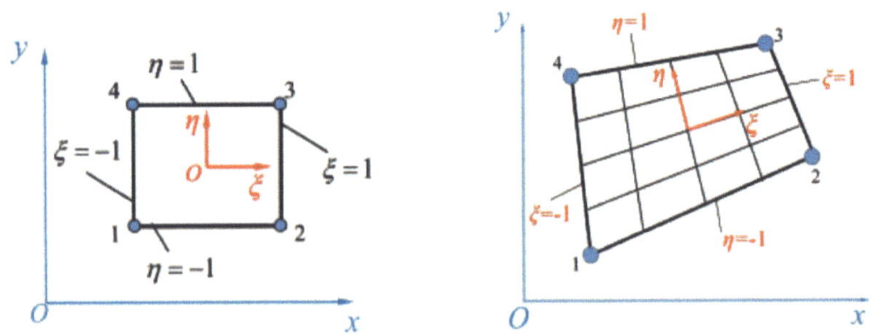

(a) Four node rectangular element (b) Quadrilateral element

Figure 7.9 Mapping of a four-node rectangular element to an isoparametric element.

interpolation function:

$$u = \sum_{i=1}^{4} \phi_i u_i$$
$$v = \sum_{i=1}^{4} \phi_i v_i. \tag{7.94}$$

ϕ_i is the shape function with the following form:

$$\phi_i = \frac{1}{4}(1 + \xi_i \xi)(1 + \eta_i \eta) \quad (i = 1\text{–}4), \tag{7.95}$$

where ξ and η are the coordinates of the point at the local coordinate frame.

A closer look at the shape functions finds that ϕ_i equals 1 at the i-th node, but 0 at others. Taking ϕ_1 as an example, at node 1, $\xi = \eta = -1$. According to Eq. (7.95), $\phi_1 = 1$. It is straightforward to check that $\phi_1 = 0$ at the other three nodes. Taking advantage of this feature, they can be used to represent the global coordinates of a point in terms of the global coordinates of the nodes so:

$$x = \sum_{i=1}^{4} \phi_i x_i$$
$$y = \sum_{i=1}^{4} \phi_i y_i, \tag{7.96}$$

where x_i and y_i are the global coordinates of the nodes; x and y are the global coordinates of a point within the element.

Through this transformation, the rectangular element is mapped to a general quadrilateral element. Now let's check if this is really the case. For the side linking nodes 1 and 2, the following equations can be obtained according to Eqs. (7.95) and (7.96):

$$x = 1 - \frac{x_1 - x_2}{2}\xi$$
$$y = 1 - \frac{y_1 - y_2}{2}\xi. \tag{7.97}$$

Obviously, this is a straight line. The other three sides can be verified similarly as straight lines, as well.

If one node is added to the midpoint of each side in the rectangular element in Fig. 7.9(a), the rectangular element would have eight nodes

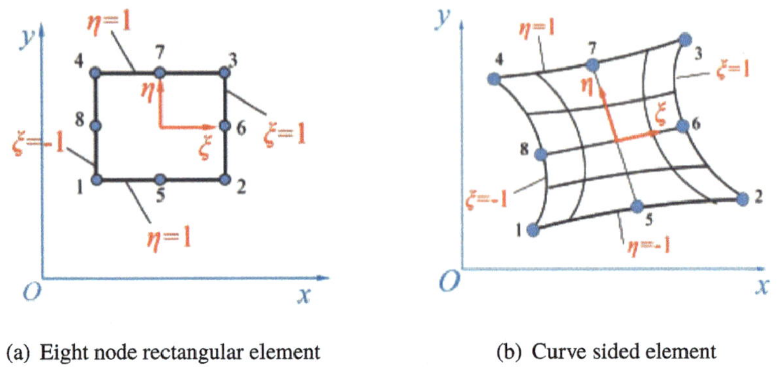

Figure 7.10 Mapping of an eight-node rectangular element to an isoparametric element.

as shown in Fig. 7.10(a). The natural coordinates of the nodes ξ_i and η_i $(i = 1\text{–}8)$ will be either 0 or ± 1.

The shape functions will be

$$
\begin{aligned}
\phi_i &= \frac{1}{4}(1+\xi_i\xi)(1+\eta_i\eta)(\xi_i\xi+\eta_i\eta-1) \quad (i=1\text{–}4) \\
\phi_i &= \frac{1}{2}(1-\xi^2)(1+\eta_i\eta) \quad (i=5,7) \\
\phi_i &= \frac{1}{2}(1-\eta^2)(1+\xi_i\xi) \quad (i=6,8).
\end{aligned}
\tag{7.98}
$$

For this type of element, the relationship between an arbitrary point coordinates and the nodal coordinates in the $\xi o \eta$ and the xoy frames are still in the form of Eqs. (7.94) and (7.96), respectively. But the upper limit of the summation should be changed from four to eight.

Let's check the side from node 1 to node 2, along which $\eta = -1$. From Eqs. (7.96) and (7.98), the following equations can be obtained:

$$
\begin{aligned}
x &= x_5 + \left(\frac{x_2 - x_1}{2}\right)\xi + \left(\frac{x_2 + x_1}{2} - x_5\right)\xi^2 \\
y &= y_5 + \left(\frac{y_2 - y_1}{2}\right)\xi + \left(\frac{y_2 + y_1}{2} - y_5\right)\xi^2.
\end{aligned}
\tag{7.99}
$$

Obviously, this is a quadratic curve; thus, the element has better suitability in meshing curved boundaries.

7.3.4 Plane problems of FEM

Any physical object is three dimensional, literally. In some special cases, however, problems can be simplified as two dimensional (planar problems). Analysis and modeling with two-dimensional elements are much easier than with the three-dimensional ones.

7.3.4.1 Plane-stress and plane-strain problems

For two-dimensional objects, there are *plane-stress* and *plane-strain* problems. These two types of problems can be illustrated with Fig. 7.11(a) and Fig. 7.11(b).

(1) Plane-stress problem

If a structure is loaded in a plane and the load is constant across the cross section, the problem can be treated as a plane-stress problem. A typical example in mechanical engineering is a plate subjected to a constant load across the cross section as shown in Fig. 7.11(a).

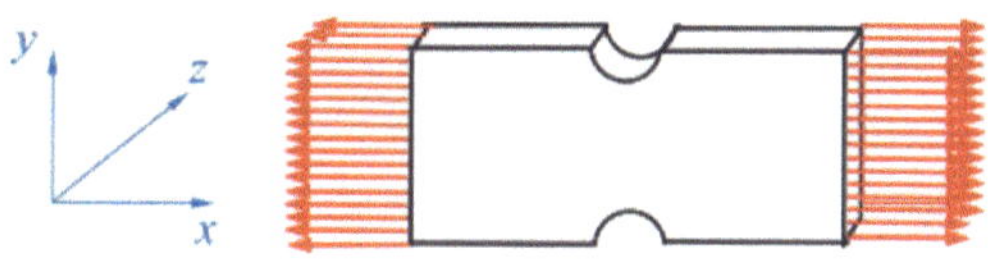

(a) Plane stress problem

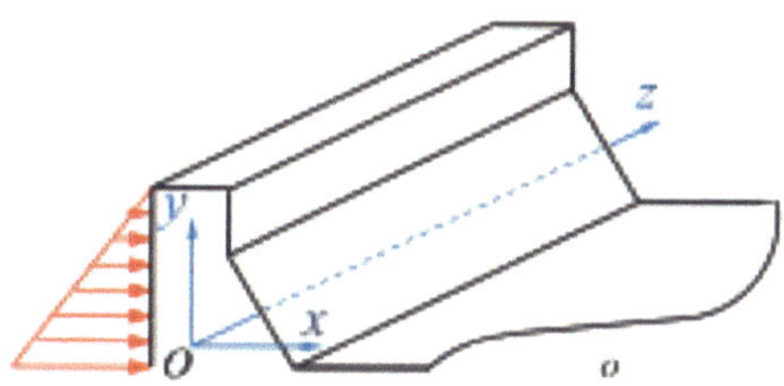

(b) Plane strain problem

Figure 7.11 Two types of planar problems.

For this plate, there is no load applied on the two free surfaces; thus, we have

$$\sigma_z = \tau_{zy} = \tau_{zx} = 0. \tag{7.100}$$

In this case, the strain and stress vectors can be represented by Eqs. (7.82) and (7.85).

(2) Plane-strain problem

If a long object is subjected to constant load along its length, such as the dam in Fig. 7.11(b), it can be treated as a plane-strain problem. For this case, we have

$$\epsilon_z = \gamma_{zy} = \gamma_{zx} = 0. \tag{7.101}$$

The $\boldsymbol{E}$ in Eq. (7.86) is replaced by

$$\boldsymbol{E} = \frac{E(1-\mu)}{(1+\mu)(1-2\mu)} \begin{bmatrix} 1 & \mu/(1-\mu) & 0 \\ 0 & 1 & 0 \\ 0 & 0 & (1-2\mu)/2(1-\mu) \end{bmatrix} \tag{7.102}$$

7.3.4.2 General rules for meshing

The process of discretization of an object is called meshing. To mesh an object, the type, number, and layout of the elements have to be carefully considered in advance. Several general rules of meshing are the following:

- Many factors affect the selection of an element, such as the type of problem you want to run, boundary conditions, geometry considerations, the results required, etc. Thus, a trade-off is required in some cases.
- Elements of higher accuracy require a smaller number of elements. Smaller element size leads to more computation. A trade-off should be considered between these factors and the computation cost.
- Critical areas should be meshed more densely. These areas include very tight radii, points of load application, abrupt changes in section, narrow sections, etc.
- If possible, the application points of concentrated loads should be chosen as nodes.

Fig. 7.12 gives some examples of meshing a spur gear tooth. In 7.12(a), linear triangular elements are used. In 7.12(b), linear triangular elements are used in some boundary areas, and rectangular elements are used for the remainder. While in 7.12(c), isoparametric elements are used. Quadrilateral elements with straight and curved sides are used for the internal and boundary areas, respectively.

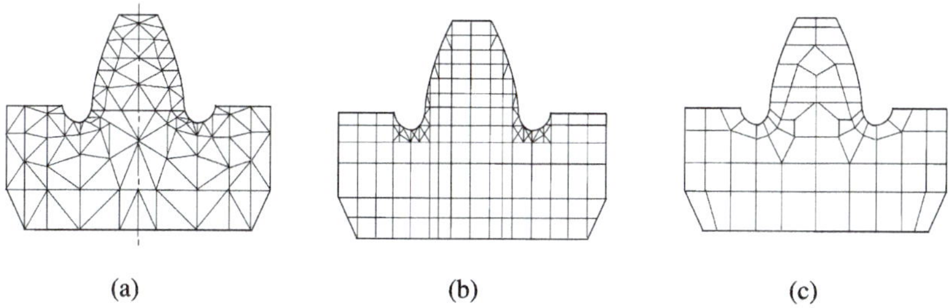

Figure 7.12 Example of meshing a tooth of a spur gear.

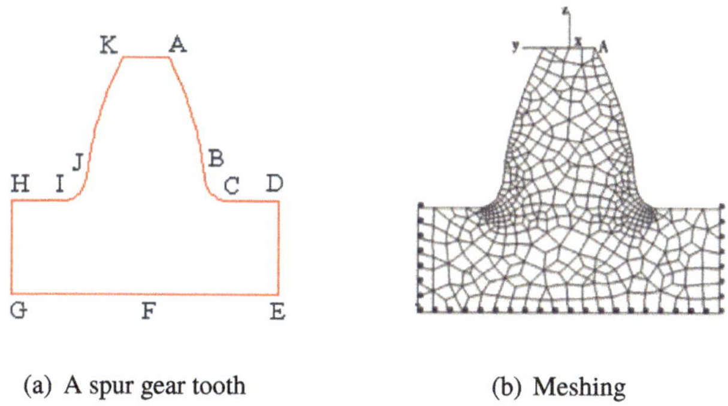

Figure 7.13 Mesh a spur gear tooth for natural frequency analysis.

7.3.4.3 An example

Natural frequencies and vibration modes of a spur gear tooth can be analyzed easily with FEM software, such as ANSYS. The effects of gear materials and gear size can be compared.

(1) Gear tooth model

For an involute spur gear, the tooth profile is the same along the thickness. In addition, the tooth is symmetrical as shown in Fig. 7.13(a). Thus, this analysis can be treated as a plane-stress problem.

In the profile, *AB* and *JK* are involute sections, while *BC* and *IJ* are the transitions at the root. To simplify the problem, only one tooth can be analyzed, and the tooth can be modeled as fix-ended with the boundary of *DEFGH*. Some dimensions involved are $DE = GH = 1.5$ m and, $EF = FG = 2.2$ m. m here is the module of the gear. The tooth is meshed as Fig. 7.13(b). Isoparametric elements of curved sides are used. The total number of element is 465 with 493 nodes.

Table 7.1 Gear and material properties.

Material	N	m **(mm)**	E **(GPa)**	ρ (N/m^2)	μ
steel	30	2.5	195	75,244	0.3
terylene 500	30	2.5	2.41	13,818	0.38

N: tooth number; m: module; E: Young's modulus; ρ: material density; μ: Poisson ratio.

Table 7.2 Order of vibration modes.

	Order of vibration modes			
	1	**2**	**3**	**4**
steel	0.781×10^5	1.631×10^5	2.203×10^5	3.401×10^5
terylene 500	0.214×10^5	0.443×10^5	0.519×10^5	0.859×10^5

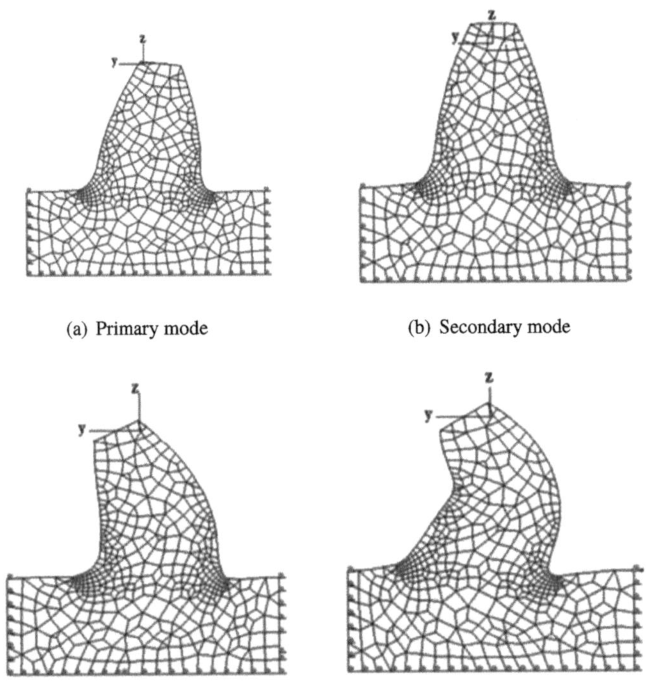

(a) Primary mode

(b) Secondary mode

(c) Third mode

(d) Fourth mode

Figure 7.14 Vibration modes of the gear tooth.

(2) Analysis results

Two types of materials, carbon steel, and terylene are analyzed. The gear parameter and material properties are given in Table 7.1.

The first four natural frequencies of the analysis are given in Table 7.2 and the corresponding vibration modes are shown in Fig. 7.14.

Table 7.3 Primary natural frequency of steel gear (Hz).

Module (mm)	**1**	**2.5**	**3**	**5**
primary natural frequency	1.97×10^5	7.81×10^4	6.51×10^4	3.94×10^4

In Table 7.3, the primary natural frequencies of a steel gear for different gear modules are given. It is clearly seen that the smaller the gear module the lower the natural frequency.

CHAPTER 8

Nonlinear vibration

8.1 Introduction

In the theory of *linear vibration* presented in Chapter 5 and 6, the equations of motion contain displacements (x) and their derivatives ($\dot{x}$ and $\ddot{x}$) only to the first degree. In other words, the inertial force, damping force, and elastic restoring force are proportional to the acceleration, velocity, and displacement, respectively. Theories of linear vibration have become substantially mature. Well-developed methods and analytic tools, such as the modal superposition, time domain, and frequency domain analysis, are available for solving linear vibration problems. Linear vibration analysis is generally adequate for machines that operate at relatively low speed and in which the deformation of components is relatively small. However, there are two conflicting trends pursued simultaneously in modern machine industry, higher speed and lighter weight. These trends lead to relatively larger deformation of the machine components. In some cases, the stress level goes beyond the proportional limit, making the Hook's law invalid. In addition, dry friction and joint clearance etc are common factors in various machine. For these problems, dynamic analysis based on the theory of *nonlinear vibration* becomes necessary to achieve better accuracy and more satisfactory results.

Compared with the linear vibration theory (Chapters 5 and 6), both the theory and responses of nonlinear vibration are much more complicated. Some unique features of nonlinear vibration differing from its linear counterpart can be are summarized as:

- The *superposition principle*, a powerful tool in linear vibration theory, is not applicable to *nonlinear problems*.
- Nonlinear systems have very rich responses, including periodic and aperiodic responses and multiple solutions to a harmonic force excitation.
- Initial conditions have apparent influence on the steady-state response.
- Exact closed-form solutions are only available for a few nonlinear vibration problems. In many cases, exact solutions are not available, and approximate solutions have to be sought with different methods.

This chapter intends to introduce the very basic theory and analytic tools in nonlinear vibration. Focus is placed on systems with nonlinear elastic

Machinery Dynamics
https://doi.org/10.1016/B978-0-12-815785-5.00018-2

restoring force, more specifically the Duffing equation, and the analytic method of multiple scales. The chapter is organized as follows:

- In Section 8.2, some typical examples of nonlinear vibration are introduced.
- Sections 8.3 and 8.4 are dedicated to the analysis of free and forced vibrations, respectively, of Duffing equation with the method of multiple scales.
- In Section 8.5, numerical solution and response features are briefly presented.

8.2 Examples of nonlinear systems

The nonlinear term may appear as any of the inertial damping and elastic restoring forces in the governing equation of motion. The most common nonlinearity in mechanical engineering, however, is in the elastic or the damping terms. Some examples of nonlinear vibration are given in the next subsection.

8.2.1 Single pendulum

A single pendulum shown in Fig. 8.1 is frequently used as an example in linear vibration texts.

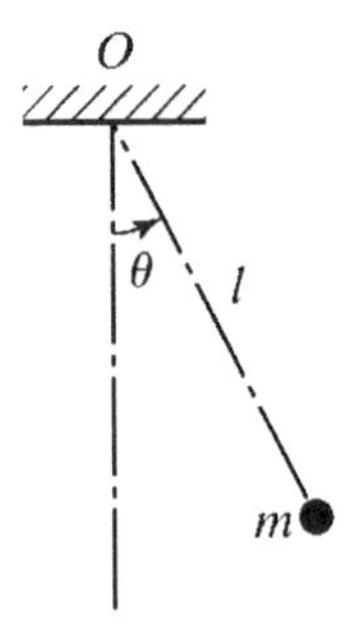

Figure 8.1 Single pendulum.

It is straightforward to write the equation with Newton's second law of motion as follows if damping is neglected:

$$ml^2\ddot{\theta} + mgl\sin\theta = 0. \tag{8.1}$$

For small amplitude vibration, $\sin\theta$ can be approximated by θ, then the previous equation is reduced to

$$\ddot{\theta} + \omega_0^2\theta = 0, \tag{8.2}$$

where

$$\omega_0 = \left(\frac{g}{l}\right)^{1/2}. \tag{8.3}$$

This is the equation found in almost all linear vibration texts. To improve the analysis accuracy, $\sin\theta$ can be approximated by retaining two terms as

$$\sin\theta \approx \theta - \frac{1}{6}\theta^3. \tag{8.4}$$

Substituting it into Eq. (8.1), the governing equation becomes

$$\ddot{\theta} + \omega_0^2\left(\theta - \frac{1}{6}\theta^3\right) = 0. \tag{8.5}$$

Obviously, the equation becomes nonlinear; the cubic nonlinearity appears in the elastic restoring force.

8.2.2 Large deformation

A solid link in a machine is often modeled as a spring, meaning that any forces applied on it causes deformation. If the elastic deformation is within the elastic proportional limit, the relationship between the force and the deformation is governed by the *Hook's law*.

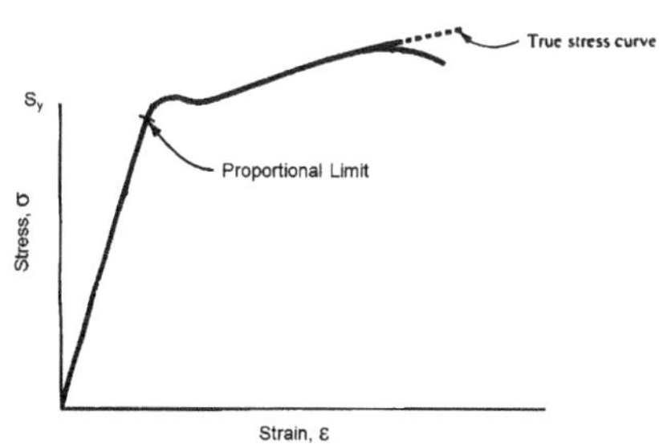

Figure 8.2 Stress strain.

However, if the deformation of a link goes beyond the *proportional limit* as shown in Fig. 8.2, the relationship between the force and the deformation will become nonlinear.

8.2.3 Joint clearance

Clearance exists in almost all joints of machines. For example, clearance is intentionally designed between engaging gear teeth for lubrication and manufacturing purpose. In a revolute joint, clearance also exists between the pin and hole for freely relative rotation.

One way to consider the clearance in dynamics is to use the model in Fig. 8.3, in which a mass is in between two springs. Two motion modes, *free flight* and *contact*, are modeled in this case. The governing equation of motion of the mass can be represented as

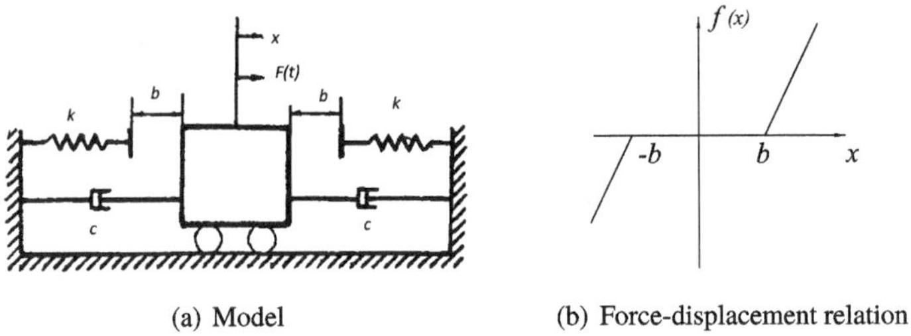

Figure 8.3 Clearance type nonlinearity.

$$m\ddot{x} + c\dot{x} + f(x) = F(t), \tag{8.6}$$

where $f(x)$ is the elastic restoring force from the spring, which can be represented as a piecewise function of x as

$$f(x) = \begin{cases} k(x-b) & \text{if } x \geq b \\ 0 & \text{if } -b < x < b \\ k(x+b) & \text{if } x \leq -b. \end{cases} \tag{8.7}$$

This is schematically illustrated in Fig. 8.3(b).

Although the spring force is linear in each segment, it is nonlinear overall.

In general, nonlinear springs are classified as *hardening springs* and *softening springs*, as shown in Fig. 8.4, depending on the relationship between the linear and nonlinear restoring forces. The solid line and dashed line in the figure represent the nonlinear and linear restoring forces, respectively.

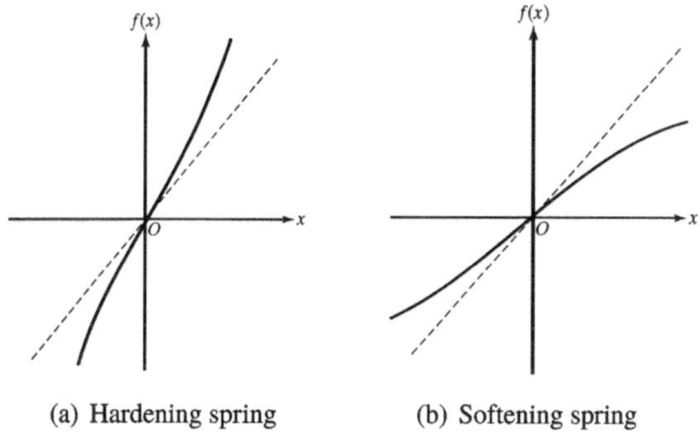

Figure 8.4 Nonlinear springs.

8.2.4 Dry friction

The nonlinearity may appear also as a damping term. Dry friction, for example, causes the so-called *stick-slip vibration*, which is found in many engineering applications. This type of vibration is generally investigated with the model of block-on-belt shown in Fig. 8.5(a). The belt is in continuous

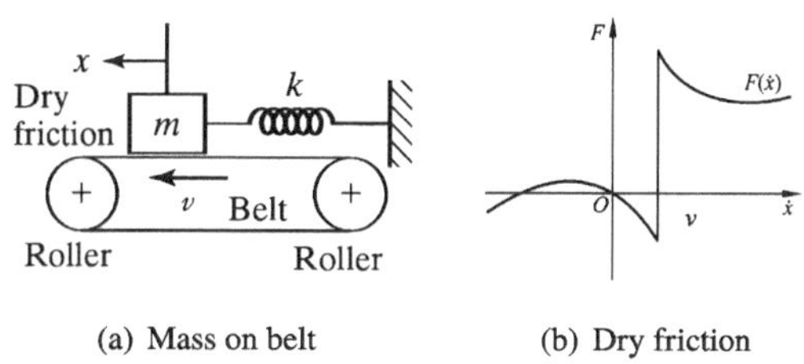

Figure 8.5 Nonlinear damping.

motion with a constant velocity v. A mass m is placed on top of the moving belt and connected to the frame through a linear spring k. Dry friction exists in between the mass m and the belt. The motion of the mass is then described by

$$m\ddot{x} - F(\dot{x} - v) + kx = 0. \tag{8.8}$$

$F(\dot{x} - v)$ is the dry friction between the mass m and the moving belt, which can be described mathematically by *static friction* and *kinetic friction*.

When the mass does not slide on the belt, the friction is static, and the maximum static friction is defined as

$$F = \mu_s N, \tag{8.9}$$

where μ_s is the coefficient of static friction.

When the mass starts sliding on the belt, the friction decreases and is referred to as kinetic friction. There are several different forms of description of the kinetic friction, one of which is schematically shown in Fig. 8.5(b).

The nonlinearity may also appear in the inertial term. Examples can be found in Rao (2010), and Nayfeh and Mook (2008). Given the need for brevity, no example is given here.

8.3 Approximate analysis of free vibration

If we consider only the nonlinear restoring spring force, a general form of equation of single DOF undamped free vibration can be represented as

$$\ddot{x} + f(x) = 0, \tag{8.10}$$

where $f(x)$ is the nonlinear restoring force. Assuming it can be expanded into series $f(x) = \sum_{i=n}^{\infty} \alpha_n x^n$, Eq. (8.10) then becomes

$$\ddot{x} + \sum_{i=n}^{\infty} \alpha_n x^n = 0. \tag{8.11}$$

They are many analytic methods available for approximately solving nonlinear vibration problems, such as *harmonic balance* method and various perturbation methods. In this chapter, we introduce only the method of *multiple scales*, a special technique of perturbation.

The basic idea of multiple scales is to express the response into different *time scales* as T_n in the following form:

$$T_n = \epsilon^n t \quad n = 0, 1, 2, \cdots, \tag{8.12}$$

where ϵ is a small parameter. The solution to Eq. (8.11) is then represented as

$$x(t) = \epsilon x_1(T_0, T_1, T_2, \cdots) + \epsilon^2 x_2(T_0, T_1, T_2, \cdots) + \cdots. \tag{8.13}$$

If we expand to the order of $O(\epsilon^3)$, then this becomes

$$x(t) \approx \epsilon x_1(T_0, T_1, T_2) + \epsilon^2 x_2(T_0, T_1, T_2) + \epsilon^3 x_3(T_0, T_1, T_2). \tag{8.14}$$

With this equation the derivatives with respect to time t can be changed to partial derivatives with respect to the newly defined time scales as

$$\begin{cases} \dfrac{\mathrm{d}}{\mathrm{d}t} = \dfrac{\mathrm{d}T_0}{\mathrm{d}t}\dfrac{\partial}{\partial T_0} + \dfrac{\mathrm{d}T_1}{\mathrm{d}t}\dfrac{\partial}{\partial T_1} + \dfrac{\mathrm{d}T_2}{\mathrm{d}t}\dfrac{\partial}{\partial T_2} \cdots = D_0 + \epsilon D_1 + \epsilon^2 D_2 + \cdots \\ \dfrac{\mathrm{d}^2}{\mathrm{d}t^2} = D_0^2 + 2\epsilon D_0 D_1 + \epsilon^2(D_1^2 + 2D_0 D_2) + \cdots, \end{cases} \tag{8.15}$$

where D_i is the operator defined as

$$D_i = \frac{\partial}{\partial T_i}. \tag{8.16}$$

Substituting Eq. (8.13) into Eq. (8.11) and equating the coefficients ϵ, ϵ^2 and ϵ^3, we have

$$D_0^2 x_1 + \omega_0^2 x_1 = 0 \tag{8.17}$$

$$D_0^2 x_2 + \omega_0^2 x_2 = -2D_0 D_1 x_1 - \alpha_2 x_1^2 \tag{8.18}$$

$$D_0^2 x_3 + \omega_0^2 x_3 = -2D_0 D_1 x_2 - D_1^2 x_1 - 2D_0 D_2 x_1 - 2\alpha_2 x_1 x_2 - \alpha_3 x_1,^3 \tag{8.19}$$

where $\omega_0^2 = \alpha_1$.

According to Eq. (5.9) (see p. 113), the homogeneous solution to Eq. (8.17) has the form

$$\begin{aligned} x_1 &= a(T_1, T_2)\cos\left[\omega_0 T_0 + \varphi(T_1, T_2)\right] \\ &= A(T_1, T_2)e^{i\omega_0 T_0} + \bar{A}e^{-i\omega_0 T_0}. \end{aligned} \tag{8.20}$$

In this equation,

- a, φ and A are functions of T_1 and T_2.
- $A = ae^{i\varphi}/2$ is the complex amplitude; $\bar{A}$ means the complex conjugate.
- The relationship $\cos\theta = (e^{i\theta} + e^{-i\theta})/2$ is applied in deriving this.

Substituting Eq. (8.20) into Eq. (8.18) gives

$$D_0^2 x_2 + \omega_0^2 x_2 = -2i\omega_0 D_1 A e^{i\omega_0 T_0} - \alpha_2\left[A^2 e^{2i\omega_0 T_0} + A\bar{A}\right] + cc, \tag{8.21}$$

where cc stands for the complex conjugates of all the proceeding terms.

Generally, only the particular solution of Eq. (8.21) is of interest. According to the superposition principle, its particular solution is the summation of responses to each individual term on the right-hand side.

Let's look at the first term specifically. According to Eqs. (5.35) and (5.39) (see p. 121), the dynamic magnification factor β is infinite because $\lambda = \omega_0/\omega_0 = 1$, meaning that the response amplitude is boundless. This term is called the *secular term*. However, actual observation indicates that the response is bounded. Thus, to get a meaningful solution, the secular term has to be diminished by letting the coefficient of this term zero. To this end, we need have

$$D_1 A = 0. \tag{8.22}$$

After the secular term is removed, one particular solution to Eq. (8.21) then is represented in the following form according to Eq. (5.34) (see p. 121):

$$x_2 = \frac{\alpha_2 A^2}{3\omega_0^2} e^{2i\omega_0 T_0} - \frac{\alpha_2}{\omega_0^2} A\bar{A} + cc. \tag{8.23}$$

Substituting x_1 and x_2 into Eq. (8.19) and taking advantage of Eq. (8.22), we have

$$\begin{aligned} D_0^2 x_3 + \omega_0^2 x_3 = & -\left[2i\omega_0 D_2 A - \frac{10\alpha_2^2 - 9\alpha_3\omega_0^2}{3\omega_0^2} A^2\bar{A}\right] e^{i\omega_0 T_0} \\ & - \frac{3\alpha_3\omega_0^2 + 2\alpha_2^2}{3\omega_0^2} A^2 e^{3i\omega_0 T_0} + cc. \end{aligned} \tag{8.24}$$

The first term on the right-hand side is the secular term. To eliminate it, we have

$$2i\omega_0 D_2 A - \frac{10\alpha_2^2 - 9\alpha_3\omega_0^2}{3\omega_0^2} A^2\bar{A} = 0. \tag{8.25}$$

Substituting $A = ae^{i\varphi}/2$ into the previous equation and letting both the real and imaginary parts equal zero yield

$$\begin{cases} D_2 a = 0 \\ D_2\varphi = -\dfrac{10\alpha_2^2 - 9\alpha_3\omega_0^2}{24\omega_0^3} a^2. \end{cases} \tag{8.26}$$

Solving this equation yields

$$\begin{cases} a = c \\ \varphi = -\dfrac{10\alpha_2^2 - 9\alpha_3\omega_0^2}{24\omega_0^3} a^2 T_2 + \varphi_0, \end{cases} \tag{8.27}$$

where c and φ_0 are constants. Substituting Eq. (8.27) into Eqs. (8.20) and (8.23) and taking advantage of Eq. (8.13), we have

$$\begin{cases} x = \epsilon a \cos(\omega t + \varphi_0) - \dfrac{\epsilon^2 a^2 \alpha_2}{2\alpha_1}\left[1 - \dfrac{1}{3}\cos(2\omega t + 2\varphi_0)\right] + O(\epsilon^3) \\ \omega = \sqrt{\alpha_1}\left[1 + \dfrac{9\alpha_3\alpha_1 - 10\alpha_2^2}{24\alpha_1^2}\epsilon^2 a^2\right] + O(\epsilon^3). \end{cases} \tag{8.28}$$

The second part of Eq. (8.28) indicates that the free vibration frequency ω depends on the vibration amplitude a. This is a fundamental difference from the linear counterpart, in which the free-vibration frequency is related to only the system parameters (k and m).

8.4 Approximate analysis of forced vibration

The *Duffing equation*, named after Georg Duffing, is a nonlinear second-order ODE with cubic nonlinearity. The general form subjected to a harmonic excitation can be represented as

$$\ddot{x} + 2\epsilon\mu\dot{x} + \omega_0^2 x + \epsilon\alpha x^3 = F_0 \cos\omega t. \tag{8.29}$$

When the excitation frequency equals to the natural frequency of the linear equation, $\omega \approx \omega_0$, *primary resonance* happens. In this case, to obtain a valid approximation the excitation needs to be placed in the order of ϵ. To this end, we let $F_0 = \epsilon f_0$, and the equation takes the following form:

$$\ddot{x} + 2\epsilon\mu\dot{x} + \omega_0^2 x + \epsilon\alpha x^3 = \epsilon f_0 \cos\omega t. \tag{8.30}$$

In the section following, the response of Duffing equation will be discussed using the method of multiple scales.

8.4.1 Primary resonance

When the excitation frequency is around ω_0, *primary resonance* happens. To represent the closeness of the two frequencies, we define

$$\omega = \omega_0 + \epsilon\sigma, \tag{8.31}$$

where σ is referred to as the *detuning parameter*. Then the solution of Eq. (8.30) is assumed to be in the form

$$x(t, \epsilon) = x_0(T_0, T_1) + \epsilon x_1(T_0, T_1) + \cdots. \tag{8.32}$$

$T_0 = t$ and $T_1 = \epsilon t$ are the two time scales to approximate the solution. Substituting Eq. (8.32) into Eq. (8.30) and equating the coefficients of ϵ^0 and ϵ^1 on both sides yields

$$D_0^2 x_0 + \omega_0^2 x_0 = 0 \tag{8.33}$$

$$D_0^2 x_1 + \omega_0^2 x_1 = -2D_1 D_0 x_0 - 2\mu D_0 x_0 - \alpha x_0^3 + f_0 \cos(\omega_0 T_0 + \sigma T_1). \tag{8.34}$$

The homogeneous solution of Eq. (8.33) is in the form of a free vibration as

$$x_0 = A(T_1)e^{i\omega_0 T_0} + \bar{A}(T_1)e^{-i\omega_0 T_0}. \tag{8.35}$$

$A(T_1)$ is the complex amplitude to be determined. It is a function of T_1, but does not rely on T_0 (see Eq. (5.9) on p. 113). Taking advantage of the relationship $\cos\theta = (e^{i\theta} + e^{-i\theta})/2$, we transform the excitation in Eq. (8.34) into the complex form

$$f_0 \cos(\omega_0 T_0 + \sigma T_1) = \frac{1}{2} f_0 \left[e^{i(\omega_0 T_0 + \sigma T_1)} + e^{-i(\omega_0 T_0 + \sigma T_1)} \right]. \tag{8.36}$$

Substituting Eqs. (8.35) and (8.36) into Eq. (8.34), Eq. (8.34) is transformed into the following form:

$$\begin{aligned} D_0^2 x_1 + \omega_0^2 x_1 = & - \left[2i\omega_0 \left(A' + \mu A \right) + 3\alpha A^2 \bar{A} \right] e^{i\omega_0 T_0} \\ & - \alpha A^3 e^{3i\omega_0 T_0} \\ & + \frac{1}{2} f_0 e^{i\sigma T_1} e^{i\omega_0 T_0} + cc, \end{aligned} \tag{8.37}$$

where cc represents the complex conjugates of all the proceeding terms. Obviously, the first and third terms on the right-hand side are the secular terms causing boundless vibration. To eliminate the secular term, we have

$$2i\omega_0 \left(A' + \mu A \right) + 3\alpha A^2 \bar{A} - \frac{1}{2} f_0 e^{i\sigma T_1} = 0. \tag{8.38}$$

Representing A in polar form

$$A = \frac{1}{2} a e^{i\varphi} \tag{8.39}$$

and substituting it into the previous equation, two equations will be formed;

$$\begin{cases} a' = -\mu a + \dfrac{f_0}{2\omega_0}\sin\beta \\ a\beta' = \sigma a - \dfrac{3\alpha a^3}{8\omega_0} + \dfrac{f_0}{2\omega_0}\cos\beta, \end{cases} \tag{8.40}$$

where

$$\beta = \sigma T_1 - \varphi. \tag{8.41}$$

For steady-state response, $a' = 0$, and $\beta' = 0$, we then have

$$\begin{cases} \mu a = \dfrac{f_0}{2\omega_0}\sin\beta \\ \sigma a - \dfrac{3\alpha a^3}{8\omega_0} = -\dfrac{f_0}{2\omega_0}\cos\beta. \end{cases} \tag{8.42}$$

Squaring these two equations and then summing up yields

$$\left[\mu^2 + \left(\sigma - \frac{3\alpha a^2}{8\omega_0}\right)^2\right]a^2 = \frac{f_0^2}{4\omega_0^2}. \tag{8.43}$$

The response to the order of $O(\epsilon)$ can be approximated as

$$x = a\cos(\omega_0 t - \varphi) = a\cos(\omega t - \beta). \tag{8.44}$$

For a given μ and f_0, the plot of a as a function of σ according to Eq. (8.43) is the so-called *amplitude–frequency curve*. Plots with parameters $\mu = 0.15, f_0 = 1.5, \omega_0 = 1.8$ are given in Fig. 8.6, from which the following observations can be made:

- In the linear case, primary resonance happens around $\sigma = 0$ (or $\omega = \omega_0$). For a given excitation, one σ corresponds to one a (unique solution) as shown in Fig. 8.6(a). This figure is found in almost all linear vibration textbooks.
- The nonlinearity bends the amplitude–frequency curve to the right when $\alpha > 0$ (hardening spring) and to the left when $\alpha < 0$ (softening spring).
- Nonlinear response has the *jump phenomenon* which a linear system does not have. Jump phenomenon means the response amplitude increases or decreases suddenly at some points when increasing or decreasing

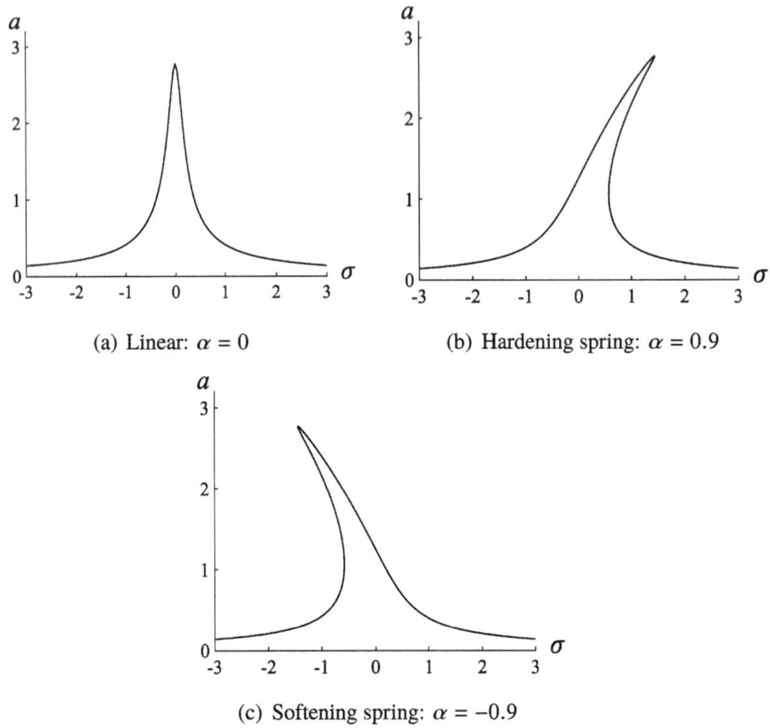

Figure 8.6 Primary resonance.

the excitation frequency gradually. For example, if we increase the excitation frequency gradually, the response amplitude will move along the path of the black arrows shown in Fig. 8.7. Clearly the amplitude suddenly drops at point A (see Fig. 8.7(a)) for a hardening spring and suddenly increases at point B (see Fig. 8.7(b)) for a softening spring. If we gradually decrease the excitation frequency, however, the amplitude will move along the path of the red arrows. Jump happens at point B in Fig. 8.7(a) for hardening springs and at point A in Fig. 8.7(b) for softening springs.

- As a result of the bending of the curve, one excitation may cause three possible solutions (multiple solutions) if the excitation frequency is in the range between points A and B. Which response the system has depends on the initial condition. This is in apparent contradiction to the linear counterpart in which initial conditions have no influence on the steady-state response.

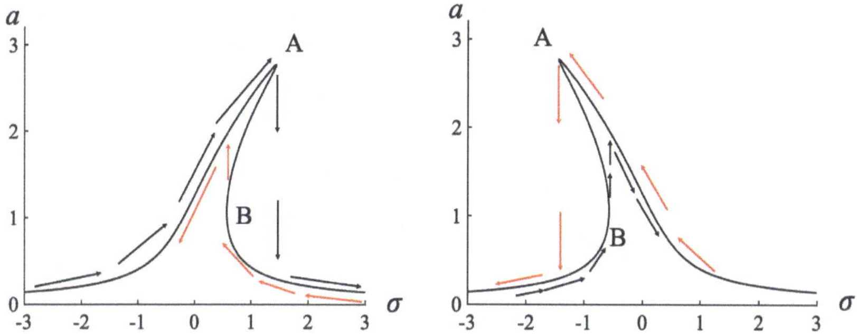

(a) Hardening spring: $\alpha = 0.9$

(b) Softening spring: $\alpha = -0.9$

Figure 8.7 Jump phenomena.

8.4.2 Nonresonant response

When the excitation frequency ω is distant from all resonant frequencies, a nonresonant response exists. In this case, we represent the nonlinear response as

$$x(t,\epsilon) = x_0(T_0, T_1) + \epsilon x_1(T_0, T_1) + \cdots. \tag{8.45}$$

Substituting Eq. (8.45) into Eq. (8.29) and equating the coefficients of ϵ^0 and ϵ^1 gives

$$D_0\ddot{x}_0 + \omega_0^2 x_0 = F_0 \cos \omega T_0 \tag{8.46}$$

$$D_0\ddot{x}_1 + \omega_0^2 x_1 = -2D_0D_1x_0 - 2\mu D_0 x_0 - \alpha x_0^3. \tag{8.47}$$

The general solution of Eq. (8.46) includes both the free and forced responses. According to Eq. (5.40) (see p. 122) it can be represented as

$$x_0 = A(T_1)e^{i\omega_0 T_0} + Be^{i\omega T_0} + cc. \tag{8.48}$$

The first and the second terms on the right-hand side represent the free vibration and forced vibration responses, respectively. A and B are the amplitudes of complex form. B is given by

$$B = \frac{F_0}{2(\omega_0^2 - \omega^2)}. \tag{8.49}$$

Substituting Eq. (8.48) into Eq. (8.47) gives

$$
\begin{aligned}
D_0\ddot{x}_1 + \omega_0^2 x_1 = & -\left[2i\omega_0(A' + \mu A) + 6\alpha AB^2 + 3\alpha A^2\bar{A}\right]e^{i\omega_0 T_0} \\
& - \alpha A^3 e^{3i\omega_0 T_0} \\
& - \alpha B^3 e^{3i\omega T_0} \\
& - 3\alpha A^2 B e^{i(\omega+2\omega_0)T_0} \\
& - 3\alpha \bar{A}^2 B e^{i(\omega-2\omega_0)T_0} \\
& - 3\alpha AB^2 e^{i(\omega_0+2\omega)T_0} \\
& - 3\alpha AB^2 e^{i(\omega_0-2\omega)T_0} \\
& - B\left[2i\mu\omega + 3\alpha B^2 + 6\alpha A\bar{A}\right]e^{i\omega T_0} \\
& + cc.
\end{aligned} \tag{8.50}
$$

This equation is the basis for analysis of the responses other than the primary resonance.

In the case of nonresonance, the first term on the right-hand side is the secular term. Eliminating the secular term leads to

$$
2i\omega_0(A' + \mu A) + 6\alpha AB^2 + 3\alpha A^2\bar{A} = 0. \tag{8.51}
$$

Substituting $A = ae^{i\varphi}/2$ into Eq. (8.51) and equating both the real and imaginary parts zero, we have

$$
a' = -\mu a \tag{8.52}
$$

$$
\omega_0 a\varphi' = 3\alpha\left(B^2 + \frac{1}{8}a^2\right)a. \tag{8.53}
$$

Integrating these gives a and φ. Substituting them into Eq. (8.48) gives the nonresonant response to the order of ϵ^1:

$$
x = a\cos(\omega_0 t + \varphi) + \frac{F_0}{\omega_0^2 - \omega^2}\cos\omega t + O(\epsilon). \tag{8.54}
$$

From Eq. (8.52), we obtain $a = a_0 e^{-\mu T_1}$. For a positive μ, a approaches zero eventually when time is large enough, meaning that the free vibration dies out and only the forced vibration is left. This is similar to the linear case.

8.4.3 Superharmonic resonances

For the Duffing system having a cubic nonlinear restoring force, experience indicates that, when the excitation frequency ω approaches $1/3\omega_0$, $1/9\omega_0$, $\cdots$, responses become very large. This is a type of resonance termed as *superharmonic resonance*.

To analyze the superharmonic resonance $\omega \approx 1/3\omega_0$, we let

$$3\omega = \omega_0 + \epsilon\sigma \tag{8.55}$$

$$3\omega T_0 = \omega_0 T_0 + \sigma T_1. \tag{8.56}$$

As stated before, ϵ and σ are the small parameter and the detuning parameter, respectively.

In this case, in addition to the first term, the term of $e^{3i\omega T_0}$ (the third term on the right-hand side) in Eq. (8.50) is also a secular term. To eliminate the secular terms, we should have

$$2i\omega_0(A' + \mu A) + 6\alpha AB^2 + 3\alpha A^2\bar{A} + \alpha B^3 e^{i\sigma T_1} = 0. \tag{8.57}$$

Substituting $A = ae^{i\varphi}/2$ into this and making both the real and imaginary parts zero, we have

$$a' = -\mu a - \frac{\alpha B^3 \sin\gamma}{\omega_0} \tag{8.58}$$

$$a\gamma' = \left(\sigma - \frac{3\alpha B^2}{\omega_0}\right)a - \frac{3\alpha a^3}{8\omega_0} - \frac{\alpha B^3 \cos\gamma}{\omega_0}, \tag{8.59}$$

where

$$\gamma = \sigma T_1 - \varphi. \tag{8.60}$$

For the steady-state response, $a' = \gamma' = 0$, then we have

$$-\mu a = \frac{\alpha B^3 \sin\gamma}{\omega_0} \tag{8.61}$$

$$\left(\sigma - \frac{3\alpha B^2}{\omega_0}\right)a - \frac{3\alpha a^3}{8\omega_0} = \frac{\alpha B^3 \cos\gamma}{\omega_0}. \tag{8.62}$$

Squaring and adding these two equations yields

$$\left[\mu^2 + \left(\sigma - \frac{3\alpha B^2}{\omega_0} - \frac{3\alpha a^2}{8\omega_0}\right)^2\right]a^2 = \frac{\alpha^2 B^6}{\omega_0^2}. \tag{8.63}$$

The response of the system is then in the form

$$x = a\cos(3\omega t - \gamma) + \frac{F_0}{\omega_0^2 - \omega^2}\cos\omega t + O(\epsilon). \tag{8.64}$$

The free vibration does not die out as the response of linear systems does. The nonlinearity, however, tunes the free vibration frequency to 3ω.

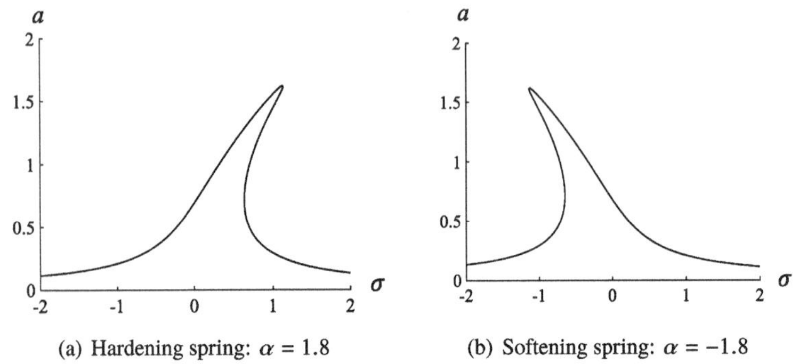

Figure 8.8 Superharmonic resonance.

The relationship between the free vibration amplitude a and the detuning parameter σ (or the excitation frequency ω) is shown in Fig. 8.8 which is drawn with parameters $\mu = 0.15$, $\omega_0 = 1.8$, $F_0 = 3.5$, $\epsilon = 0.1$. Similar to the case of primary resonance, the nonlinearity bends the amplitude–frequency curves.

8.4.4 Subharmonic resonance

If the excitation frequency is close to three times ω_0, *subharmonic resonance* happens. To represent the closeness of ω to $3\omega_0$, we let

$$\omega = 3\omega_0 + \epsilon\sigma. \tag{8.65}$$

From Eq. (8.50), we find that terms of $e^{i\omega_0 T_0}$ and $e^{i(\omega-2\omega_0)T_0}$ (the first and the fifth terms on the right-hand side) are the secular terms causing the response to grow infinitely. Thus, to eliminate the secular terms, we shall have

$$2i\omega_0(A' + \mu A) + 6\alpha AB^2 + 3\alpha A^2\bar{A} + 3\alpha B\bar{A}^2 e^{i\sigma T_1} = 0. \tag{8.66}$$

Substituting $A = ae^{i\varphi}/2$ into this and making both the real and imaginary parts zero, we have

$$a' = -\mu a - \frac{3\alpha Ba^2}{4\omega_0}\sin\gamma \tag{8.67}$$

$$a\gamma' = \left(\sigma - \frac{9\alpha B^2}{\omega_0}\right)a - \frac{9\alpha a^3}{8\omega_0} - \frac{9\alpha Ba^2}{4\omega_0}\cos\gamma, \tag{8.68}$$

where

$$\gamma = \sigma T_1 - 3\varphi. \tag{8.69}$$

For the steady-state response, $a' = \gamma' = 0$. This leads to

$$\begin{cases} -\mu a = \dfrac{3\alpha B a^2}{4\omega_0}\sin\gamma \\ \left(\sigma - \dfrac{9\alpha B^2}{\omega_0}\right)a - \dfrac{9\alpha a^3}{8\omega_0} = \dfrac{9\alpha B a^2}{4\omega_0}\cos\gamma. \end{cases} \tag{8.70}$$

From Eq. (8.70), we have

$$\left[9\mu^2 + \left(\sigma - \frac{9\alpha B^2}{\omega_0} - \frac{9\alpha a^2}{8\omega_0}\right)^2\right]a^2 = \frac{81\alpha^2 B^2}{16\omega_0^2}a^4. \tag{8.71}$$

This leads to either

$$a = 0 \tag{8.72}$$

or

$$9\mu^2 + \left(\sigma - \frac{9\alpha B^2}{\omega_0} - \frac{9\alpha a^2}{8\omega_0}\right)^2 = \frac{81\alpha^2 B^2}{16\omega_0^2}a^2. \tag{8.73}$$

A plot of Eq. (8.73) under a set of parameters $\mu = 0.15$, $\omega_0 = 1.8$, $F_0 = 3.5$, $\alpha = 1$ and $\epsilon = 0.1$ is given in Fig. 8.9.

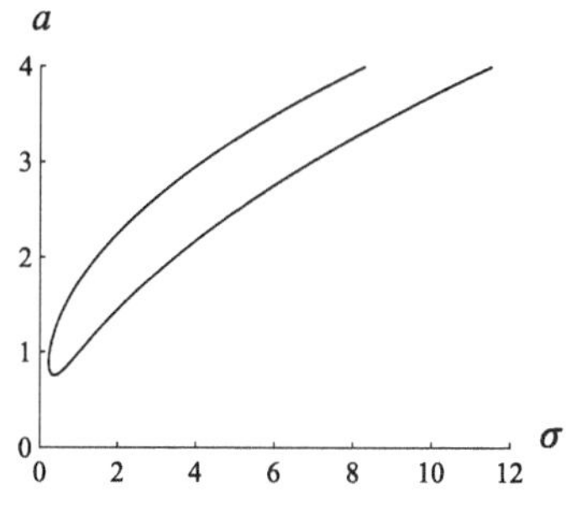

Figure 8.9 Subharmonic resonance.

It is worth pointing out that nontrivial solutions to Eq. (8.73) exist only when specific conditions are met. Given the page limit, these conditions are not discussed here; readers who are interested in this topic are advised to consult the monograph by Nayfeh and Mook (2008).

Once a is obtained, the response of the subharmonic resonance is represented as

$$x = a\cos\left[\frac{1}{3}(\omega t - \gamma)\right] + \frac{F_0}{\omega_0^2 - \omega^2}\cos\omega t + O(\epsilon). \tag{8.74}$$

It can be seen that the free vibration does not decay as in the linear cases despite the existence of damping. Similar to the superharmonic response, the nonlinearity adjusts the free-vibration frequency to $1/3\omega$.

8.5 Numerical analysis

Various analytical methods are suitable for finding the periodic responses of a nonlinear vibration problem. However, a nonlinear system may have very rich responses, including periodic and aperiodic, such as chaos. In some cases, numerical methods are used to capture a more complete picture of the response. In addition, numerical methods may be the only option for multiple DOF systems. In this section, we will present an example of numerical solution of the Duffing equation, and some typical responses.

The Duffing equation under a harmonic force excitation can be represented in the following general for:

$$\ddot{x} + 2\mu\dot{x} + \omega_0^2 x + \alpha x^3 = F_0\cos\omega t. \tag{8.75}$$

To solve this numerically, we first define

$$\mathbf{X} = \begin{bmatrix} x_1 \\ x_2 \end{bmatrix} = \begin{bmatrix} x \\ \dot{x} \end{bmatrix}, \tag{8.76}$$

and then transform Eq. (8.75) into the form of first order as follows:

$$\dot{x}_1 = x_2 \tag{8.77}$$

$$\dot{x}_2 = -2\mu x_2 - \omega_0^2 x_1 - \alpha x_1^3 + F_0\cos\omega t. \tag{8.78}$$

This equation can then be solved with various numerical methods, such as the well known Runge–Kutta method.

As an example, the following Duffing equation is solved with Matlab® to illustrate the procedures of numerical solution and some features of the responses (Wagg and Neild, 2016):

$$\ddot{x} + 0.4\dot{x} - x + x^3 = F_0\cos(1.8t). \tag{8.79}$$

The first step is to transform it into a set of first-order ODEs:

$$\begin{cases} \dot{x}_1 = x_2 \\ \dot{x}_2 = -0.4x_2 + x_1 - x_1^3 + F_0 \cos(1.8t). \end{cases} \tag{8.80}$$

This equation is solved by the ODE solver *ode*45, the Runge–Kutta method, in Matlab. The result of time histories for $F_0 = 0.6, 0.73, 1.51,$ and 1.8 are presented in Fig. 8.10.

It can be seen that Fig. 8.10(a) is periodic, but may or may not be harmonic. The response in 8.10(b) and 8.10(c) are periodic, but clearly not harmonic; while Fig. 8.10(d) does not even look like periodic.

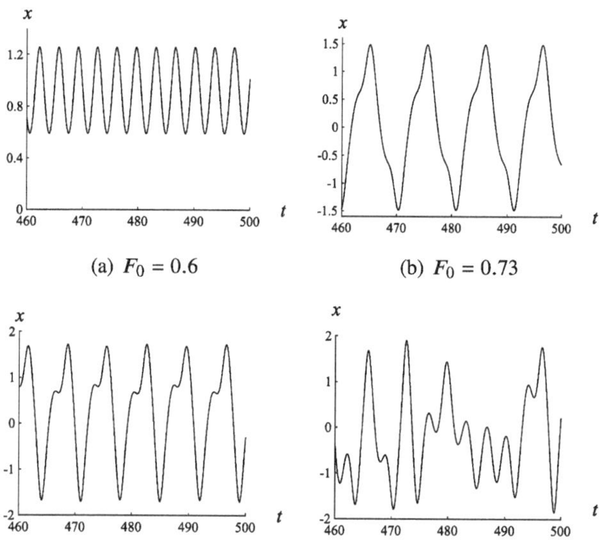

(a) $F_0 = 0.6$ (b) $F_0 = 0.73$

(c) $F_0 = 1.51$ (d) $F_0 = 1.8$

Figure 8.10 Time histories.

The plane defined by $(x, \dot{x})$ is referred to as *phase plane*. The response drawn in the phase plane is called the trajectory. Trajectories are often used in *qualitative analysis* of nonlinear vibration and are capable of providing unique information about the response.

The trajectories for the above four cases are presented in Fig. 8.11. One can see that in Fig. 8.11(a) the trajectory curve is closed, meaning that the response repeats after some time, i.e., it is periodic. It is also seen from Fig. 8.10(a) that the response frequency in this case seems the same

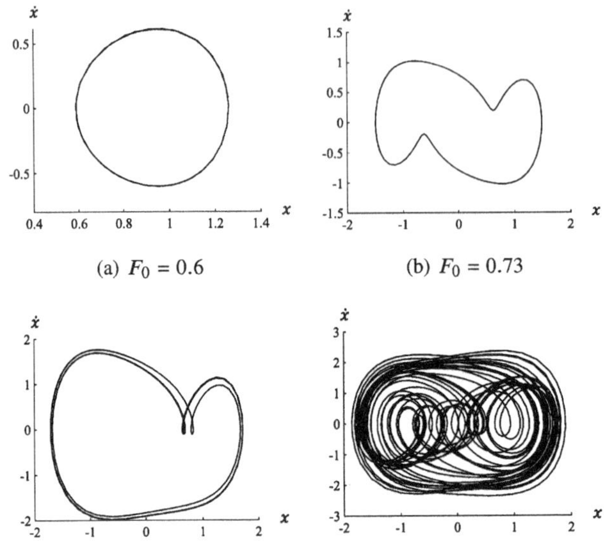

Figure 8.11 Phase plane trajectories.

of the excitation frequency. In Fig. 8.11(b) the closed trajectory means it is a periodic response, as well. The response repeats after three excitation periods. Fig. 8.11(c) indicates the response in this case is periodic, i.e., the response repeats after two excitation periods. In the case of Fig. 8.11(d), the trajectory is a curve that never repeats, indicating that the response is aperiodic. This type of response is referred to as *chaos*.

The above analysis is qualitative. To verify this analysis, the time history can be easily analyzed with the *fast Fourier transform* (FFT) in Matlab, which gives the frequencies in each response. Here we directly give the result by Wagg and Neild (2016) as follows without presenting further details:

- When $F_0 = 0.6$ (Figs. 8.10(a) and 8.11(a)), the excitation frequency is dominant in the response. In addition, the second and thirds are also evident.
- When $F_0 = 0.73$ (Figs. 8.10(b) and 8.11(b)), the excitation frequency, the *third* subharmonic and the third harmonic are obvious. In addition, some non-integer harmonic components, including 5/3, 7/3, 11/3, and 13/3, are also seen.
- When $F_0 = 1.51$ (Figs. 8.10(c) and 8.11(c)), all the integer harmonic from the first to the fifth are obvious. In addition, 1/2 subharmonic,

and non-integer harmonics, including 3/2, 5/2 7/2 and 9/2, are seen as well.

- When $F_0 = 1.8$ (Figs. 8.10(d) and 8.11(d)), the spectrum is continuous in a range with several peaks at the first, third, and fifth harmonics. The continuous spectrum indicates that the response is chaotic.

PART 3

Elasto-dynamics

After WWI, light weight and high speed became two prominent trends of development for some machines. Faster speed has two effects on machine dynamics. First, it causes a greater inertial load. Second, it excites the system with higher frequencies. On the other hand, lighter weight generally means more flexibility and lower natural frequency. These two trends may cause a series of harmful impacts on the machine, such as vibration and noise, degradation of motion accuracy, and the acceleration of the onset of fatigue. In addition, the increased deformation of components may make the rigid assumption in dynamic analysis inappropriate in some cases. In view of this fact, elastic deformation first was started to be considered in machine dynamics around the mid-20th century. In the 1970s, the term *elasto-dynamics* was coined for the dynamic study of elastic linkages (Zhang and Yang, 2020). This word perfectly describes the essence of this level of dynamics by including the two keywords, *elasticity* and *dynamics*. For this reason, we categorize as elasto-dynamics dynamics combined with the elasticity of any mechanism in this book.

It should be pointed out that, in the literature, the use of the term is not consistent. In gear and cam dynamics, the term has not been commonly used even though elasticity is considered. In shaft vibration, the term is not used either. The reason for this, at least partially, is that the development of theory in these fields occurred earlier than its use in linkages for which the term was initially invented.

The inclusion of component elasticity is a critical step forwards in the refinement of dynamic modeling. The other important progress in dynamic modeling is the consideration of clearances in kinematic pairs (joints).

The inclusion of either elasticity or clearance in dynamics makes the analysis much more complicated, necessitating much higher computational power than that in the rigid-body dynamics. The problems are often nonlinear, and numerical solutions are required. Thus, modern computers and the vibration theory are the technical and theoretical basis of elastodynamics.

CHAPTER 9

Vibration of shafts and shaft systems

9.1 Introduction

Shafts are widely used in various machines mainly for transmitting motion and power. Generally shafts are relatively long and, in most cases, have circular cross sections. Due to the requirement of light weight, shafts nowadays are more flexible than they used to be, shifting the natural frequency downward. On the other hand, the operation speed of machines is alway in on the increase. These two effects make severe vibration possible in shaft systems. Vibration of shafts attracted researchers attention in the mid-19th century. In 1869, Rankine published the first paper on dynamics of rotating shafts (Rankine, 1869). Since the 1930s, the number of publications on shaft dynamics and vibration has increased dramatically. A thorough literature review on shaft dynamics and vibration can be found in the monograph (Loewy and Piarulli, 1970).

Two vibration modes, *lateral vibration* and *torsional vibration*, may happen on a shaft.

Lateral vibration: When a shaft is in rotation, the center of mass may deviate from the rotating axis due to various reasons, such as uneven material properties, manufacturing errors, and dynamic elastic deflection of the shaft. The periodic inertial force caused by the deviation of the rotating axis's center of mass causes lateral vibration in a shaft. The frequency of the excitation or the inertial centrifugal force is coincident with the shaft's rotating frequency. When this frequency equals the natural frequency of lateral vibration of the shaft, resonance happens. The operation speeds causing lateral resonance of a shaft are referred to as *critical speeds*. If machines operate at speeds around the critical frequency, severe vibration happens, which deteriorates the working conditions and can even damage the machine. To avoid this situation, the critical speed needs to be calculated accurately, and the shaft needs to be designed to operate far from the rotating frequency.

Torsional vibration: Shafts in a machine are subjected to various torques, which generally periodically change with time. For example, the output torque of an internal combustion engine is periodic. The cutting

Machinery Dynamics
https://doi.org/10.1016/B978-0-12-815785-5.00020-0

resistance torque in some metal-cutting operations, such as drilling and milling, also periodically changes with time. In addition, impact torque may exist in some machines. One example is the impact happening between engaging gears. These torques, impacts, or time changing causes torsional vibration of the shaft. To avoid severe torsional vibration, the natural frequency of torsional vibration needs to be predicted.

Natural frequency is critical for both the lateral and torsional vibrations of shafts. Two methods are commonly used to calculate a shaft's natural frequencies. The first is by solving the eigenvalue problem introduced in Chapter 6. The other is the transfer matrix method, which is especially suitable for various chain-like structures. Given that closed-form solutions are only available for very simple continuous vibration problems, discrete models, including the lumped-parameter model and the finite element (FE) model, are more often used when dealing with shaft-vibration problems.

This chapter will introduce both lumped-parameter model and the FE model, as well as the eigenvalue problem related to natural frequencies and vibrational modes. The lumped-mass model and transfer matrix method are presented for both lateral and torsional vibrations. The FE model is briefly introduced for lateral vibration.

The chapter is organized as follows:

- Section 9.1 presents an introduction.
- Sections 9.2 and 9.3 deal with the natural frequency of the torsional vibration and the transfer matrix method for critical speeds, respectively.
- Section 9.4 focuses on the finite element method, while a brief review on rotor dynamics is presented in Section 9.5.

9.2 Natural frequency of torsional vibration

Calculation of the natural frequency of shaft systems is of primary interest in the study of shaft vibration. Thus, this section focuses on the calculation of torsional natural frequency of a shaft.

9.2.1 Dynamic model of torsional vibration

As stated in the introduction, the lumped-mass model can be used to study the torsional vibration of shaft systems. Power-transmission components, such as gears, sprockets and pulleys etc., are often installed on For a general shaft system, the shaft itself is often modeled as a massless torsional spring, and the installed components are treated as lumped mass/inertia.

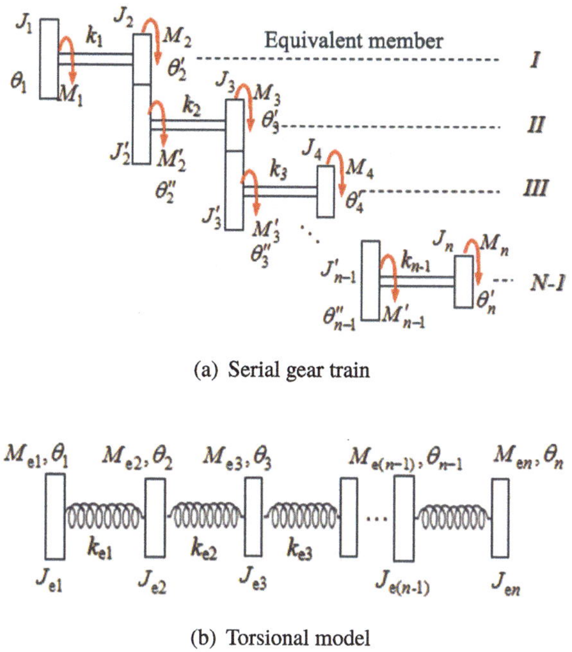

Figure 9.1 Equivalent torsional model.

Take the serial gear train in Fig. 9.1(a) as an example. Based on the just mentioned rule, it can be represented by a lumped-mass model in Fig. 9.1(b).

To find the parameters in the model of Fig. 9.1(b), including the equivalent spring stiffness k_{ei}, the equivalent moment of inertia J_{ei}, and the equivalent torque M_{ei}, each segment of the shaft in Fig. 9.1(a) is treated with the method introduced in Chapter 3. Specifically, the equivalent moment of inertia J_{ei} is calculated based on the principle of keeping the kinetic energy unchanged, while the calculation of equivalent torque M_{ei} is based on the principle of unchanged virtual work caused by the loads. The calculation of the equivalent stiffness, k_{ei}, although not covered in Chapter 3, can be calculated on the basis of unchanged elastic potential energy, which is illustrated next.

9.2.1.1 Equivalent torsional stiffness

The detailed calculation varies depending on geometry of the shaft.

1. Uniform shafts

If the shaft has a uniform cross section, the torsional stiffness is calculated by

$$k = \frac{GI}{l}, \tag{9.1}$$

where k is the torsional stiffness, G is the shear modulus of elasticity of the shaft material, l is the length of the shaft, and I is the polar moment of inertia of the cross section. For the most common circular shaft, I is calculated as

$$I = \frac{\pi d^4}{32}, \tag{9.2}$$

where d is the diameter of the shaft.

2. Stepped shafts

A stepped shaft can be treated as multiple uniform segments connected in series. For the example shown in Fig. 9.2(a), each segment with diameter d_1 or d_2 is treated as a uniform shaft. Two equivalent springs of k_1 and k_2 are calculated as uniform shafts by Eq. (9.1). The two springs are connected in series as in Fig. 9.2(b). The overall equivalent

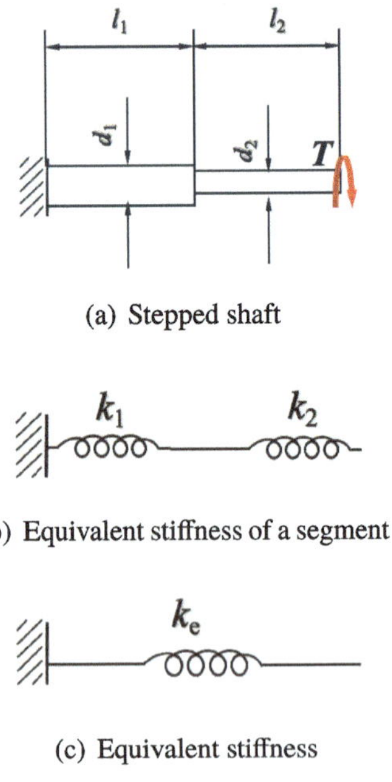

Figure 9.2 Equivalent stiffness of stepped shafts.

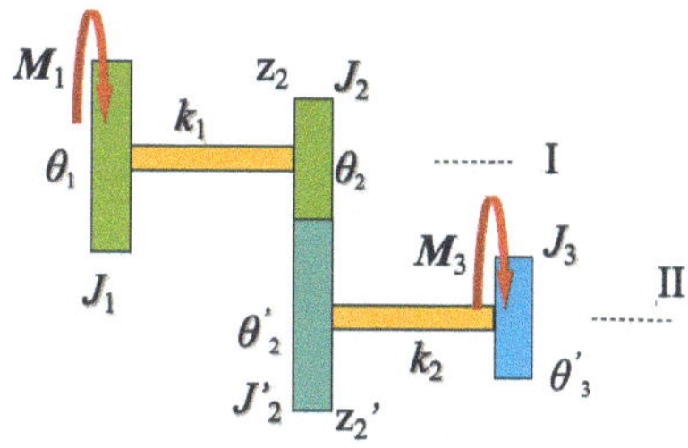

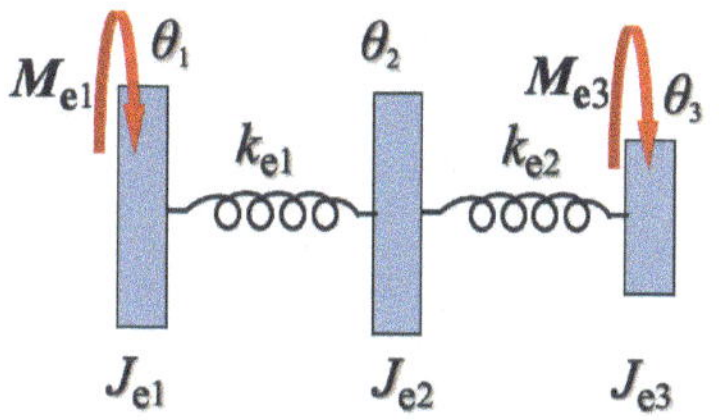

Figure 9.3 Equivalent stiffness to a different shaft.

stiffness, k_e, is then calculated as

$$\frac{1}{k_e} = \frac{1}{k_1} + \frac{1}{k_2}$$

or more conveniently

$$k_e = \frac{k_1 k_2}{k_1 + k_2}. \tag{9.3}$$

3. Equivalent stiffness applied to a different shaft
Sometimes, the equivalency needs to be applied to a different shaft. For example, in the gear and shaft system shown in Fig. 9.3(a) we want to find the equivalent parameters of shaft *II* on shaft *I*. The gear ratio of the intermediate engaging gears is

$$i = \frac{\dot{\theta}_2'}{\dot{\theta}_2} = -\frac{z_2}{z_2'}, \tag{9.4}$$

where z_2' and z_2 are the tooth numbers of the driving and the driven gear, respectively.

To transform the shaft *II* and its gears to shaft *I*, we first transform the rotation angles at the gear z_2' and the gear z_3 to equivalent rotation angles on shaft *I* based on the following relationship:

$$\begin{cases} \theta_2 = \dfrac{\theta_2'}{i} \\ \theta_3 = \dfrac{\theta_3'}{i} \\ \dot{\theta}_2 = \dfrac{\dot{\theta}_2'}{i} \\ \dot{\theta}_3 = \dfrac{\dot{\theta}_3'}{i}. \end{cases} \tag{9.5}$$

Then a model shown in Fig. 9.3(b) can be obtained, in which the equivalent parameters are

$$\begin{cases} J_{e1} = J_1 \\ J_{e2} = J_2 + J_2' i^2 \\ J_{e3} = J_3 i^2 \end{cases} \tag{9.6}$$

and

$$\begin{cases} M_{e1} = M_1 \\ M_{e3} = M_3 i. \end{cases} \tag{9.7}$$

The equivalent stiffness is calculated based on the principle of unchanged elastic potential energy. If we assume the equivalent stiffness transformed from shaft *II* to shaft *I* is k_{e2}, then the following is true:

$$\frac{1}{2} k_2 \left(\theta_3' - \theta_2'\right)^2 = \frac{1}{2} k_{e2} \left(\theta_3 - \theta_2\right)^2. \tag{9.8}$$

Substituting Eq. (9.5) into Eq. (9.7) gives

$$k_{e2} = k_2 i^2. \tag{9.9}$$

Eqs. (9.6) and (9.9) simply indicate that the equivalent moment of inertia and the equivalent stiffness transformed to a faster shaft from a slower one become smaller.

9.2.1.2 Dynamic model of serial systems

The previous procedures can be applied to more complicated serial systems. For the one in Fig. 9.1(a), we can first calculate the gear ratios from each of the shafts from II to $N-1$

$$\begin{cases} i_{II,I} = \dfrac{\omega_2''}{\omega_2'} = -\dfrac{z_2}{z_2'} \\ i_{III,I} = \dfrac{\omega_3''\omega_2''}{\omega_3'\omega_2'} = (-1)^2\dfrac{z_2 z_3}{z_2' z_3'} \\ \cdots \\ i_{N-1,I} = (-1)^p \dfrac{z_{n-1}z_{n-2}\cdots z_3 z_2}{z_{n-1}' z_{n-2}' \cdots z_3' z_2'}, \end{cases} \tag{9.10}$$

where p is the total number of engaging gear pairs in between. Then the equivalent rotation angle transformed to shaft I are

$$\begin{cases} \theta_1 = \theta_1' \\ \theta_2 = \theta_2' = \dfrac{\theta_2''}{i_{II,I}} \\ \cdots \\ \theta_n = \dfrac{\theta_n'}{i_{N-1,I}}. \end{cases} \tag{9.11}$$

The equivalent moments of inertia are

$$\begin{cases} J_{e1} = J_1 \\ J_{e2} = J_2 + J_2' i_{II,I}^2 \\ J_{e3} = J_3 i_{II,I}^2 + J_3' i_{III,I}^2 \\ \cdots \\ J_{en} = J_n i_{N-1,I}^2. \end{cases} \tag{9.12}$$

The equivalent stiffnesses are

$$\begin{cases} k_{e1} = k_1 \\ k_{e2} = k_2 i_{II,I}^2 \\ \cdots \\ k_{e(n-1)} = k_{n-1} i_{N-1,I}^2. \end{cases} \tag{9.13}$$

The equivalent torques are

$$\begin{cases} M_{e1} = M_1 \\ M_{e2} = M_2 + M_2' i_{II,I} \\ \cdots \\ M_{en} = M_n i_{N-1,I}. \end{cases} \tag{9.14}$$

After these substitutions, an equivalent dynamic model as shown in Fig. 9.1(b) is obtained, which has n generalized coordinates $\theta_1 - \theta_n$.

The selection of generalized coordinates is not unique. If the rotation of the system is split into a rigid motion of angular velocity $\dot{\theta}_1$ superimposed by an elastic rotation and the $n-1$ elastic rotation angles are

$$\begin{cases} q_1 = \theta_2 - \theta_1 \\ q_2 = \theta_3 - \theta_2 \\ \cdots \\ q_{n-1} = \theta_n - \theta_{n-1}, \end{cases} \tag{9.15}$$

then the generalized coordinates can be taken as $\theta_1, q_1, q_2, \cdots, q_{n-1}$.

9.2.2 Transfer matrix method for torsional vibration

Once the dynamic model as shown in Fig. 9.1(b) is established, the natural frequencies of the torsional vibration of the shaft system can be obtained by solving the eigenvalue problem introduced in Chapter 6.

In this section, we introduce another method, *transfer matrix*, which is a very efficient for calculating the natural frequencies of chain-like structures, such as a shaft system. The method is first proposed by Holzer (1921) and now has been widely accepted in shaft and beam vibration problems.

9.2.2.1 State vector

A shaft system can be simplified into a chain-like structure as shown in Fig. 9.4(a) through the equivalent procedures introduced in the previous section. For the sake of brevity, the subscript e is dropped from the parameters; however, we should keep in mind that they are the equivalent parameters. The structure consists of lumped masses J_i, which may be transformed from gears, sprockets, or pulleys, etc., in a real shaft or shaft segments. The location of the lumped masses are referred to as *stations* or mass (disk) elements, while the shaft segments are treated as massless, and

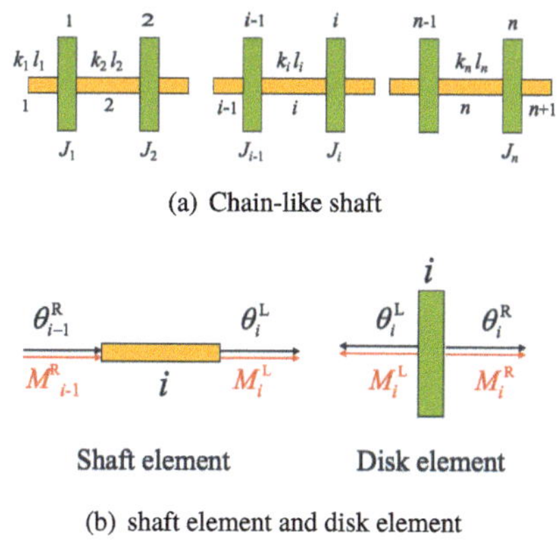

Figure 9.4 Transfer matrix.

are called *fields* or shaft elements. An individual shaft element and a disk element are shown in Fig. 9.4(b). The disks (stations) and the shaft segments (fields) are connected through nodes. In this section, the subscript i means the node of the i-th mass, and the superscripts L and R represent the left and right of the lumped masses, respectively.

At each node, the internal force, torque M, and rotation angle θ are used to represent the *state* of the node; $\{\theta \quad M\}^T$ is referred to as the *state vector*. The positive directions of θ and M are defined based on the common right-hand rule, and are shown in Fig. 9.4(b).

9.2.2.2 Point transfer matrix

For an individual mass i, the rotation of the mass is governed by Euler's second law:

$$J_i\ddot{\theta}_i = M_i^R - M_i^L, \tag{9.16}$$

where M_i^R and M_i^L are the torque on the right-hand and left-hand side of the mass. J_i is the moment of inertia. Assuming the mass moves harmonically with frequency p, we have

$$\theta_i = \vartheta_i \sin(pt - \alpha) \tag{9.17}$$

$$\ddot{\theta}_i = -p^2\vartheta_i \sin(pt - \alpha) = -p^2\theta_i. \tag{9.18}$$

Substituting Eqs. (9.17) and (9.18) into Eq. (9.16) gives

$$M_i^R = M_i^L - J_i p^2 \theta_i. \tag{9.19}$$

The lumped mass is an inertial element, having no elastic deformation. Thus, we have

$$\theta_i^{\mathrm{L}} = \theta_i^{\mathrm{R}} = \theta_i. \tag{9.20}$$

Combining Eqs. (9.19) and (9.20) into matrix form, we have

$$\begin{Bmatrix} \theta \\ M \end{Bmatrix}_i^{\mathrm{R}} = \begin{bmatrix} 1 & 0 \\ -p^2 J & 1 \end{bmatrix}_i \begin{Bmatrix} \theta \\ M \end{Bmatrix}_i^{\mathrm{L}}, \tag{9.21}$$

where

$$\begin{Bmatrix} \theta \\ M \end{Bmatrix}_i^{\mathrm{R}} \quad \text{and} \quad \begin{Bmatrix} \theta \\ M \end{Bmatrix}_i^{\mathrm{L}}$$

are the state vector of the right node and left node of the lumped mass. Clearly, Eq. (9.21) constitutes the transference of the state vector over a mass. Letting

$$\boldsymbol{T}_{\mathrm{p}i} = \begin{bmatrix} 1 & 0 \\ -p^2 J & 1 \end{bmatrix}, \tag{9.22}$$

it is called the transfer matrix over the station, or *point transfer matrix*.

9.2.2.3 Field transfer matrix

Now let's look at a shaft segment (field) shown in Fig. 9.4(b). As stated before, the shaft segments are treated as massless; thus, we have

$$M_i^{\mathrm{L}} = M_{i-1}^{\mathrm{R}}. \tag{9.23}$$

The elastic rotation angle between the two ends of the segment under the torque is

$$\theta_i^{\mathrm{L}} - \theta_{i-1}^{\mathrm{R}} = \frac{M_{i-1}^{\mathrm{R}}}{k_i}, \tag{9.24}$$

where k_i is the stiffness of the i-th shaft element. Combining Eqs. (9.23) and (9.24), we have

$$\begin{Bmatrix} \theta \\ M \end{Bmatrix}_i^{\mathrm{L}} = \begin{bmatrix} 1 & 1/k \\ 0 & 1 \end{bmatrix}_i \begin{Bmatrix} \theta \\ M \end{Bmatrix}_{i-1}^{\mathrm{R}}. \tag{9.25}$$

This represents the state vector transference from the left end to the right end of a shaft segment. Letting

$$\boldsymbol{T}_{\mathrm{f}i} = \begin{bmatrix} 1 & 1/k \\ 0 & 1 \end{bmatrix}_i, \tag{9.26}$$

this is called the transfer matrix for the i-th shaft segment or *field transfer matrix*.

9.2.2.4 Frequency equation

With the point and field transfer matrices, we can write the transferring relationship of the state vector from the left to the right end of the whole shaft as

$$\begin{Bmatrix} \theta \\ M \end{Bmatrix}_{n+1}^{\mathrm{L}} = \boldsymbol{T} \begin{Bmatrix} \theta \\ M \end{Bmatrix}_0^{\mathrm{R}}, \tag{9.27}$$

where $\boldsymbol{T}$ is the global transfer matrix. It is in the following form:

$$\boldsymbol{T} = \boldsymbol{T}_{\mathrm{f}n+1}\boldsymbol{T}_{\mathrm{p}n}\boldsymbol{T}_{\mathrm{f}n}\cdots\boldsymbol{T}_{\mathrm{p}2}\boldsymbol{T}_{\mathrm{f}2}\boldsymbol{T}_{\mathrm{p}1}\boldsymbol{T}_{\mathrm{f}1}. \tag{9.28}$$

Eq. (9.27) represents the transferring of the state vector from the 0-th disk to the $(n+1)$-th disk.

By incorporating the boundary conditions, the *frequency equation* can be obtained. The boundary conditions can be summarized as:

- **Free end:** $M=0$;
- **Fixed end:** $\theta=0$.

Take the shaft in Fig. 9.3 in which both ends are free as an example. Thus, we have

$$M_{n+1}^{\mathrm{L}} = 0, \quad M_0^{\mathrm{R}} = 0. \tag{9.29}$$

The second row of Eq. (9.27) is

$$M_{n+1}^{\mathrm{L}} = T_{21}\theta_0^{\mathrm{R}} + T_{22}M_0^{\mathrm{R}}. \tag{9.30}$$

From Eq. (9.29) we have

$$T_{21} = 0. \tag{9.31}$$

This equation contains the natural frequency p so it is called the *frequency equation*.

9.2.2.5 Solving the frequency equation

Solving the frequency equation Eq. (9.31) gives the natural frequency p. It can be seen that p^2 appears in the point transfer matrix of Eq. (9.22). If the system has n DOFs, the frequency equation Eq. (9.31) will be a *second*–order polynomial. Generally, numerical methods have to be relied on for its solution. Many numerical schemes are available for solving the polynomial equation, such as the bracketing methods and open methods (Chapra and Canale, 2011). One method is to start from a guessed p and gradually find an approximate solution. The flow chart is given in Fig. 9.5.

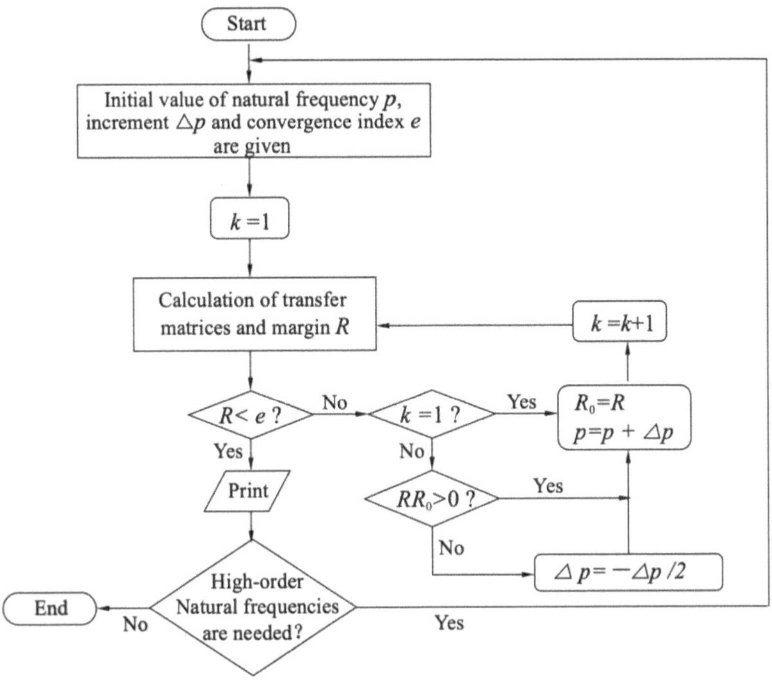

Figure 9.5 Flow chart.

9.3 Transfer matrix method for critical speeds

The lateral natural frequencies of a shaft are called *critical speeds*, at which the shaft experiences resonance and very large deflection. In the following, the transfer matrix method to determine the critical speeds of a shaft is introduced. Fig. 9.6 shows a typical shaft with some disk-like components installed, such as gears, sprockets, and pulleys, etc. The shaft is supported

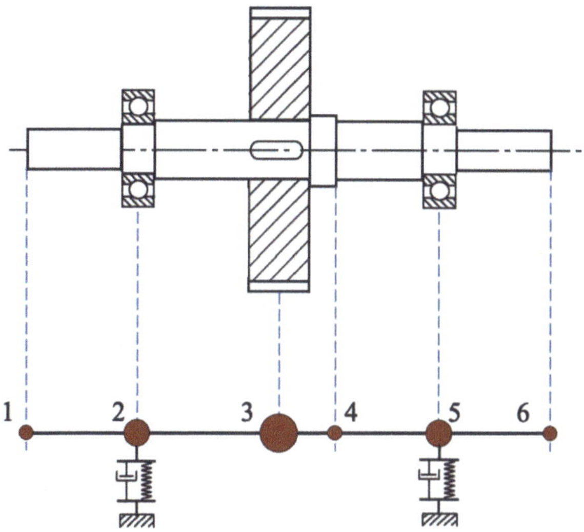

Figure 9.6 Lumped-parameter model for gears and shafts.

by two bearings. We first divide the shaft into lumped masses (stations) and massless beam segments (fields) as shown. In doing this, the masses of the beam segments may be transformed to the lumped masses at the two ends for accuracy.

9.3.1 Point and field transfer matrix

There are only masses at some stations, such as the stations 1, 3, 4, and 6. While at some other stations, there are masses and supports, such as stations 2 and 5, in this case, there are two types of point transfer matrices, namely point mass transfer matrix and point support matrix. In the following, we derive the point matrices and the field matrix of the lateral vibration.

9.3.1.1 Point transfer matrix of masses

The lumped masses and relevant beam segments without supports are shown in Fig. 9.7(a). Individual mass and beam segment along with nodal internal forces and deformation are given in Fig. 9.7(b). Similar to the previous section, we use the superscripts L and R to represent the quantities to the left and right of a station. In the theory of beam bending, deflection $\tilde{y}$ and slope angle $\tilde{\theta}$ are used to represent the displacements of a cross section. The internal force include the moment $\tilde{M}$ and shear force $\tilde{Q}$. With

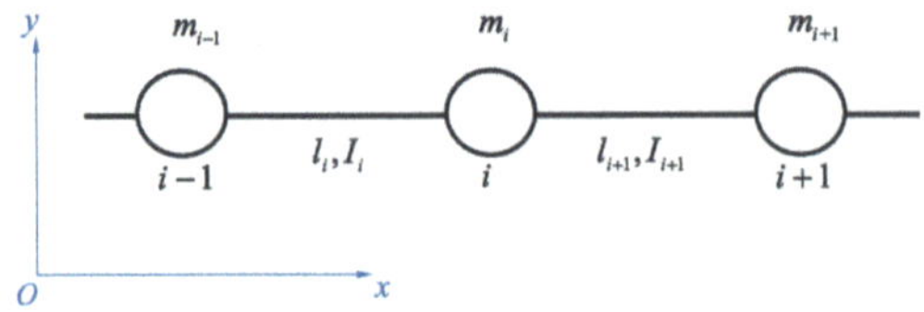

(a) Shaft model

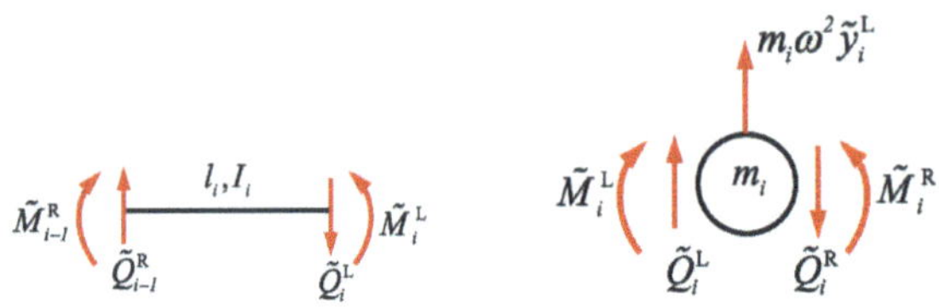

(b) Beam element (c) Mass element

Figure 9.7 Mass and beam elements in transfer matrix.

reference to Fig. 9.7(c), we have the following relationship:

$$\begin{cases} \tilde{y}_i^{R} = \tilde{y}_i^{L} \\ \tilde{\theta}_i^{R} = \tilde{\theta}_i^{L} \\ \tilde{M}_i^{R} = \tilde{M}_i^{L} + J_i\ddot{\tilde{\theta}}_i \\ \tilde{Q}_i^{R} = \tilde{Q}_i^{L} - m_i\ddot{\tilde{y}}. \end{cases} \tag{9.32}$$

Assuming harmonic vibration with frequency p, we have

$$\begin{cases} \tilde{y}_i^{R} = y_i^{R}\sin(pt-\alpha), & \tilde{y}_i^{L} = y_i^{L}\sin(pt-\alpha) \\ \tilde{\theta}_i^{R} = \theta_i^{R}\sin(pt-\alpha), & \tilde{\theta}_i^{L} = \theta_i^{L}\sin(pt-\alpha) \\ \tilde{M}_i^{R} = M_i^{R}\sin(pt-\alpha), & \tilde{M}_i^{L} = M_i^{L}\sin(pt-\alpha) \\ \tilde{Q}_i^{R} = Q_i^{R}\sin(pt-\alpha), & \tilde{Q}_i^{L} = Q_i^{L}\sin(pt-\alpha). \end{cases} \tag{9.33}$$

Substitution of Eq. (9.33) into Eq. (9.32) gives

$$\begin{cases} y_i^{R} = y_i^{L} \\ \theta_i^{R} = \theta_i^{L} \\ M_i^{R} = M_i^{L} + J_i\ddot{\theta}_i \\ Q_i^{R} = Q_i^{L} + m_ip^2y_i. \end{cases} \tag{9.34}$$

Letting the state vectors be

$$\boldsymbol{Z}_i^{\mathrm{L}} = \begin{Bmatrix} y \\ \theta \\ M \\ Q \end{Bmatrix}_i^{\mathrm{L}} \quad \text{and} \quad \boldsymbol{Z}_i^{\mathrm{R}} = \begin{Bmatrix} y \\ \theta \\ M \\ Q \end{Bmatrix}_i^{\mathrm{R}}, \tag{9.35}$$

Eq. (9.34) can be represented as

$$\boldsymbol{Z}_i^{\mathrm{R}} = \boldsymbol{T}_{\mathrm{p}i}\boldsymbol{Z}_i^{\mathrm{L}}. \tag{9.36}$$

$\boldsymbol{T}_{\mathrm{p}i}$ is the point transfer matrix for a mass. It has the following form:

$$\boldsymbol{T}_{\mathrm{p}i} = \begin{bmatrix} 1 & 0 & 0 & 0 \\ 0 & 1 & 0 & 0 \\ 0 & -p^2 J & 1 & 0 \\ p^2 m & 0 & 0 & 1 \end{bmatrix}. \tag{9.37}$$

9.3.1.2 Field transfer matrix

For the beam segment shown in Fig. 9.7(b), the following relationships are true:

$$\begin{cases} \tilde{M}_i^{\mathrm{L}} = \tilde{M}_{i-1}^{\mathrm{R}} + \tilde{Q}_{i-1}^{\mathrm{R}} l_i \\ \tilde{Q}_i^{\mathrm{L}} = \tilde{Q}_{i-1}^{\mathrm{R}}, \end{cases} \tag{9.38}$$

where l_i is the length of the segment. From mechanics of materials, we have

$$\tilde{\theta}_i^{\mathrm{L}} = \tilde{\theta}_{i-1}^{\mathrm{R}} + \frac{1}{EI_i}\int_0^{l_i} \tilde{M}_i^{\mathrm{L}}\,\mathrm{d}x = \tilde{\theta}_{i-1}^{\mathrm{R}} + \frac{\tilde{M}_{i-1}^{\mathrm{R}} l_i}{EI_i} + \frac{\tilde{Q}_{i-1}^{\mathrm{R}} l_i^2}{2EI_i} \tag{9.39}$$

$$\tilde{y}_i^{\mathrm{L}} = \tilde{y}_{i-1}^{\mathrm{R}} + \int_0^{l_i} \tilde{\theta}_i^{\mathrm{L}}\,\mathrm{d}x = \tilde{y}_{i-1}^{\mathrm{R}} + \frac{\tilde{M}_{i-1}^{\mathrm{R}} l_i^2}{2EI_i} + \frac{\tilde{Q}_{i-1}^{\mathrm{R}} l_i^3}{6EI_i}. \tag{9.40}$$

Combining Eqs. (9.38) through (9.40) yields

$$\begin{cases} y_i^{\mathrm{L}} = y_{i-1}^{\mathrm{R}} + \int_0^{l_i} \theta_i^{\mathrm{L}}\,\mathrm{d}x = y_{i-1}^{\mathrm{R}} + \dfrac{M_{i-1}^{\mathrm{R}} l_i^2}{2EI_i} + \dfrac{Q_{i-1}^{\mathrm{R}} l_i^3}{6EI_i} \\ \theta_i^{\mathrm{L}} = \theta_{i-1}^{\mathrm{R}} + \dfrac{1}{EI_i}\int_0^{l_i} M_i^{\mathrm{L}}\,\mathrm{d}x = \theta_{i-1}^{\mathrm{R}} + \dfrac{M_{i-1}^{\mathrm{R}} l_i}{EI_i} + \dfrac{Q_{i-1}^{\mathrm{R}} l_i^2}{2EI_i} \\ M_i^{\mathrm{L}} = M_{i-1}^{\mathrm{R}} + Q_{i-1}^{\mathrm{R}} l_i \\ Q_i^{\mathrm{L}} = Q_{i-1}^{\mathrm{R}}. \end{cases} \tag{9.41}$$

It can be recast into more compact matrix form as

$$\boldsymbol{Z}_i^{\mathrm{L}} = \boldsymbol{T}_{\mathrm{f}i}\boldsymbol{Z}_{i-1}^{\mathrm{R}}, \tag{9.42}$$

where $\boldsymbol{T}_{\mathrm{f}i}$ is the field transfer matrix having the following for:

$$\boldsymbol{T}_{\mathrm{f}i} = \begin{bmatrix} 1 & l & \frac{l^2}{2EI} & \frac{l^3}{6EI} \\ 0 & 1 & \frac{l}{EI} & \frac{l^2}{2EI} \\ 0 & 0 & 1 & l \\ 0 & 0 & 0 & 1 \end{bmatrix}. \tag{9.43}$$

Given that a point mass is always connected to a beam segment, it often combines a point mass and a beam segment to form a unit. The transferring relationship of the state vector over the unit and the transfer matrix are represented as

$$\boldsymbol{Z}_i^{\mathrm{R}} = \boldsymbol{T}_{\mathrm{p}i}\boldsymbol{Z}_i^{\mathrm{L}} = \boldsymbol{T}_{\mathrm{p}i}\boldsymbol{T}_{\mathrm{f}i}\boldsymbol{Z}_{i-1}^{\mathrm{R}} \tag{9.44}$$

$$\boldsymbol{T}_i = \boldsymbol{T}_{\mathrm{p}i}\boldsymbol{T}_{\mathrm{f}i} = \begin{bmatrix} 1 & l & \dfrac{l^2}{2EI} & \dfrac{l^3}{6EI} \\ 0 & 1 & \dfrac{l}{EI} & \dfrac{l^2}{2EI} \\ 0 & -p^2J & 1-\dfrac{p^2Jl}{EI} & l-\dfrac{p^2Jl^2}{2EI} \\ p^2m & p^2ml & \dfrac{p^2ml^2}{2EI} & 1+\dfrac{p^2ml^3}{6EI} \end{bmatrix}. \tag{9.45}$$

9.3.1.3 Point transfer matrix of supports

A support is schematically shown in Fig. 9.8. The state variables across a support have the following relationship:

$$\begin{cases} \tilde{y}_i^{\mathrm{R}} = \tilde{y}_i^{\mathrm{L}} \\ \tilde{\theta}_i^{\mathrm{R}} = \tilde{\theta}_i^{\mathrm{L}} \\ \tilde{M}_i^{\mathrm{R}} = \tilde{M}_i^{\mathrm{L}} \\ \tilde{Q}_i^{\mathrm{R}} = \tilde{Q}_i^{\mathrm{L}} - m_i\ddot{\tilde{y}}_i - c_i\dot{\tilde{y}}_i + k_i\tilde{y}_i, \end{cases} \tag{9.46}$$

where m_i, k_i, and c_i are the equivalent mass, equivalent stiffness, and equivalent damping coefficient of the support. If the damping term is neglected and harmonic vibration is assumed, the transferring relation of Eq. (9.46) can be represented as

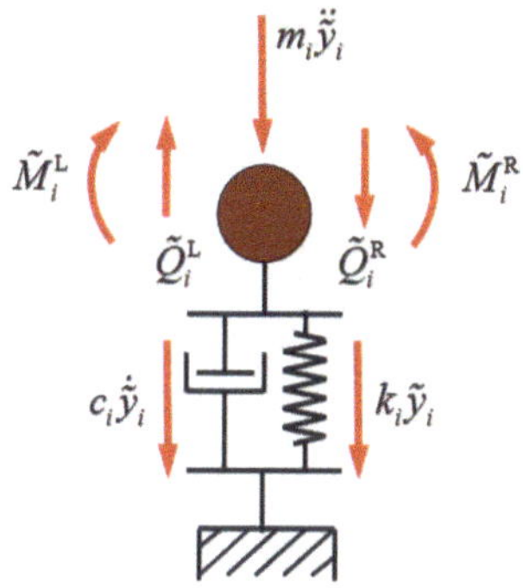

Figure 9.8 Support element in transfer matrix.

$$\boldsymbol{Z}_i^{\mathrm{R}} = \boldsymbol{T}_{\mathrm{s}i}\boldsymbol{Z}_i^{\mathrm{L}}. \tag{9.47}$$

$\boldsymbol{T}_{\mathrm{s}i}$ is the point transfer matrix for the support. It has the following form:

$$\boldsymbol{T}_{\mathrm{s}i} = \begin{bmatrix} 1 & 0 & 0 & 0 \\ 0 & 1 & 0 & 0 \\ 0 & 0 & 1 & 0 \\ (p^2m - k) & 0 & 0 & 1 \end{bmatrix}_i. \tag{9.48}$$

9.3.2 Global transfer matrix

For the shaft in Fig. 9.6, the nodes are shown and numbered as in Fig. 9.9. In addition, the fields and stations are grouped as elements, which are numbered as shown and organized as follows:

- Element 1: mass
- Elements 2 and 6: beam segments
- Elements 3 and 7: supports
- Elements 4, 5, and 8: combination of a beam segment and a mass.

The subscript of the point and field transfer matrices indicate the element number.

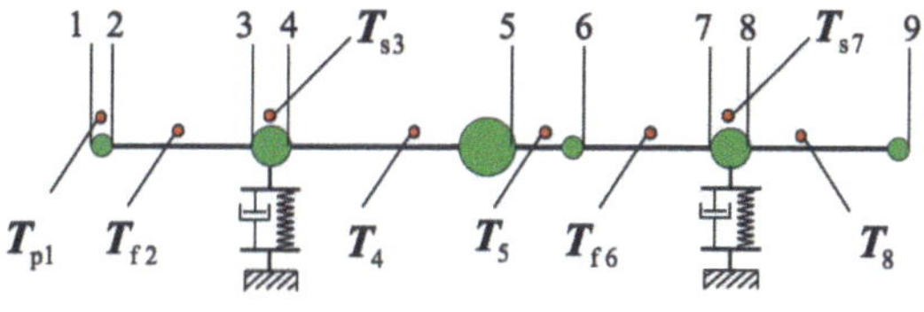

Figure 9.9 Shaft division and element transfer matrix.

According to Eqs. (9.36), (9.42), (9.44), and (9.47), we have

$$\begin{aligned} \boldsymbol{Z}_9 &= \boldsymbol{T}_8\boldsymbol{Z}_8 \\ &= \boldsymbol{T}_8\boldsymbol{T}_{s7}\boldsymbol{Z}_7 \\ &= \boldsymbol{T}_8\boldsymbol{T}_{s7}\boldsymbol{T}_{f6}\boldsymbol{Z}_6 \\ &= \ldots \\ &= \boldsymbol{T}_8\boldsymbol{T}_{s7}\boldsymbol{T}_{f6}\boldsymbol{T}_5\boldsymbol{T}_4\boldsymbol{T}_{s3}\boldsymbol{T}_{f2}\boldsymbol{T}_{p1}\boldsymbol{Z}_1. \end{aligned} \tag{9.49}$$

This can be shortened to

$$\boldsymbol{Z}_9 = \boldsymbol{T}\boldsymbol{Z}_1, \tag{9.50}$$

where $\boldsymbol{T}$ is the global transfer matrix in the following form:

$$\boldsymbol{T} = \boldsymbol{T}_8\boldsymbol{T}_{s7}\boldsymbol{T}_{f6}\boldsymbol{T}_5\boldsymbol{T}_4\boldsymbol{T}_{s3}\boldsymbol{T}_{f2}\boldsymbol{T}_{p1}. \tag{9.51}$$

$\boldsymbol{T}$ is a 4×4 matrix.

9.3.3 Frequency equation and solution

It can be seen that the transfer matrix contains the natural frequency matrix p. The natural frequency equation can be obtained with the boundary conditions. The frequency equation is generally a polynomial of p, which, in most cases, has to be solved with numerical methods. The boundary conditions of the shaft can be summarized as follows:

1. **Both ends free**

   Boundary conditions:

$$\begin{aligned} M_1 &= 0, & Q_1 &= 0 \\ M_n &= 0, & Q_n &= 0 \end{aligned}$$

   Frequency equation:

$$f(p) = \begin{vmatrix} T_{31} & T_{32} \\ T_{41} & T_{42} \end{vmatrix} = 0.$$

2. **Both ends simply supported**

   Boundary conditions:

$$\begin{aligned} M_1 &= 0, & y_1 &= 0 \\ M_n &= 0, & y_n &= 0 \end{aligned}$$

Frequency equation:

$$f(p) = \begin{vmatrix} T_{12} & T_{14} \\ T_{32} & T_{34} \end{vmatrix} = 0.$$

3. **Element 1 simply supported, element n free**
Boundary conditions:

$$M_1 = 0, \qquad y_1 = 0$$
$$M_n = 0, \qquad Q_n = 0$$

Frequency equation:

$$f(p) = \begin{vmatrix} T_{32} & T_{34} \\ T_{42} & T_{44} \end{vmatrix} = 0.$$

4. **Element 1 fixed, element n free**
Boundary conditions:

$$\theta_1 = 0, \qquad y_1 = 0$$
$$M_n = 0, \qquad Q_n = 0$$

Frequency equation:

$$f(p) = \begin{vmatrix} T_{33} & T_{34} \\ T_{43} & T_{44} \end{vmatrix} = 0.$$

We will show the discretization process of a shaft next through the example shown in Fig. 9.10(a), which is the spindle of a general-purpose lathe. ①, ②, and ③ are radial bearings; ④ and ⑤ are axial bearings. ⑥ and ⑦ are the chuck and the workpiece respectively. ⑧ is a gear.

The shaft can be represented by a lumped-parameter model shown in Fig. 9.10(b) in which 20 elements, including nine lumped masses, eight beam segments, and three supports coincident with the locations of the three radial bearings. In the discretization process, the shaft is divided first into segments of various diameters. The mass of each segment is split into two masses located at both ends based on the principle of keeping the center of mass unchanged. The three bearing supports are treated as springs of equivalent stiffness. In real calculations, the number of shaft segments to

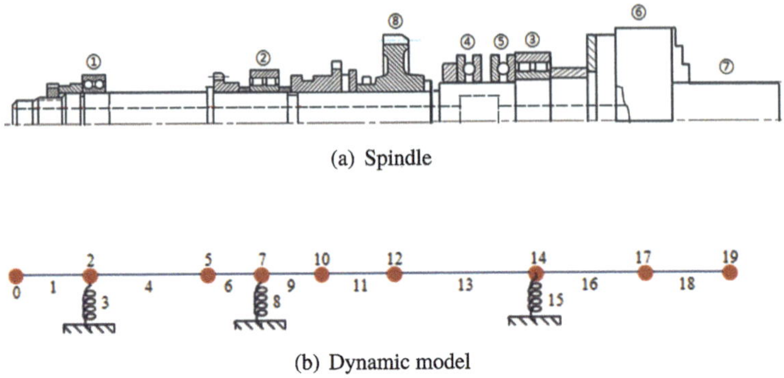

Figure 9.10 Spindle system of machine tool.

divide also depends on the required accuracy. If higher accuracy is required, the number of segments may be made higher.

9.4 Finite element method for critical speeds

The critical speeds can also be calculated with the finite element (FE) method. The procedures with the FE method are outlined here:

- Discretize the shaft to build the FE model.
- Develop element stiffness and mass matrices, $\boldsymbol{k}$ and $\boldsymbol{m}$, based on which the global stiffness and mass matrices $\boldsymbol{K}$ and $\boldsymbol{M}$ are obtained.
- Solve the eigenvalue problem to obtain the natural frequencies (critical speeds).

9.4.1 Finite element model

The procedures to develop the FE model are illustrated through a shaft with a disk installed shown in Fig. 9.11(a).

9.4.1.1 Discretization of shaft

For a stepped shaft, here are some general rules for discretization:

1) Shaft shoulders are always taken as nodes. Short shaft segments can be treated as one element. Long segments are further divided into multiple elements.
2) The disk installed on the shaft, such as gears, sprockets, or pulleys, are always treated as individual elements.

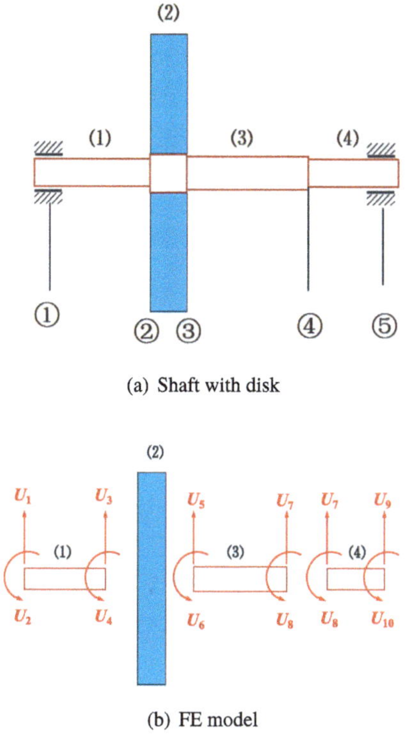

Figure 9.11 FE discretization of shaft.

3) Supports, more specifically the midpoint of the bearing width, are always treated as nodes.

If the number of element is designated as N_e, the number of nodes N_n is

$$N_n = N_e + 1. \tag{9.52}$$

Numbering of the elements and nodes is shown in Fig. 9.11(a). For differentiation, the element number and node number are put in parenthesis and circles, respectively. For each node, two coordinates, the lateral displacement and the rotation angle, are chosen. For the i-th node, the two generalized coordinates are U_{2i-1} and U_{2i}. The total number of generalized coordinates are

$$N_u = 2N_n. \tag{9.53}$$

The shaft in Fig. 9.11(a) is divided into four elements, including three shaft elements, one disk element, and five nodes. Therefore, there are 10 generalized coordinates in total. The parameters needed for the calculation include the number of element N_e, the length of element l_i, shaft internal and external diameters D_{1i} and D_{2i}, the disk's moment of inertia, material's density ρ, and the modulus of elasticity E.

9.4.1.2 Element matrices

The shaft element stiffness and mass matrices $\boldsymbol{k}$ and $\boldsymbol{m}$ are calculated according to Eqs. (7.43) and (7.46).

The mass matrix of the disk element contains two parts, namely the part caused by the laterally linear displacement and the part caused by the angular displacement θ as shown in Fig. 9.12.

$$\boldsymbol{m} = \boldsymbol{m}_y + \boldsymbol{m}_\theta, \tag{9.54}$$

where $\boldsymbol{m}_y$ is the mass matrix caused by the linear displacement and $\boldsymbol{m}_\theta$ is the part caused by the angular displacement θ. $\boldsymbol{m}_y$ is calculated by Eq. (7.46). The calculation of $\boldsymbol{m}_\theta$ is outlined subsequently.

The kinetic energy caused by θ is

$$E_\theta = \frac{1}{2}\int_0^l J_\mathrm{d}\left[\frac{\mathrm{d}\theta(x,t)}{\mathrm{d}t}\right]^2 \mathrm{d}x, \tag{9.55}$$

where J_d is the moment of inertia of the disk in unit length about its axis.

$\theta(x, t)$ is calculated by

$$\theta(x,t) = \frac{\partial u(x,t)}{\partial x}, \tag{9.56}$$

where $u(x, t)$ is the lateral displacement at x. According to Eq. (7.40), $\theta(x, t)$

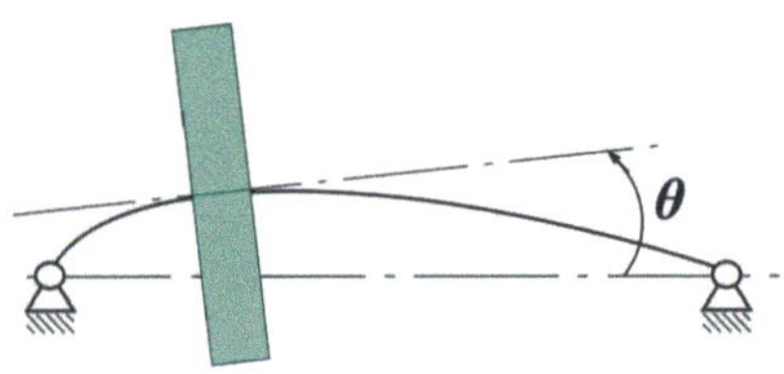

Figure 9.12 Rotation of disk.

is formulated as

$$\theta(x,t) = \sum_{i=1}^{4} \phi_i'(x) u_i(t). \tag{9.57}$$

Substituting Eq. (9.57) into Eq. (9.55) gives

$$E_\theta = \frac{1}{2}\sum_{i=1}^{4}\sum_{j=1}^{4}\dot{u}_i(t)\left[\int_0^l J_d\phi_i'(x)\phi_j'(x)\,dx\right]\dot{u}_j(t) = \frac{1}{2}\dot{\boldsymbol{u}}^T\boldsymbol{m}_\theta\dot{\boldsymbol{u}}. \tag{9.58}$$

The entries of $\boldsymbol{m}_\theta$ is calculated by

$$(m_\theta)_{ij} = \int_0^l J_d\phi_i'(x)\phi_j'(x)\,dx. \tag{9.59}$$

Substituting the parameters in and completing the integration, we have

$$\boldsymbol{m}_\theta = \frac{J_d}{30}\begin{bmatrix} 36 & 3l & -36 & 3l \\ & 4l^2 & -3l & -l^2 \\ & symm. & 36 & -3l \\ & & & 4l^2 \end{bmatrix}. \tag{9.60}$$

$\boldsymbol{m}_\theta$ is a 4 × 4 symmetrical matrix.

When the disk-shaped part is installed on the shaft, the fit between the shaft and the hole has some effect on the stiffness of the assembly. This effect can be taken into account roughly as follows:

- For clearance fit: The shaft diameter is used in the calculation.
- For interference fit: The external diameter of the hub of the disk-shaped part is used in the calculation.

9.4.1.3 Global matrices

To combine the element matrices into global ones, the procedures introduced in Chapter 7 shall be followed. For the example in Fig. 9.11(a), the layout of the four element-mass matrices in the global is shown in Fig. 9.13.

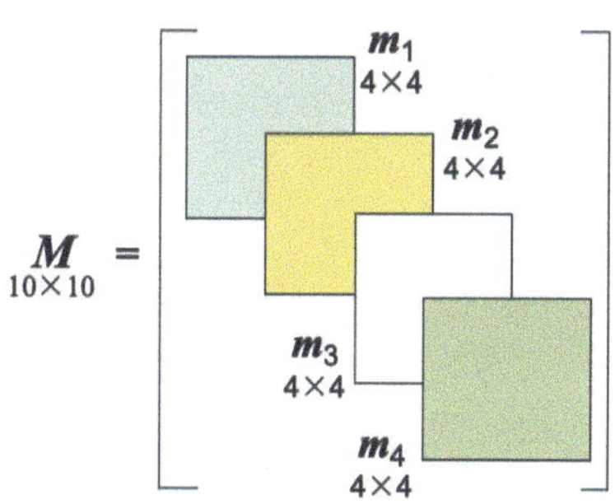

Figure 9.13 Global matrix.

For the i-th element, its four displacements $\{u_1, u_2, u_3, u_4\}^T$ are located at the global coordinates from U_{2i-1} through U_{2i+2}. Thus, the element matrix is simply placed in the global mass matrix covering the $2i-1$ to $2i+2$ rows and the $2i-1$ to $2i+2$ columns.

9.4.1.4 Supports

As a general rule, supports of the shaft are always made coincident with the nodes in the discretization. Given that a radial bearing does not restrict the motion in the axial direction, the stiffness matrix is singular, causing difficulty in the computation. This issue can be overcome through proper treatment of the support as follows.

- **Pinned support:** If the supports are treated as pinned support, the axial motion of the supporting node is actually removed. For the j-th support, simply crossing out the $2j-1$ row and $2j-1$ column in the $\boldsymbol{M}$ and $\boldsymbol{K}$, the singularity problem can be solved.
- **Elastic support:** If the support is modeled as a spring k_j, then adding k_j to the global stiffness matrix, can solve the singularity problem. Specifically, if the support at the j-th node is modeled as a spring, then replace the $K_{2j-1,2j-1}$ by $K_{2j-1,2j-1} + k_j$.

9.4.2 Critical speed

Once the stiffness and mass matrices $\boldsymbol{K}$ and $\boldsymbol{M}$ are obtained, solving the eigenvalue problem

$$\left(\boldsymbol{K} - \omega^2 \boldsymbol{M}\right) \boldsymbol{\Phi} = \boldsymbol{0} \tag{9.61}$$

gives the natural frequencies, ω_i $(i = 1, 2, \cdots)$. In practice, only the several lower frequencies, especially the primary natural frequency, are important.

Conventionally, the critical speed is given in rpm. The natural frequencies are needed to transform this into rpm units. For the primary critical speed, this is

$$n_c = \frac{30}{\pi}\omega_1. \tag{9.62}$$

The real operation speed of the shaft should be very distant from the critical speed. If the operation speed cannot be changed for some reason, n_c can be changed by changing the shaft design parameters.

9.5 Introduction to rotor dynamics

A rotating shaft, together with the disks installed on it, is referred to as a *rotor*. The rotor, plus with the bearings and the casing/housing, is referred to as a *rotor system*. With the development of various rotating machines, from electric motors to large-scale steam turbines, vibration of rotor systems becomes a challenging problem. In addition, the pursuit of higher power, higher speed, and higher efficiency is changing the design philosophy of rotors, making the vibration problem more prominent. *Rotor dynamics* is a scientific subject specialized in the study of vibration of rotors and the supporting structures. In the following, a brief discussion of rotor dynamics is presented; readers interested in more detailed information of this topic are advised to consult more specialized monographs or texts, such as the one by Matsushita et al. (2017).

In the early days, machine speed was relatively low. The main cause behind rotor vibration is unbalanced masses of the disks on the rotor. Static balancing is effective in reducing the imbalance. For rotors having multiple disks separated from each other, dynamic balancing can be used to control the vibration. Static and dynamic balancing techniques are generally covered in most undergraduate curriculum. However, for a slender rotor, severe vibration can still exist even when well-balanced. The speeds at which lateral resonances happen are termed the critical speeds.

In many rotating machines, the shaft speed is lower than the primary critical speed. Rotors of this kind are referred to as *rigid rotors*. Rotors in most electric motors belong to this category. For some other machines, such as large steam turbine generators, the operation speed is above the primary critical speed. During the start-up, the machine quickly exceeds the critical speed with proper acceleration to avoid severe vibration. This type of rotor are is termed *flexible rotors*. Generally, flexible rotors are more slender and have larger flexure deflection during operation, which makes the dynamic analysis and dynamic balancing of flexible rotors much more complicated than for rigid ones. Large-scale steam turbine generators in power plants generally have flexible rotors. Unexpected vibration and failures have severe economic and safety consequences.

In addition to the forced vibration caused by the unbalanced masses, *whirling* is also a possible vibration mode to high-speed rotors. In forced vibration, the excitation frequency is the rotation speed of the rotor. Whirling is a type of self-excited vibration, which is related to many factors, such as friction, the material's internal damping, and the oil film in bearings.

Whirling may lead to instability of the rotor, causing fatigue and lubrication issues. Stability is an important topic in rotor dynamics.

In some cases, the vibration of supporting structure is also incorporated into rotor dynamics. This is especially true for large rotors in steam turbines.

Some other important developments in rotor dynamics include operation monitoring and fault diagnosis, as well as integrating the knowledge of fracture mechanics into the dynamic analysis of cracked rotors.

CHAPTER 10

Dynamics of cam mechanisms

10.1 Introduction

Cam mechanisms are widely used in various machines. The valve cam mechanism in an internal combustion engine and the indexing cam mechanism in various automatic machines are two typical examples. With the increase of operation speed, dynamic loads on components become significant, causing a series of detrimental effects on machine performance. Wear, fatigue and noise are the most common problems. More importantly, increased dynamic load often results in more severe vibrations and larger motion errors. These would be eventually transmitted to higher indexing errors for an indexing cam.

Conventionally, analysis and design of cam mechanisms are based on two basic assumptions:

1) The cam mechanism is considered as a rigid system;
2) The cam rotates at a constant speed.

In this case, the follower would move exactly the same as described by the motion analysis. For a cam mechanism operating at relatively low speed, these assumptions are acceptable.

It used to be common practice that the follower's motion was chosen based on only kinematics. In special cases, when dynamics has to be considered, the maximum acceleration might be checked. The follower's motion of *constant acceleration*, which has a relatively low peak acceleration value, was viewed as a good option in the past. By the 1930s, however, with the increase of operation speed of internal combustion engines, malfunction of cam mechanisms became a prominent problem in automobile and airplanes. This problem couldn't be explained with the rigid-body assumption until Hrones et al. (1948) built the first vibration model by incorporating a factor for elasticity into the equation. He revealed theoretically that the follower motion of constant acceleration performed poorly at high speeds. Shortly later, Mitchell (1950) validated Hrones's finding experimentally. These two breakthrough works are landmarks of the dynamics of high-speed cam mechanisms. Thereafter, studies on dynamics of cam mechanisms entered a new period.

Machinery Dynamics
https://doi.org/10.1016/B978-0-12-815785-5.00021-2

The term *high speed* is not clearly defined. Historically, various different criteria were proposed. Now, a consensus has been reached that the evaluation of the speed should be made in terms of not only the cam's physical speed, but also the system's dynamic parameters, such as stiffness and mass. From the viewpoint of vibration, the closeness of the excitation frequency (the cam rotation speed) to the natural frequency is the critical factor. In some publications (Volmer, 1976), a time ratio, called the *period ratio*, is defined as

$$\lambda = \frac{t_h}{t_0}, \tag{10.1}$$

where t_h is the time of the follower's rising, and t_0 is the natural period corresponding to the primary natural frequency.

Based on the value of λ, the following classification was established:

- $\lambda > 15$: Is low speed; the conventional analysis based on rigid assumption is valid.
- $6 < \lambda < 15$: Is medium speed. The mechanism can be treated as rigid, but the follower's motion should be selected with careful consideration of the dynamic effect.
- $\lambda < 6$: Is high speed. The elastic deformation of parts needs to be considered and a thorough dynamic analysis should be conducted.

According to this classification, although the physical speed may not be very high in some cases, it is still considered high speed if λ has a low value. Thus, systematic dynamic analysis is required.

This chapter deals with the dynamics of high-speed cam mechanisms. The chapter is organized as follows:

- In Section 10.2, types of follower's motion for high-speed cam mechanisms are presented.
- Section 10.3 focuses on the development of dynamic model for high-speed cam mechanisms, while the analysis of the dynamic models is presented in Section 10.4.
- Section 10.5 presents some important design considerations for high-speed cam mechanisms.

10.2 Follower motions for high-speed cam mechanisms

10.2.1 Two types of motion constraints

There are generally two types of motion constraints in cam mechanisms, namely, *critical extreme position* (CEP) and *critical path motion* (CPM) (Norton, 2004).

1) **Critical extreme position (CEP):** In this type of constraint, the starting and ending positions of the follower are the only requirement. The motion in between the two limiting positions is not considered critical. Most cam mechanisms in automatic machines fall into this category. Designers have more room in this case to choose the follower's motion.
2) **Critical path motion (CPM):** In some cases, the follower motion in between the two limiting positions is also essential. For example, the feeding mechanism of a cutting tool in an automated machine tool must meet the requirements not only on the limit positions, but also on the constant velocity of motion. Obviously, this case is more tightly constrained, and the designer has much less freedom in design compared with the CEP case.

This chapter deals with mechanisms having only CEP constraints.

10.2.2 Normalization of motion parameters

When selecting follower's motion, we need to first analyze and compare the properties of different motions, such as the rotation speed of the camshaft, the rising angle, and the lift of follower. These properties may be very different from one mechanism to another. For comparison purpose, normalization (nondimensionalization) of parameters is often done in advance.

The displacement of the follower is generally represented as $s = s(t)$. Starting from $t = 0$, the follower is lifted h during the time t_h of the *rising* stage. The time is often normalized with respect to the rising time t_h as

$$T = \frac{t}{t_h}, \tag{10.2}$$

obviously during the rising stage, $0 \le T \le 1$.

As a convention in this book, a lower-case letter represents an absolute quantity, while a capital letter represents the normalized quantity. Specifically, s, v, a and j represent the absolute displacement, velocity, acceleration and jerk of the follower, while S, V, A and J represent the corresponding normalized parameters, which are defined as:

$$S = \frac{s}{h} \tag{10.3}$$

$$V = \frac{dS}{dT} = \frac{d(s/h)}{d(t/t_h)} = \frac{vt_h}{h} \tag{10.4}$$

$$A = \frac{d^2(S)}{dT^2} = \frac{at_h^2}{h} \tag{10.5}$$

$$J = \frac{d^3(S)}{dT^3} = \frac{jt_h^3}{h}. \tag{10.6}$$

Common practice is to draw these normalized parameters together with time t or rotation angle θ as the abscissa. For convenience, they are vertically aligned as shown in Fig. 10.1.

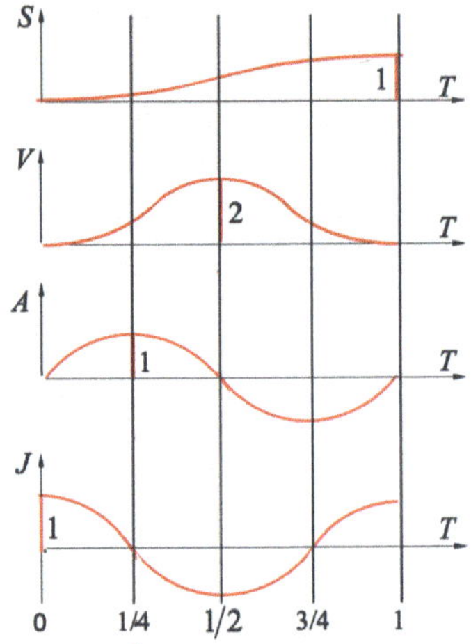

Figure 10.1 SVAJ diagrams.

Either time t or rotation angle θ is the abscissa. It makes no difference as long as the cam rotation speed remains constant.

10.2.3 Characteristic quantities

Selection of the follower motion is the most important step in high-speed cam mechanism design. Different follower motions lead to different dimensions of the cam, different loads and vibration levels. To roughly evaluate the follower motion, some normalized kinematic or dynamic quantities are used as the characteristic quantities. By comparing these quantities, the mechanism's dynamic performance can be roughly judged. Several commonly used such quantities are:

1. **Maximum velocity** V_m

   If the base circles of cams are the same, larger V_m means the larger pressure angle or vice visa. On the other hand, if the allowable pressure angle is given, a smaller V_m value means a smaller radius of base circle can be adopted. If dynamics is not a concern, such as in low-speed cam mechanisms, the follower motion can be selected based on V_m value.

In addition, larger V_m means larger momentum for a certain mass. In the case of emergency braking, higher braking torque is needed. Thus, smaller V_m means better safety performance. This is especially important for a system with large inertia (mass).

2. **Maximum acceleration** A_m
Inertia load becomes dominant at high speed. A large inertia force deteriorates lubrication and accelerates wear of contact surfaces and kinematic pairs. For this reason A_m has long been used as a criterion for selection of follower motion. However, it should be noted that the relationship between vibration response and follower motions in a cam mechanism is very complicated; vibration response is influenced by many other factors besides A_m.
3. **Maximum** $(AV)_m$
The driving torque required for a high-speed cam mechanism is calculated by Eq. (1.36). For convenience it is repeated here:

$$M_d \approx \frac{m}{\omega}\dddot{s}\dot{s}, \tag{10.7}$$

where $\dot{s}$ and $\ddot{s}$ are the follower's velocity and acceleration . Considering the normalized quantities given in Eqs. (10.2) through (10.6), it can be recast as

$$M_d \approx \frac{m}{\omega}\left(A\frac{h}{t_h^2}\right)\left(V\frac{h}{t_h}\right) = AV\frac{mh^2}{\omega t_h^3}. \tag{10.8}$$

Obviously a larger $(AV)_m$ value means a larger driving torque is required to maintain the expected motion. On the other hand, for a certain driving torque, a smaller $(AV)_m$ value means a narrower range of speed fluctuation.
4. **Continuity order** p and **maximum jerk** J_m
In the vibration model of a cam system, the cam profile is the excitation (Section 10.3). Thus, the smoothness of the cam profile has a direct effect on the vibration response. The *order of continuity p* is generally used as an indicator of the cam's profile smoothness (Koster, 1974b). It is defined as the order of the derivative of the motion in which discontinuities appear. For the follower motion with a constant velocity as shown in Fig. 10.2, for instance, the first discontinuity appears in V; thus, $p=1$. For the constant acceleration motion in Fig. 10.3, the first discontinuity appears in A; thus, $p=2$.
The maximum jerk J_m has long been used as an indicator of vibration character. However, this is mainly based on common sense, rather than

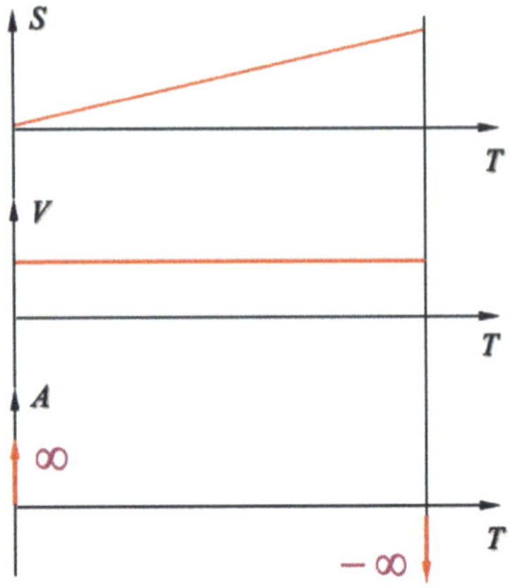

Figure 10.2 Constant velocity motion.

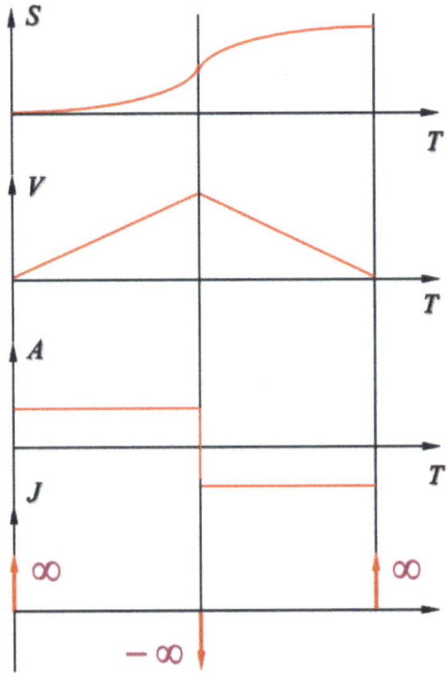

Figure 10.3 Constant acceleration motion.

rigorous scientific validation. The actual relationship between vibration response and the follower motion in a cam mechanism is much more complicated.

As a general design rule, p must be at least three, and J_m must be finite for general cam mechanisms. This rule is treated as the *fundamental law of cam design* by Norton (2004).

However, later studies found exceptions to the general rule. For example, the motions optimized with the least residual vibration in Wiederrich and Roth (1978); Chew and Chuang (1995) have discontinuous A. Yao and Zhang (1998) systematically investigated this issue and gave a rigorous explanation on the relationship between residual vibration and the continuity of the motion curves.

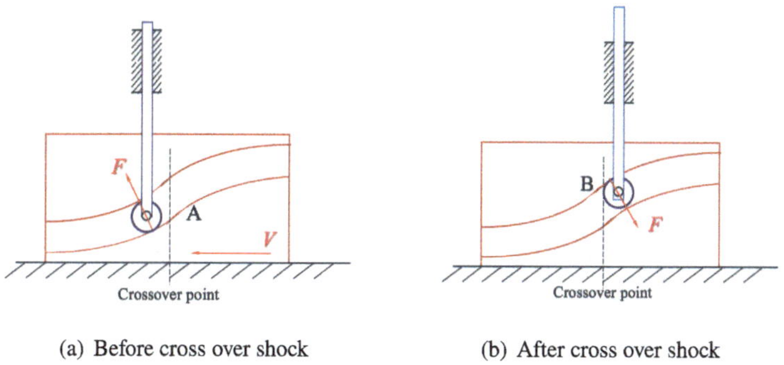

Figure 10.4 Cross over shock.

5. **Crossover shock** $\sqrt[3]{J_{imp}}$ For form-closed cams, impact may happen when the acceleration changes sign. This impact is termed *crossover shock*, which is illustrated with the rising stage in Fig. 10.4. Initially, the acceleration is +, and the roller is in contact with the lower side of the groove at A as shown in Fig. 10.4(a). At some point, the acceleration becomes − (deceleration); the contact of the groove with the roller shifts from the lower side to the upper side at B as shown in Fig. 10.4(b). At the moment when the acceleration changes sign, the velocity reaches its maximum, and the contact changes from one side to the other. This point is referred to as the *crossover point*. If there is a clearance between the groove and the roller, the change of contact creates a shock or impact, which causes the cam mechanism to vibrate. In high-speed cam mechanisms, the crossover shock is one of the main limiting factors for increased speed. A detailed analysis of the crossover shock is given in Appendix A, in which two main conclusions are drawn: 1) Greater clearance leads to greater shock; 2) the crossover shock is proportional to $\sqrt[3]{J_{imp}}$. J_{imp} is the jerk at the crossover point. In view of the second conclusion, $\sqrt[3]{J_{imp}}$ is widely used as a characteristic indicator of the crossover shock by some researchers. Some other researchers use the change rate of $(AV)_m$ with respect to time (desig-

Table 10.1 Characteristic quantities of some follower motions.

Motion	V_m	A_m	J_m	p	$(AV)_m$	$\sqrt[3]{J_{imp}}$	τ_m
1	1.0	∞	–	1	–	–	–
2	2.0	4.0	∞	2	8.0	$-\infty$	–
3	2.0	6.283	39.48	3	8.162	3.405	44.14
4	1.76	5.528	69.47	3	5.435	2.850	34.17
5	2.0	4.888	61.43	3	8.048	3.946	26.71
6	1.875	5.773	60	3	6.694	3.107	36.72
7	2.188	7.511	52.5	4	10.75	3.745	64.25
8	1.275	8.013	201.4	3	5.671	4.064	63.30

1: constant velocity; 2: constant acceleration; 3: sinusoidal acceleration; 4: modified sinusoidal acceleration; 5: modified trapezoidal acceleration; 6: 3–4–5 polynomial; 7: 4–5–6–7 polynomial; 8: modified constant velocity.

nated as τ_m) as the crossover shock indicator. The two indicators do not agree sometimes. Table 10.1 gives a comparison of some follower motions evaluated for the two indicators.

10.2.4 Follower motions

After Hrones discovered that the motion of constant acceleration performed poorly at high speed, many research studies have been conducted to find follower motions with better dynamic performance. In the next section, several commonly used follower motions for high-speed application are presented. For the sake of brevity, analysis is conducted on the rising stage only. These follower motions can be classified into two groups: *combined harmonic motions* and *polynomial motions.*

10.2.4.1 Combined harmonic motions

In automatic machines, the follower motions with an acceleration consisting of simple harmonic segments and horizontal straight-line segments are widely accepted.

1. **Combined harmonic motions**

   Several most commonly used motions of this type are as follows:

   1) **Sinusoidal acceleration:** The motion shown in Fig. 10.1 has a sinusoidal acceleration. Its displacement is cycloidal; thus, it is also referred to as *cycloidal displacement.*

   2) **Modified trapezoidal acceleration:** This motion modifies the constant acceleration at the starting and ending points with a quarter of the sinusoidal curve, as shown in Fig. 10.5(a). This motion

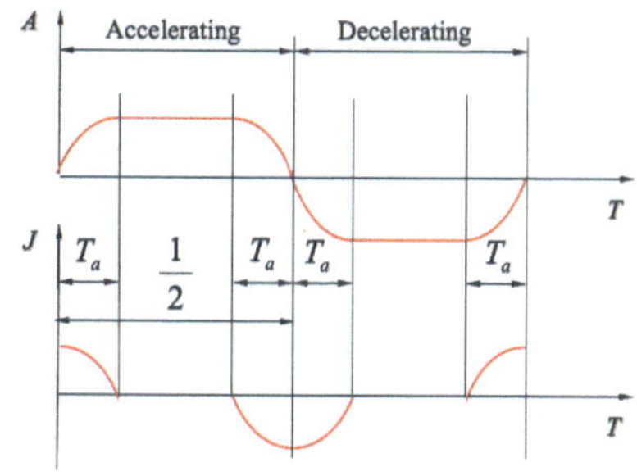

(a) Modified trapezoidal acceleration

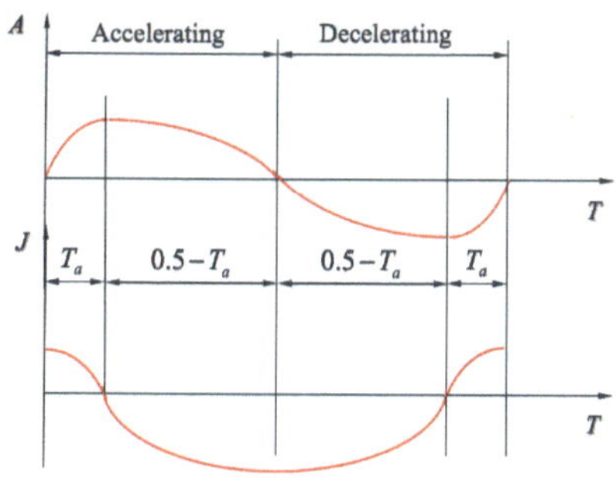

(b) Modified sinusoidal acceleration

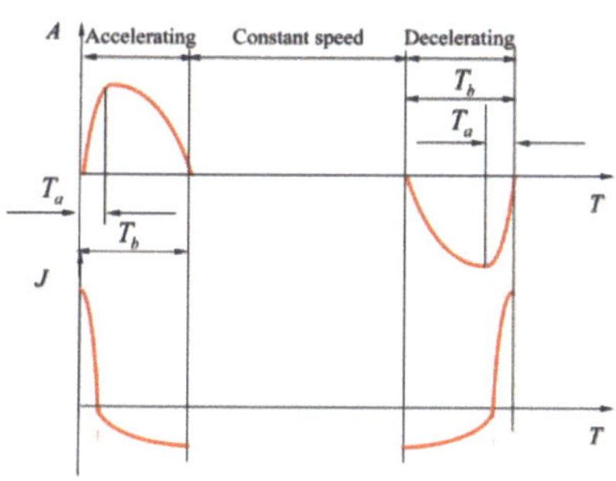

(c) Modified constant velocity

Figure 10.5 Several combined harmonic motions.

keeps the relatively low magnitude of A_m of the constant acceleration motion, but has a much smoother jerk J due to the modification.

3) **Modified sinusoidal acceleration:** In this motion, the acceleration curve consists of three segments as shown in Fig. 10.5(b). The

starting segment, $0 \le T \le T_a$, and the ending segment, $(1 - T_a) \le T \le 1$, are one quarter of the shorter sinusoidal waves, while the intermediate segment, $T_a \le T \le (1 - T_a)$, is a half of the longer sinusoidal wave.

4) **Modified constant velocity:** In this motion, the follower keeps a constant velocity the majority of the time, but modification is made to the starting and ending parts to smoothen out the acceleration by sinusoidal segments as shown in Fig. 10.5(c).

A comparison of the characteristic quantities for these motions are given in Table 10.1. In all the three motions given in Fig. 10.5, the accelerations consist of multiple segments. The displacement can be obtained by twice integrating the accelerations. Their mathematical equations are given in Appendix B.

2. **A generic combined harmonic motion**

It can be seen from Table 10.1 that all the four types of combined harmonic motions have a value of $p = 3$. Thus, they all have continuous accelerations, but discontinuous jerks. The discontinuous jerk is a limiting factor to any further increase of operation speed. It is expected that the operation speed can be increased if the jerk is made continuous.

Zhang et al. (1995) proposed a general combined harmonic motion having continuous jerk. This is shown in Fig. 10.6. The motion consists of 11 segments, each of which is numbered with a circled number. All segments with exception of ③, ⑥ and ⑨ are parts of harmonic

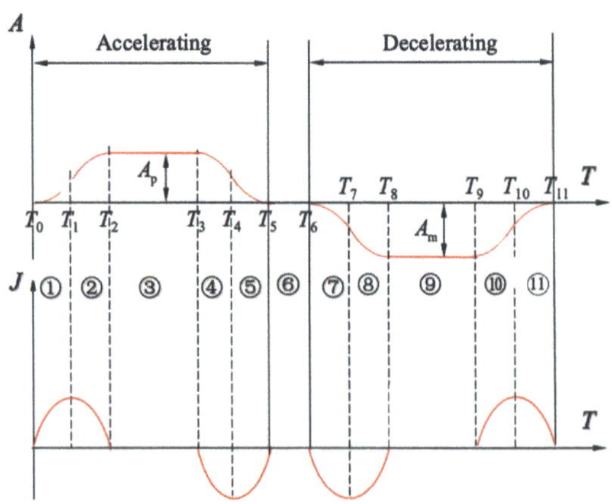

Figure 10.6 General combined harmonic motion.

curves. Segments ③ and ⑨ have constant accelerations and segment ⑥ has constant velocity, thus, zero acceleration. The nodes forming the segments are from T_1 through T_{11}. This combined motion has the following features:

1) **Several commonly used motions can be derived from it:** By choosing the values of T_i, several commonly used follower motions can be constructed. This is illustrated in Table 10.2.

Table 10.2 Motions derived from the general combined harmonic motion.

Motion	T_0	T_1	T_2	T_3	T_4	T_5	T_6	T_7	T_8	T_9	T_{10}	T_{11}
1	0	0	$\frac{1}{4}$	$\frac{1}{4}$	$\frac{1}{2}$	$\frac{1}{2}$	$\frac{1}{2}$	$\frac{1}{2}$	$\frac{3}{4}$	$\frac{3}{4}$	1	1
2	0	0	$\frac{1}{4}$	$\frac{1}{4}$	$\frac{1}{2}$	$\frac{1}{2}$	$\frac{1}{2}$	$\frac{1}{2}$	$\frac{3}{4}$	$\frac{3}{4}$	1	1
3	0	0	$\frac{1}{4}$	$\frac{1}{4}$	$\frac{1}{2}$	$\frac{1}{2}$	$\frac{1}{2}$	$\frac{1}{2}$	$\frac{3}{4}$	$\frac{3}{4}$	1	1
4	0	0	$\frac{1}{16}$	$\frac{1}{16}$	$\frac{1}{4}$	$\frac{1}{4}$	$\frac{3}{4}$	$\frac{3}{4}$	$\frac{15}{16}$	$\frac{15}{16}$	1	1
5	0	0	$\frac{1}{8}$	$\frac{3}{8}$	$\frac{1}{2}$	$\frac{1}{2}$	$\frac{1}{2}$	$\frac{1}{2}$	$\frac{5}{8}$	$\frac{7}{8}$	1	1
6	0	0	$\frac{1}{8}$	$\frac{1}{8}$	$\frac{1}{2}$	$\frac{1}{2}$	$\frac{1}{2}$	$\frac{1}{2}$	$\frac{7}{8}$	$\frac{7}{8}$	1	1
7	0	0	$\frac{1}{10}$	$\frac{3}{10}$	$\frac{2}{5}$	$\frac{2}{5}$	$\frac{2}{5}$	$\frac{2}{5}$	$\frac{11}{20}$	$\frac{17}{20}$	1	1
8	0	0	$\frac{1}{8}$	$\frac{3}{8}$	$\frac{1}{2}$	$\frac{1}{2}$	$\frac{1}{2}$	$\frac{1}{2}$	$\frac{5}{8}$	$\frac{7}{8}$	1	1

1–constant velocity; 2–sinusoidal; 3–cycloidal; 4–modified constant velocity; 5–modified trapezoidal; 6–modified sinusoidal; 7–skewed trapezoidal; 8–trapezoidal cycloidal.

2) **More complicated motions can be derived from it:** More complicated follower motions can be constructed based on it to meet some special requirements. For example, to reduce the residual vibration at dwells, one measure is to reduce the inertial impact during deceleration. To this end, a skewed acceleration curve with a shorter acceleration but longer deceleration can be constructed. If T_5 and T_6 are made less than 1/2, this requirement can be satisfied. For form-closed cam mechanisms, to suppress the crossover shock, the jerk at the crossover point should be zero or, at least, minimized. This requirement can be met by retaining the segments of ⑤ and ⑦.

   Table 10.3 gives the values of parameters for a skewed motion with continuous jerk (JUMT). In this motion the time ratio of the acceleration and deceleration is 4 : 6. It features zero jerk at the starting point, the ending point and the crossover point of the rising. Fig. 10.7 compares the residual vibration (maximum acceleration) of JUMT against that of two commonly used follower motions.

Table 10.3 JUMT motion.

T_0	T_1	T_2	T_3	T_4	T_5	T_6	T_7
0	0.05	0.14	0.22	0.33	0.40	0.40	0.50
T_8	T_9	T_{10}	T_{11}		V_m	A_m	J_{imp}
0.58	0.83	0.92	1		1.93	7.79	0.00

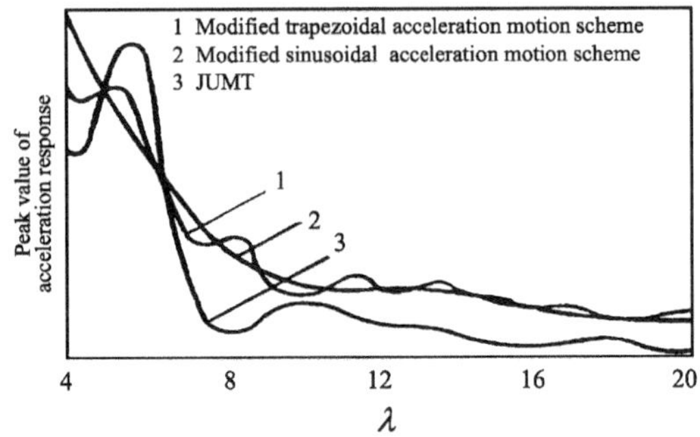

Figure 10.7 Residual vibration comparison.

In the following the derivation of this generic combined motion is presented. We start with the acceleration, which can be represented as

$$A_i = \begin{cases} A_p\,(\alpha_i + \beta_i \sin P_i) & (i = 1 - 6) \\ A_m\,(\alpha_i + \beta_i \sin P_i) & (i = 7 - 11), \end{cases} \tag{10.9}$$

where

$$P_i = \frac{T - T_{i-1}}{F_i} + \theta_i \tag{10.10}$$

$$F_i = \frac{2(T - T_{i-1})}{\pi}. \tag{10.11}$$

The values of α_i, β_i and θ_i are given in Table 10.4. The jerk is obtained by

Table 10.4 Parameters in the generic combined motion.

i	α_i	β_i	θ_i	i	α_i	β_i	θ_i
1	T_1/T_2	$-\alpha_1$	$\pi/2$	7	$-(T_7 - T_6)/(T_8 - T_6)$	$-\alpha_7$	$\pi/2$
2	α_1	$1 - \alpha_1$	0	8	α_7	$-1 - \alpha_7$	0
3	1	0	0	9	-1	0	0
4	$(T_5 - T_4)/(T_5 - T_3)$	$1 - \alpha_4$	$\pi/2$	10	$-(1 - T_{10})/(1 - T_9)$	$1 - \alpha_{10}$	$\pi/2$
5	α_4	$-\alpha_4$	0	11	α_{10}	$-\alpha_{10}$	0
6	0	0	0				

differentiating Eq. (10.9) as

$$J_i = \begin{cases} \dfrac{A_p \beta_i \cos P_i}{F_i} & (i = 1 - 6) \\ \dfrac{A_m \beta_i \cos P_i}{F_i} & (i = 7 - 11), \end{cases} \tag{10.12}$$

and the velocity and displacement are obtained by integration of Eq. (10.9) once and twice, respectively:

$$V_i = \begin{cases} C_i + A_p(\alpha_i T - \beta_i F_i \cos P_i) & (i = 1 - 6) \\ C_i + A_m(\alpha_i T - \beta_i F_i \cos P_i) & (i = 7 - 11) \end{cases} \tag{10.13}$$

$$S_i = \begin{cases} C_i T + D_i + A_p\left(\dfrac{1}{2}\alpha_i T^2 - \beta_i F_i^2 \sin P_i\right) & (i = 1 - 6) \\ C_i T + D_i + A_m\left(\dfrac{1}{2}\alpha_i T^2 - \beta_i F_i^2 \sin P_i\right) & (i = 7 - 11). \end{cases} \tag{10.14}$$

There are 24 constants in Eqs. (10.13) and (10.14), including C_i, D_i, A_p and A_m $(i = 1 - 11)$. They can be determined by using the boundary conditions and continuity conditions at intermediate nodes as follows:

1) $V_0 = 0$, $S_0 = 0$ at the left end.
2) $V_{11} = 0$, $S_{11} = 1$ at the right end.
3) $V_i = V_{i+1}$, $S_i = S_{i+1}$ at intermediate nodes.

From these conditions, this algebraic equation can be formed:

$$\boldsymbol{BX} = \boldsymbol{E}, \tag{10.15}$$

where $\boldsymbol{X}$ is the unknown vector:

$$\boldsymbol{X} = \begin{Bmatrix} C_1 & C_2 & \cdots & C_{11} & D_1 & D_2 & \cdots & D_{11} & A_p & A_m \end{Bmatrix}^{\mathrm{T}}. \tag{10.16}$$

$\boldsymbol{B}$ and $\boldsymbol{E}$ contain α_i, β_i and θ_i, which further depend on T_i. For a given T_i, $\boldsymbol{X}$ can be obtained through solving Eq. (10.15).

10.2.4.2 Polynomial motions

Polynomial motions are suitable for high-speed applications. The increase of operation speed of internal combustion engines was a major motivational force for the invention of polynomial motions. Polynomial functions are continuous, and their derivatives are also polynomials of lower order. It is easy to get a higher order of continuity (p) through designing the order of the polynomials. For example, if $p=3$ is desired, we have the following six boundary conditions.

- $S=V=A=0$ at $T=0$;
- $S=1$, $V=A=0$ at $T=1$.

Thus, these can be satisfied by the following polynomial firth order:

$$S=a_0+a_1T+a_2T^2+a_3T^3+a_4T^4+a_5T^5. \tag{10.17}$$

Differentiating this twice gives

$$V=a_1+2a_2T+3a_3T^2+4a_4T^3+5a_5T^4 \tag{10.18}$$
$$A=2a_2+6a_3T+12a_4T^2+20a_5T^3. \tag{10.19}$$

Applying the boundary conditions, the constants a_i can be obtained. Substituting them into Eq. (10.17) gives the polynomial.

$$S=10T^3-15T^4+6T^5. \tag{10.20}$$

Eq. (10.20) is often referred to as 3–4–5 polynomial motion. The V, A and J diagrams of this motion are shown as dashed line in Fig. 10.8.

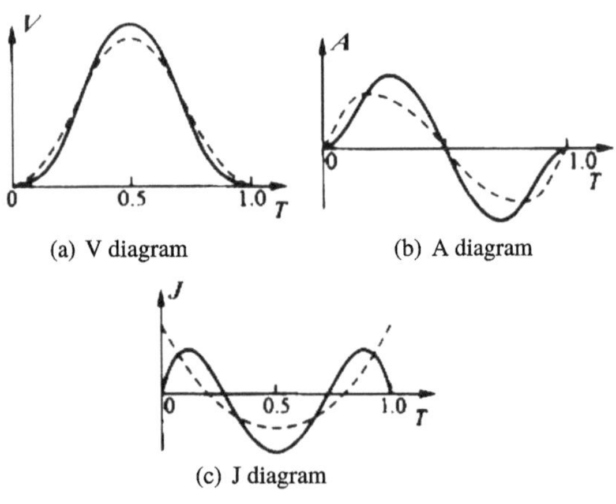

Figure 10.8 Two types of polynomials.

If $p=4$ is required, two boundary conditions are added in addition to that for $p=3$.

- $J=0$ at $T=0$.
- $J=0$ at $T=1$

In this case, eight boundary conditions are available; thus, a polynomial of seventh order can be designed as:

$$S = 35T^4 - 84T^5 + 70T^6 - 20T^7. \tag{10.21}$$

This is called the 4–5–6–7 polynomial motion. Its V, A and J are drawn in Fig. 10.8 as a solid line. It can be seen that this motion has continuous J, but higher magnitudes of A and V.

Theoretically, polynomials of any order can be designed. It is reported that in some special application a polynomial motion of 50th order was used. In addition to the boundary conditions, requirements at any intermediate moments can also be used to determine the polynomials and constants.

Example 10.1 (Design of a polynomial motion). A polynomial motion needs to be designed for a high-speed indexing cam mechanism. To minimize the residual vibration, it is required that: 1) A is continuous in the whole cycle; 2) the time ratio between acceleration and deceleration is 4 : 6; and 3) jerk J is zero at the end of indexing.

Solution. According to the first and third requirements, we have

$$S(0) = V(0) = A(0) \tag{a}$$

$$S(1) = 1 \tag{b}$$

$$V(1) = A(1) = J(1) = 0. \tag{c}$$

According to the second requirement, we have

$$A(0.4) = 0. \tag{d}$$

There are eight conditions in total to determine the polynomial constants; thus, we can assume a polynomial of seventh order:

$$S = a_0 + a_1T + a_2T^2 + a_3T^3 + a_4T^4 + a_5T^5 + a_6T^6 + a_7T^7. \tag{e}$$

From this we have V, A and J as follows through differentiation:

$$V = a_1 + 2a_2T + 3a_3T^2 + 4a_4T^3 + 5a_5T^4 + 6a_6T^5 + 7a_7T^6 \tag{f}$$

$$A = 2a_2 + 6a_3T + 12a_4T^2 + 20a_5T^3 + 30a_6T^4 + 42a_7T^5 \tag{g}$$

$$J = 6a_3 + 24a_4T + 60a_5T^2 + 120a_6T^3 + 210a_7T^4. \tag{h}$$

Substituting (a) into (f) through (h) gives

$$a_0 = a_1 = a_2 = 0. \tag{i}$$

Substituting (b), (c) and (i) into (f) through (h), we obtain the following algebraic equation set:

$$\begin{cases} a_3 + a_4 + a_5 + a_6 + a_7 = 1 \\ 3a_3 + 4a_4 + 5a_5 + 6a_6 + 7a_7 = 0 \\ 6a_3 + 12a_4 + 20a_5 + 30a_6 + 42a_7 = 0 \\ 6a_3 + 24a_4 + 60a_5 + 120a_6 + 210a_7 = 0 \\ 6a_3(0.4) + 12a_4(0.4)^2 + 20a_5(0.4)^3 + 30a_6(0.4)^4 + 42a_7(0.4)^5 = 0. \end{cases} \tag{j}$$

Solving this equation gives the constants a_i. The final form of the polynomial is then obtained as

$$S = 20T^3 - 45T^4 + 36T^5 - 10T^6. \tag{k}$$

The A and J diagrams are presented in Fig. 10.9. It is skewed with shorter

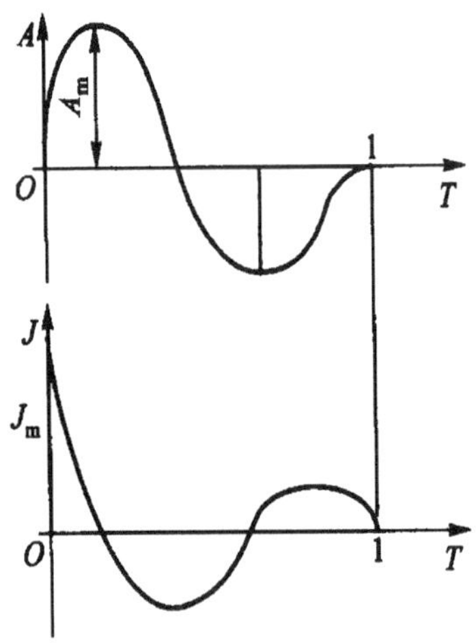

Figure 10.9 *A* and *J* diagrams.

acceleration and longer deceleration; the maximum acceleration A_m equals 8.1345. The longer deceleration makes the jerk J zero at the end; thus, it is expected to have less residual vibration.

10.3 Dynamic models of cam mechanisms

The first step to study dynamics is to develop a mathematic model. For cam mechanisms, lumped-parameter models are generally used. Components with significant flexibility but negligible mass are modeled as massless springs, while some others with significant inertia but negligible flexibility are modeled as lumped masses. If neither flexibility nor inertia is negligible, equivalent masses and equivalent springs are found through the principle of unchanged elastic potential energy and unchanged kinetic energy. In the following, a cam mechanism used in an internal combustion engine to activate the poppet valve is used to illustrate the procedures of dynamic model development.

10.3.1 Dynamic model

If the camshaft is included in the model, the whole system is divided into two subsystems, namely the cam-pushrod subsystem and the camshaft-cam subsystem.

10.3.1.1 Cam-pushrod subsystem

First, we obtain the lumped masses through the following steps:

1) **The pushrod** AB**:** The pushrod AB is replaced with two lumped masses at ends A and B based on the principle of an unchanged location of center of mass. Given that it is uniform, we have

$$m_{A2} = m_{B2} = \frac{m_2}{2}. \tag{10.22}$$

2) **The rocker** BC**:** The motion of B and C can be treated as vertical straight lines because the rotation angle of the rocker BC around O_2 is very small. We can split its mass into B and C by keeping the moment of inertia unchanged. Thus, we have

$$\begin{cases} m_{B3} = \dfrac{J_{31}}{a^2} \\ m_{C3} = \dfrac{J_{32}}{b^2}, \end{cases} \tag{10.23}$$

where J_{31} and J_{32} are the moments of inertia of the left and right parts of the rocker to point O_2. a and b are the dimensions as shown.

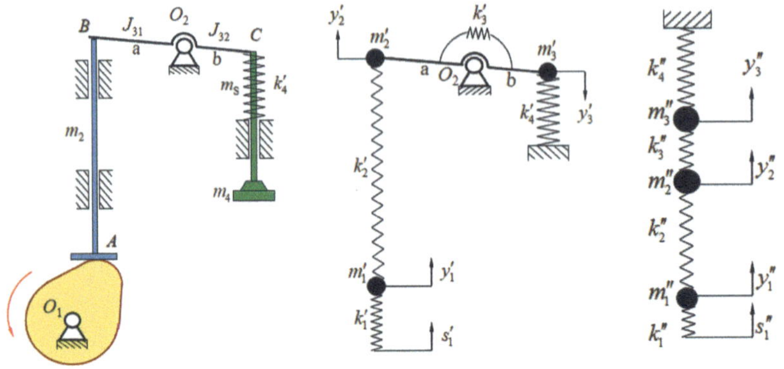

(a) Cam mechanism (b) Model I (c) Model II

Figure 10.10 Cam mechanism in internal combustion engines.

3) **The valve:** It is relatively stiff, so we neglect its flexibility and treat it as a lumped mass at C so:

$$m_{C4} = m_4 + \frac{m_S}{3}, \tag{10.24}$$

where m_4 is the mass of the valve and m_S is the mass of the spring. This is added to the valve mass according to the result in Chapter 5 (see Eq. (5.16) on p. 117).

After these steps, we obtain model I shown in Fig. 10.10(b). The parameters in this figure are:

$$\begin{cases} m_1' = m_{A2} \\ m_2' = m_{B2} + m_{B3} \\ m_3' = m_{C3} + m_{C4}. \end{cases} \tag{10.25}$$

k_1' is the contact stiffness between the pushrod AB and the cam; k_2' is the equivalent stiffness of pushrod AB in axial direction; k_3' is the bending stiffness of rocker BC; and k_4' is the stiffness of the spring. s' is the displacement of the cam motion curve.

To obtain a meaningful model, we need further work on model I by transforming the entire right part to the left. This involves the following transformation:

$$s'' = s', \quad y_1'' = y_1', \quad y_2'' = y_2', \quad y_3'' = (a/b)y_3' \tag{10.26}$$

$$k_1'' = k_1', \quad k_2'' = k_2', \quad k_3'' = k_3', \quad k_4'' = (b/a)^2 k_4' \tag{10.27}$$

$$m_1'' = m_1', \quad m_2'' = m_2', \quad m_3'' = (b/a)^2 m_3'. \tag{10.28}$$

After all these steps, the model II in Fig. 10.10(c) is obtained.

10.3.1.2 Camshaft–cam subsystem

This subsystem is as shown in Fig. 10.11(a). J_1' and J_2' are calculated by

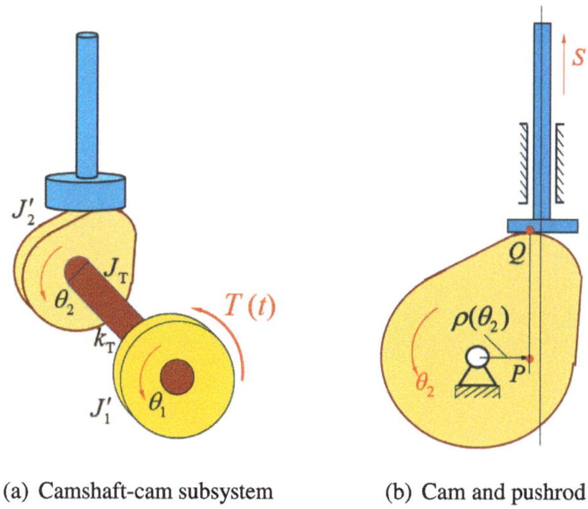

Figure 10.11 Camshaft–cam subsystem.

$$\begin{cases} J_1' = J_1 + J_{T1} \\ J_2' = J_2 + J_{T2}, \end{cases} \tag{10.29}$$

where J_1 and J_2 are the moments of inertia of the driven gear and the cam. J_{T1} and J_{T2} are the equivalent moments of inertia transformed to the gear and the cam, respectively, from the shaft.

Now, we need to transform them into the pushrod to obtain the equivalent masses and stiffness. First, the relationship between the cam rotation and the followers displacement is

$$ds = \rho(\theta_2)d\theta_2. \tag{10.30}$$

$\rho(\theta_2)$ is the distance from the center of rotation of the cam to the instant center of velocity P as shown in Fig. 10.11(b).

Torque T can be transformed into an equivalent force F_e:

$$F_e = \frac{T}{\rho(\theta_2)}. \tag{10.31}$$

Rotation angles θ_1 and θ_2 are transformed into equivalent linear displacements as:

$$\begin{cases} y_{\theta_1} = \theta_1 \rho(\theta_2) \\ y_{\theta_2} = \theta_2 \rho(\theta_2) = s. \end{cases} \tag{10.32}$$

Moments of inertia J_1 and J_2 are transformed into equivalent masses:

$$\begin{cases} m_{J1} = \dfrac{J_1}{[\rho(\theta_2)]^2} \\ m_{J2} = \dfrac{J_2}{[\rho(\theta_2)]^2}. \end{cases} \tag{10.33}$$

After these steps, we obtain a model shown in Fig. 10.12, which has five lumped masses, thus, five DOFs.

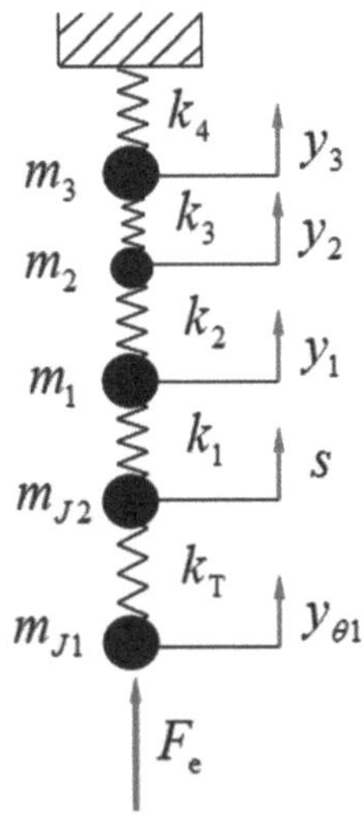

Figure 10.12 Dynamic model of cam system.

10.3.1.3 Governing equations

With the model shown in Fig. 10.12, it is straightforward to derive the undamped vibration equation as:

$$\boldsymbol{M}\ddot{\boldsymbol{U}} + \boldsymbol{K}\boldsymbol{U} = \boldsymbol{F}, \tag{10.34}$$

where

$$\boldsymbol{U} = \left\{ y_{\theta_1} \quad s \quad y_1 \quad y_2 \quad y_3 \right\}^{\mathrm{T}} \tag{10.35}$$

$$\boldsymbol{F} = \left\{ F_e \quad 0 \quad 0 \quad 0 \quad 0 \right\}^{\mathrm{T}}. \tag{10.36}$$

$\boldsymbol{M}$ and $\boldsymbol{K}$ are 5 × 5 mass-and-stiffness matrices. They vary with the rotation angle θ_2. Thus, Eq. (10.34) is a parametric vibration system.

Obviously, the process just described and the model are pretty awkward and complicated. It is justified only when requiring high analytical accuracy. If accuracy is not a critical concern, the model can be greatly simplified. In the following, we will discuss the procedures for model reduction.

10.3.2 Reduction of model

If the camshaft has very large stiffness, it can be removed from the vibration model. A model with three DOFs as shown in Fig. 10.10(c) can then be reached. In addition to a reduction of DOF, this model also removes the time-varying parameter from the equation, making the solution and analysis much more straightforward. If needed, the model can be further reduced to a single/DOF model, which is required in some cases.

According to the modal superposition principle introduced in Chapter 6, the response of a multiple DOF system can be calculated by summing up the contribution from each principal vibration mode. More often, modal truncation is used by summing the contribution from only the several lower natural-frequency vibration modes. In some cases, only the primary vibration mode can be used, as long as the primary and the secondary natural frequencies are well separated. In a real cam mechanism, the excitation frequency is generally well below the primary natural frequency; thus, the primary vibration mode makes the most contribution to the response. In the following we will present the procedure to construct an equivalent dynamic model of a cam mechanism having a single DOF.

This equivalent single-DOF model is constructed with the following steps:

1) **Natural frequency:** The natural frequency of the equivalent model equals the primary natural frequency of the original model. $\omega_e = \omega_1$.

2) **Equivalent stiffness:** The equivalent stiffness k_e of the model in Fig. 10.10(c) is calculated by

$$\frac{1}{k_e} = \frac{1}{k_1''} + \frac{1}{k_2''} + \frac{1}{k_3''}. \tag{10.37}$$

3) **Equivalent mass:** The equivalent mass m_e can be calculated as:

$$m_e = \frac{k_e}{\omega_e^2} = \frac{k_e}{\omega_1^2}. \tag{10.38}$$

It should be pointed out that a single-DOF model can also be directly derived from the physical system. The details of the direct method can be found in Rothbart (2004).

10.4 Dynamic analysis of cam mechanisms

Vibration inflicts a series of harmful effects on cam mechanisms, such as deterioration of motion accuracy, wear, and fatigue, as well as the radiation of noise. Dynamic analysis is the basic technique to improve the performance of mechanisms.

For simplicity, we analyze only single-DOF systems in the following.

10.4.1 Analysis of single-DOF systems

A equivalent single DOF model, shown in Fig. 10.13, is obtained after the procedures in the previous section. For the sake of brevity, the subscript e is already dropped off. k_s is the stiffness of the closure spring; y and s are the displacements of the follower's top end and bottom end (contact with the cam), respectively. The free-body diagram (FBD) of m is shown in Fig. 10.13(b), in which F_P is the preload of the closure spring and G is the external load. According to Newton's second law of motion, we have

$$m\ddot{y} = k(s-y) - k_s y - F_P - G. \tag{10.39}$$

F_P and G only cause static deformation; thus, they can be dropped off the vibration equation. In addition, k_s can be neglected because generally $k_s << k$. With these factors considered, Eq. (10.39) is reduced to

$$m\ddot{y} + ky = ks. \tag{10.40}$$

Clearly the cam profile s is a displacement excitation to the system. If the follower is very stiff, the inertial load, $m\ddot{y}$, can be neglected. Then we have

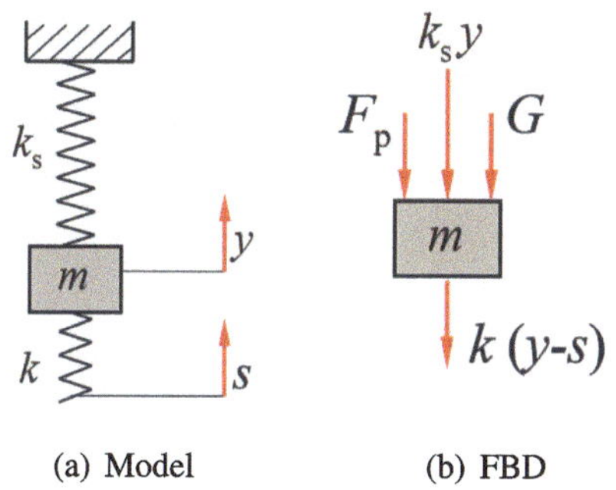

Figure 10.13 Single-DOF model.

$y = s$. This is the rigid-body case covered in the undergraduate mechanism textbooks. However, if the elasticity is considered, y will be different from s.

In the following, the analysis is focused on the general dynamic performance of various cam-follower motions without dealing with specific mechanisms. To this end, we first nondimensionalize the variables as

$$Y = \frac{y}{h} \tag{10.41}$$

$$T = \frac{t}{t_h} \tag{10.42}$$

$$S = \frac{s}{h}, \tag{10.43}$$

where h and t_h are the total rise displacement and the time of rise, respectively. With these dimensionless quantities, we have the following by using chain rule of differentiation:

$$\ddot{y} = \frac{d^2y}{dt^2} = \frac{h}{t_h^2} \cdot \frac{d^2Y}{dT^2} = \frac{h}{t_h^2} Y''. \tag{10.44}$$

Here $'$ and $''$ represent the first- and second-order derivatives with respect to the dimensionless time T. Substituting the previous equation into Eq. (10.40) gives

$$Y'' + \omega_n^2 t_h^2 Y = \omega_n^2 t_h^2 S, \tag{10.45}$$

where ω_n is the natural frequency calculated by

$$\omega_n = \sqrt{\frac{k}{m}}. \tag{10.46}$$

Recall that we introduced a dimensionless λ in Eq. (10.1). λ plays an important role in dynamics of a cam mechanism. Then, we rewrite Eq. (10.45) in the following form:

$$Y'' + (2\pi\lambda)^2 Y = (2\pi\lambda)^2 S. \tag{10.47}$$

This is the dimensionless equation we are going to analyze in the following. We will analyze and compare the response to three followers' motions, including: 1) cycloidal motion, 2) constant acceleration, and 3) the generic combined harmonic motion.

10.4.2 Analysis of cycloidal motion

The dimensionless form of cycloidal motion is

$$S = T - \frac{1}{2\pi}\sin(2\pi T). \tag{10.48}$$

Substituting Eq. (10.48) into Eq. (10.47) gives

$$Y'' + (2\pi\lambda)^2 Y = (2\pi\lambda)^2\left[T - \frac{1}{2\pi}\sin(2\pi T)\right]. \tag{10.49}$$

Its solution is in the following form:

$$Y(T) = C_1\sin(2\pi\lambda T) + C_2\cos(2\pi\lambda T) + \left[T - \frac{1}{2\pi}\frac{1}{\lambda^2 - 1}\sin(2\pi T)\right]. \tag{10.50}$$

C_1 and C_2 are to be determined from the initial conditions. If the cam is initially at rest, we have zero conditions, $Y(0) = 0,\ Y'(0) = 0$. Solving for C_1 and C_2 and substituting back into Eq. (10.50) gives the response as

$$Y(T) = T - \frac{1}{2\pi}\frac{1}{\lambda^2 - 1}\left[\lambda^2\sin(2\pi T) - \frac{1}{\lambda}\sin(2\pi\lambda T)\right]. \tag{10.51}$$

The velocity and acceleration can be obtained by taking the derivative with respect to the dimensionless time T once and twice, respectively:

$$Y'(T) = 1 - \frac{1}{\lambda^2 - 1}\left[\lambda^2\cos(2\pi T) - \cos(2\pi\lambda T)\right] \tag{10.52}$$

$$Y''(T) = 2\pi\frac{\lambda}{\lambda^2 - 1}\left[\lambda\sin(2\pi T) - \sin(2\pi\lambda T)\right]. \tag{10.53}$$

Eqs. (10.51) through (10.53) are the forced vibration responses to the excitation of $S(T)$ in the rising period $0 \le t \le t_h$ or $0 \le T \le 1$. The vibration in the rising stage is referred to as the *principal vibration*. $Y(T)$, $Y'(T)$ and $Y''(T)$ represent the actual displacement, velocity and acceleration of the upper end of the follower, respectively.

The rise ends at $t = t_h$, and the excitation $S(T)$ disappears. Thus, the vibration equation of Eq. (10.40) is reduced to

$$m\ddot{y}_r + ky_r = 0. \tag{10.54}$$

Its dimensionless form is derived from Eq. (10.47):

$$\ddot{Y}_r + (2\pi\lambda)^2 Y_r = 0. \tag{10.55}$$

This is free vibration because of no external excitation. Given that this vibration happens after the principal vibration of the rise stage, it is called *residual vibration*. To distinguish it from the principal vibration, subscript *r* is already used in Eq. (10.54) and (10.55). Thus, we have the following relationships:

$$y_r = y - h \tag{10.56}$$

$$Y_r = Y - 1. \tag{10.57}$$

The general solution of Eq. (10.55) is

$$Y_r = C_1 \sin(2\pi\lambda T) + C_2 \cos(2\pi\lambda T). \tag{10.58}$$

C_1 and C_2 are determined by the initial conditions. These are obtained through the following steps: 1) $Y(1)$ and $Y'(1)$ are calculated by Eq. (10.51) and (10.52), which are the displacement and velocity at $T = 1$, and 2) from Eq. (10.57), we know that $Y_r(0) = Y(1) - 1$; $Y_r'(0) = Y'(1)$.

Once C_1 and C_2 are determined, the response is obtained as follows:

$$Y_r(T) = \frac{1}{\pi}\frac{1}{\lambda^3 - \lambda}\sin(\pi\lambda)\cos\left[\pi\lambda(2T - 1)\right]. \tag{10.59}$$

The acceleration is obtained through differentiating twice with respect to T.

$$Y_r''(T) = -\frac{4\pi\lambda}{\lambda^2 - 1}\sin(\pi\lambda)\cos\left[\pi\lambda(2T - 1)\right]. \tag{10.60}$$

For cam mechanisms with CPM constraints, $\delta = y - s$ is the dynamic error between the theoretical and the actual displacements. In most automatic machines, the cam mechanisms are of the category of CEP constraints, and

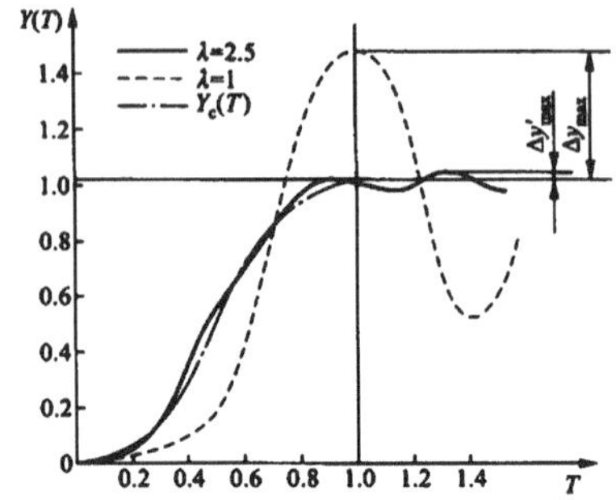

Figure 10.14 Displacement response of cycloidal motion.

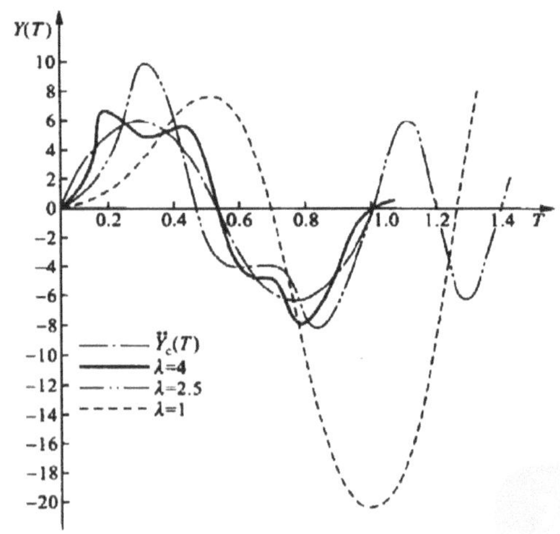

Figure 10.15 Acceleration response of cycloidal motion.

the error at the ends are of more interest. So $(y_r)_{max}$ represents the dynamic accuracy.

Fig. 10.14 shows the responses of $Y(T)$ and $Y_r(T)$, while Fig. 10.15 gives the response of $Y''(T)$ and $Y_r''(T)$. $T = 1$ is the dividing point between the principal vibration and residual vibration responses.

10.4.3 Analysis of constant acceleration motion

The constant acceleration motion is shown in Fig. 10.3. Its dimensionless displacements in the acceleration and deceleration periods are

$$S_1 = 2T^2 \qquad \text{acceleration} \qquad (10.61)$$

$$S_2 = -2T^2 + 4T - 1 \qquad \text{deceleration.} \qquad (10.62)$$

Substituting Eq. (10.61) into Eq. (10.47) gives the equation in the acceleration period:

$$\ddot{Y}_1 + (2\pi\lambda)^2 Y_1 = (2\pi\lambda)^2(2T^2). \tag{10.63}$$

The complete solution of this equation is

$$Y_1 = C_1\cos(2\pi\lambda T) + C_2\sin(2\pi\lambda T) + 2T^2 - \frac{1}{(\pi\lambda)^2}, \tag{10.64}$$

where C_1 and C_2 are constants.

Taking advantage of the initial conditions $Y_1(0) = 0$; $Y_1'(0) = 0$, C_1 and C_2 can be determined. Then the response is represented as

$$Y_1 = \frac{4}{(2\pi\lambda)^2}\cos(2\pi\lambda T) + 2T^2 - \frac{1}{(\pi\lambda)^2}. \tag{10.65}$$

For the deceleration period, the governing equation can be obtained similarly by substituting Eq. (10.62) into Eq. (10.47):

$$\ddot{Y}_2 + (2\pi\lambda)^2 Y_2 = (2\pi\lambda)^2(-2T^2 + 4T - 1). \tag{10.66}$$

Its complete solution is

$$Y_2 = C_3\cos(2\pi\lambda T) + C_4\sin(2\pi\lambda T) - 2T^2 + 4T - 1 + \frac{1}{(\pi\lambda)^2}. \tag{10.67}$$

C_3 and C_4 are constants that are determined through the continuity condition of the crossover point, $Y(1/2) = Y_2(1/2)$; $Y_1'(1/2) = Y_2'(1/2)$. Then the response is represented as

$$\begin{aligned} Y_2 = {} & \frac{1 - 2\cos\pi\lambda}{(\pi\lambda)^2}\cos(2\pi\lambda T) - \frac{2\sin\pi\lambda}{(\pi\lambda)^2}\sin(2\pi\lambda T) \\ & - 2T^2 + 4T - 1 + \frac{1}{(\pi\lambda)^2}. \end{aligned} \tag{10.68}$$

The velocity and acceleration are obtained through differentiating the displacements with respect to T, as

$$Y_1'(T) = \frac{-2}{\pi\lambda}\sin(2\pi\lambda T) + 4T \tag{10.69}$$

$$Y_1''(T) = -4\cos(2\pi\lambda T) + 4 \tag{10.70}$$

and

$$Y_2'(T) = \frac{2(2\cos\pi\lambda - 1)}{\pi\lambda}\sin(2\pi\lambda T) - \frac{4\sin(\pi\lambda - 1)}{\pi\lambda}\cos(2\pi\lambda T) - 4T + 4 \tag{10.71}$$

$$Y_2''(T) = 4[2\cos(\pi\lambda) - 1]\cos(2\pi\lambda T) + 8\sin(\pi\lambda)\sin(2\pi\lambda T) - 4. \tag{10.72}$$

The equation for the residual vibration and its solution are the same with Eq. (10.55). For convenience, we repeat it here:

$$Y_r = C_5\cos(2\pi\lambda T) + C_6\sin(2\pi\lambda T). \tag{10.73}$$

C_5 and C_6 are to be determined through the initial conditions, $Y_r(0) = Y_2(1) - 1$, $Y_r' = Y_2'(1)$. They are

$$C_5 = \frac{4[1 - 2\cos(\pi\lambda)]\cos(2\pi\lambda)}{(2\pi\lambda)^2} - \frac{8\sin(\pi\lambda)\sin(2\pi\lambda)}{(2\pi\lambda)^2} + \frac{4}{(2\pi\lambda)^2} \tag{10.74}$$

$$C_6 = \frac{4[2\cos(\pi\lambda) - 1]\sin(2\pi\lambda)}{(2\pi\lambda)^2} - \frac{8\sin(\pi\lambda)\cos(2\pi\lambda)}{(2\pi\lambda)^2} + \frac{4}{(2\pi\lambda)^2}. \tag{10.75}$$

Using the trigonometric identities, they can be shortened to

$$C_5 = \frac{8\cos(\pi\lambda)[\cos(\pi\lambda) - 1]}{(2\pi\lambda)^2} \tag{10.76}$$

$$C_6 = \frac{8\sin(\pi\lambda)[1 - \cos(\pi\lambda)]}{(2\pi\lambda)^2}. \tag{10.77}$$

The residual acceleration response is then:

$$Y_r'' = -C_5(2\pi\lambda)^2\cos(2\pi\lambda T) - C_6(2\pi\lambda)^2\sin(2\pi\lambda T). \tag{10.78}$$

10.4.4 Analysis of generic combined harmonic motion

As stated earlier, a few commonly used motions can be derived from the generic combined harmonic motion. Thus, the dynamics of analysis of this motion is of more interest than analysis of individual motions. For convenience, we introduce viscous damping into Eq. (10.45):

$$\ddot{Y} + 2\omega_n t_h \xi Y' + \omega_n^2 t_h^2 Y = \omega_n^2 t_h^2 S, \tag{10.79}$$

where ξ is the damping ratio. Substituting the S in Eq. (10.17) into the previous equation, we have its particular solution as:

$$Y_i^* = G_i\cos P_i + H_i\sin P_i + L_i T^2 + M_i T + N_i \quad (i = 1, 2, \cdots, 11). \tag{10.80}$$

The coefficients are represented for the segments as:

$$\begin{cases} H_i = \dfrac{F_i^4 (t_\mathrm{h}\omega_\mathrm{n})^2 \left[1-(F_i t_\mathrm{h}\omega_\mathrm{n})^2\right]}{\left[1-(F_i t_\mathrm{h}\omega_\mathrm{n})^2\right] + 0.5\xi\, t_\mathrm{h}\omega_\mathrm{n} F_i^2} \left(A_\mathrm{p}\beta_i + A_\mathrm{m}\mu_i\right) \\ G_i = \dfrac{0.5\xi\, t_\mathrm{h}\omega_\mathrm{n} F_i^2}{1-(F_i t_\mathrm{h}\omega_\mathrm{n})^2} H_i \\ L_i = 0.5\left(A_\mathrm{p}\alpha_i + A_\mathrm{m}\mu_i\right) \\ M_i = C_i - \dfrac{t_\mathrm{h}\xi}{t_\mathrm{h}^2\omega_\mathrm{n}} L_i \\ N_i = D_i - \dfrac{2L_i + 0.5 t_\mathrm{h}\xi M_i}{(t_\mathrm{h}\omega_\mathrm{n})^2}, \end{cases} \tag{10.81}$$

where F_i, P_i, C_i, D_i, α_i, β_i, A_p, and A_m are already given in Section 10.2. Based on these, the response of this combined harmonic motion can be represented as

$$\begin{aligned} Y_i = {} & e^{-t_\mathrm{h}\xi T}\left[e_i\cos(qT) + f_i\sin(qT)\right] + G_i\cos P_i + H_i\sin P_i \\ & + L_i T^2 + M_i T + N_i \quad (i = 1, 2, \cdots, 11), \end{aligned} \tag{10.82}$$

where

$$q = t_\mathrm{h}\sqrt{\omega_\mathrm{n}^2 - \xi^2}. \tag{10.83}$$

e_i and f_i are determined by the boundary conditions and the continuity conditions at the intermediate nodes, $Y_i = Y_{i+1}$, $Y_i' = Y_{i+1}'$.

Differentiating Eq. (10.82) with respect to T once and twice gives the dimensionless velocity and acceleration, respectively.

The equation of the residual vibration is then:

$$\ddot{Y}_\mathrm{r} + 2\omega_\mathrm{n} t_\mathrm{h}\xi Y_\mathrm{r}' + \omega_\mathrm{n}^2 t_\mathrm{h}^2 Y_\mathrm{r} = 0. \tag{10.84}$$

Its general solution is:

$$Y_\mathrm{r} = e^{-t_\mathrm{h}\xi T}\left[e_\mathrm{r}\cos(qT) + f_\mathrm{r}\sin(qT)\right], \tag{10.85}$$

where e_r and f_r are determined by the initial conditions.

10.4.5 Effect of λ and dynamic response spectrum

It is clear from Figs. 10.14 and 10.15 that the period ratio λ has a significant effect on the dynamic response. The effect can be studied by the *dynamic response spectrum* (DRS).

For example, the residual acceleration response of the constant acceleration motion is given by Eq. (10.78). It is a sinusoidal wave with an amplitude of

$$(Y_r'')_{\max} = (2\pi\lambda)^2\sqrt{C_5^2 + C_6^2} = 8[1 - \cos(\pi\lambda)]. \qquad (10.86)$$

Clearly, when λ is an odd integer, $(Y_r'')_{\max} = 16$; when λ is an even integer, $(Y_r'')_{\max} = 0$. The variation of $(Y_r'')_{\max}$ with λ is shown as the dashed line in Fig. 10.16(a). Similarly, we can draw the amplitude of the principal ac-

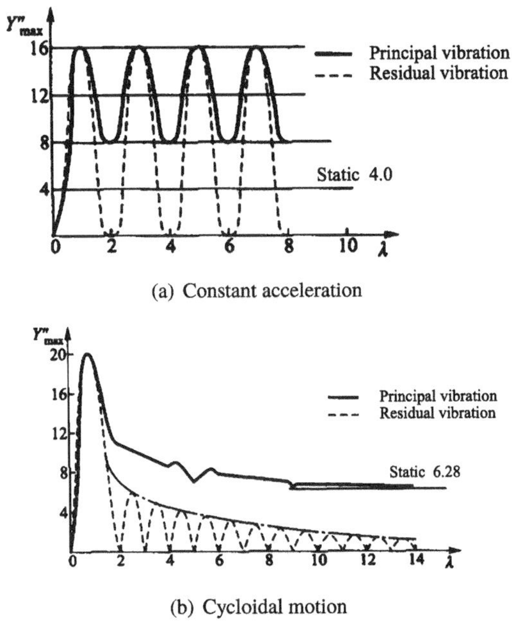

Figure 10.16 Dynamic response spectrum.

celeration response against λ, which is shown as a solid line. The principal and residual acceleration responses for the cycloidal motion are given in Fig. 10.16(b).

Fig. 10.16 depicts the relationship between the acceleration amplitude and the period ratio λ; thus, they are referred to as the *acceleration spectrum*. In a similar way, we can have the *displacement spectrum* and *velocity spectrum* if needed. From Fig. 10.16 we clearly see that the constant velocity motion is not good in terms of dynamic acceleration response.

When λ is at some specific values, $(Y_r'')_{\max}$ may be zero, meaning that the free vibration is not excited. However, this has very limited practical significance because it is almost impossible to make λ meet the requirement due to the inevitable uncertainties. In practice the envelope of the DRS, e.g., the dash-dotted line in Fig. 10.16(b), is more meaningful. The DRS

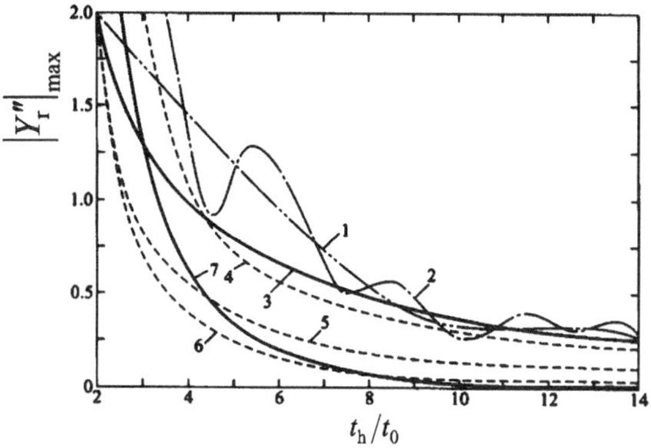

1: modified sinusoida–l; 2: modified trapezoidal; 3: 3–4–5 polynomial; 4: cycloidal; 5: 3–4–5–6–7 polynomial; 6: 4–5–6–7 polynomial; 7: polynomial of 11 degrees

Figure 10.17 Envelops of residual vibration spectra.

envelope of residual vibration for several commonly used follower motions are given in Fig. 10.17 in which 1, 2, 3 and 4 have continuous acceleration, but discontinuous jerk ($p = 3$); 5 and 6 have continuous jerk ($p = 4$); 7 has higher-order continuity. It can be seen that, in general, follower motions with higher p values have lower residual vibration when λ is relatively large. It is also noted that motions 6 and 7, although having continuous jerk, have even higher response when λ takes a low value. These phenomena simply reflect the inadequacy of the characteristic quantities and the general rules in the selection of follower motion for high-speed applications. Thus, thorough analysis of dynamics is necessary for high-speed applications.

10.5 Dynamic design of cam mechanisms

10.5.1 Introduction to design of high-speed cam mechanisms

From the analysis in previous section, we know that the actual displacement of the follower y is different from the theoretical follower motion s. At high speed, this deviation may be significant, posing a challenge to the design of high-speed cam mechanisms. Generally, two routes are commonly taken in designing high-speed cam mechanisms.

1) **Static design, dynamic verification:** This method chooses the follower motion from a kinematics point of view. Then dynamic analysis is conducted to verify if the chosen motion meets the dynamic requirement.
2) **Dynamic design:** This method establishes a dynamic model first. Then dynamic synthesis is conducted in which the follower motion s is derived from y.

The period ratio, λ, has a significant effect on dynamics. As a rule of thumb, cam design follows the following guidelines.

- When $6 < \lambda < 15$, route 1) can be followed;
- When $\lambda < 6$, route 2) is followed.

Many methods have been developed for the dynamic design of cam mechanisms in the past half century. In this section, one method based on the polydyne cams is introduced.

10.5.2 Polydyne cams

A dynamic design method, *polydyne cams*, was proposed by Stoddart (1953). This method is based on a polynomial to find the cam profile input s with a desired actual follower motion y.

10.5.2.1 Rationale

In Section 10.3, the single-DOF dynamic model of a cam mechanism, Eq. (10.39), is presented. We rearrange it into the following form:

$$m\ddot{y} + (k + k_s)y = ks - F_P - G. \tag{10.87}$$

In dynamic analysis, we solve this equation to find y with a given s.

In dynamic synthesis, however, we will use this equation the other way around. For a given y we find the s.

10.5.2.2 Polynomial dynamic cam curve

We first recast Eq. (10.87) into

$$s = \frac{F_P + G}{k} + \frac{k + k_s}{k}y + \frac{m}{k}\frac{d^2 y}{dt^2}. \tag{10.88}$$

Given that the follower motion can be given as a function of the cam rotation angle φ, we make the following variable transformation:

$$\begin{cases} \dfrac{dy}{d\varphi} = \dfrac{dy}{\omega dt} = \dfrac{1}{\omega}\dfrac{dy}{dt} \\ \dfrac{d^2y}{d^2\varphi} = \dfrac{1}{\omega^2}\dfrac{d^2y}{d^2t}. \end{cases} \tag{10.89}$$

Substituting this into Eq. (10.88), we have

$$s(\varphi) = \frac{F_P + G}{k} + \frac{k + k_s}{k}y + \frac{m\omega^2}{k}\frac{d^2y}{d\varphi^2}. \tag{10.90}$$

Differentiating it twice with respective to φ gives

$$\frac{d^2s}{d^2\varphi} = \frac{k + k_s}{k}\frac{d^2y}{d\varphi^2} + \frac{m\omega^2}{k}\frac{d^4y}{d\varphi^4}. \tag{10.91}$$

$d^2s/d^2\varphi$ is the curvature of the cam profile and also an indicator of the input acceleration. For high-speed cams, the smoothness of the cam profile is critical, requiring continuous $d^2s/d^2\varphi$. In the other words, y is required to be continuous up to the order of four.

We assume the output y is a polynomial as;

$$y(\tau) = h(a_0 + a_1\tau + a_2\tau^2 + \cdots + a_n\tau^n), \tag{10.92}$$

where a_i $(i = 1, \cdots, n)$ are constants. τ is the dimensionless rotation angle.

$$\tau = \frac{\varphi}{\varphi_0} = T \tag{10.93}$$

φ_0 is the rotation angle of rising. T is the dimensionless time. a_i can be determined by the boundary conditions. To make y continuous up to the order of four, we have ten boundary conditions available:

$$\begin{cases} y(0) = 0 \\ \dfrac{dy(0)}{d\tau} = \dfrac{d^2y(0)}{d\tau^2} = \dfrac{d^3y(0)}{d\tau^3} = \dfrac{d^4y(0)}{d\tau^4} = 0 \\ y(1) = h \\ \dfrac{dy(1)}{d\tau} = \dfrac{d^2y(1)}{d\tau^2} = \dfrac{d^3y(1)}{d\tau^3} = \dfrac{d^4y(1)}{d\tau^4} = 0. \end{cases} \tag{10.94}$$

With the ten boundary conditions, the ten constants, a_i ($i = 0, 1, \cdots, 9$), can be determined. Substituting the constants into Eq. (10.92) gives

$$y(\tau) = h(126\tau^5 - 420\tau^6 + 540\tau^7 - 135\tau^8 + 70\tau^9). \tag{10.95}$$

This is referred to as the 5–6–7–8–9 polynomial motion. It guarantees no impact at the input because of the fourth order continuity of y.

If the boundary conditions, $\tau = 0$ and 1, $d^4y/d\tau^4 = 0$, are removed, a 4–5–6–7 polynomial will be obtained that has tiny impacts at $\tau = 0$ and 1.

With $y(\tau)$ known, the required $s(\varphi)$ can be obtained by Eqs. (10.89) and (10.90).

10.5.2.3 Discussion

The previous method was proposed by Stoddart (1953); it requires that the equivalent mass m, equivalent stiffness k, and rotation speed ω are known in advance. This means that the expected dynamic performance can only be accomplished at the specific values of m, k and ω. On the other hand, it is not easy to estimate m and k accurately until the completion of the detailed design. This dilemma is the main drawback of this method. Experience indicates that cams designed with this method are very sensitive to the variation of parameters. It is reported that a 10% change of ω can significantly worsen the dynamic performance.

Although the sensitivity largely limits the application of the method, its real significance lies in the breakthrough in design philosophy, changing from static design to dynamic design. Later, Tesar and Matthew (1976) made effort to improve the robustness of this method. The basic idea is summarized here:

1) For a given ω, find the required s from y using Stoddart's method.
2) To analyze the parametric sensitivity of dynamic performance to speed variation, find the s having better performance in a specific speed range.

10.5.3 General rules for design of high-seed cams

Although the performance of high-speed cam mechanisms is affected by many factors, some general rules can be followed in design:

1) Stiffer and lighter followers lead to higher primary natural frequency; thus, higher λ. In view of this, if permitted, higher k and smaller m are always desirable.

2) The manufacturing quality of the cam profile is critical for the dynamics. Cams expected to work at high speed should be machined on CNC machine-tools to guarantee a high quality of profile surface.
3) Attention should be paid to balancing so that centrifugal excitation is minimized.

In recent years, some researchers refined the dynamic model by getting rid of the constant rotation speed (Yan et al., 1996).

10.6 High-speed indexing cam mechanisms

In many automatic machines, such as packaging machines, printing machines and CNC machine tools, continuous rotating input needs to be transformed into intermittent output. Ratchet and Geneva mechanisms are conventional solutions to this type of applications. *Indexing cams*, however, are gaining more and more of the market share. Two applications of indexing cams are given in Fig. 10.18.

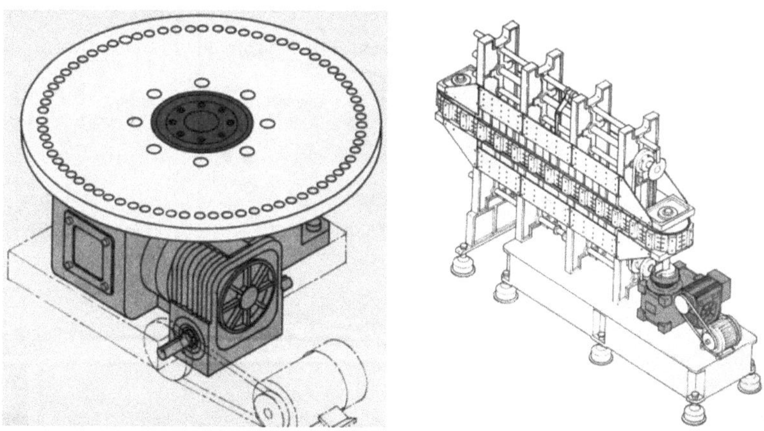

(a) Rotary table (b) Flexible chains

Figure 10.18 Application of indexing cams.

There are various indexing cams, of which three types are shown in Fig. 10.19, including 1) grooved concave globoidal cams, 2) cylindrical cams, and 3) parallel cams, are the most common. Indexing-cam mechanisms have the following features:

1) **Form-closed:** Indexing cam mechanisms are all form-closed. Thus, crossover shock is a concern at high speed.

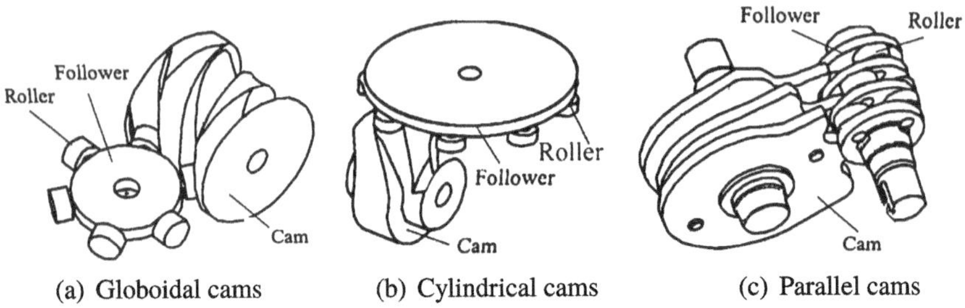

Figure 10.19 Types of indexing cams.

2) **Varying speed ratio:** The speed ratio of indexing-cam mechanism is not a constant; it varies with the position of engagement or time t.
3) **Coupled torsional and lateral vibrations:** Inertial loads cause both torsional and lateral vibrations. Generally, these two vibrations are coupled.
4) **Interaction:** An indexing-cam mechanism, in most cases, is only a subsystem of a whole machine. Interaction between the cam and other subsystems of the machine is generally significant and has to be considered.
5) **Parametric excitation:** In the working process, the contact point and the tooth number in meshing change with time, forming time-varying parameters or parametric excitations.

In the following, the dynamic model and analysis of a rotary table and flexible chains are briefly presented.

10.6.1 Rotary table driven by globoidal cam

One application of the globoidal indexing cam is to drive a rotary table shown in Fig. 10.18(a). In this type of cam mechanisms, the input and the output axes are perpendicular to each other and located in different planes. The distance between the two axes is called the *center distance* designated by a. The cam on the input shaft is very similar to a worm gear with a varying helix angle. The follower or the indexing wheel has rollers installed radially.

10.6.1.1 Dynamic model

Without inclusion of the driver and the subsequent parts, the cam is assumed to rotate at a constant speed. Also the clearance between the roller

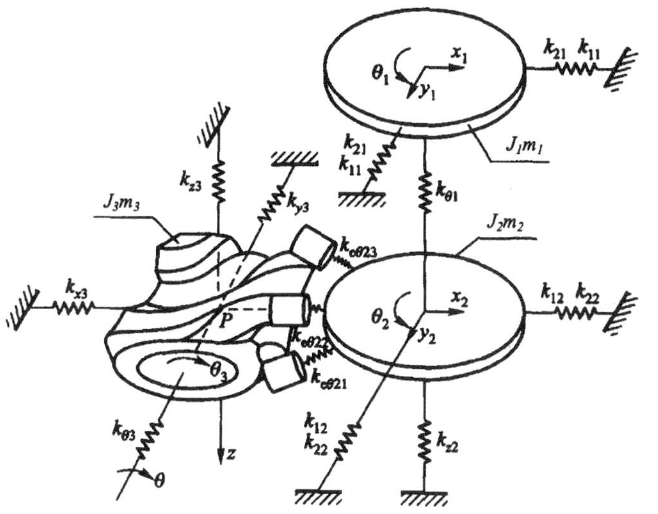

Figure 10.20 Dynamic model.

and the groove of cam is not considered. A dynamic model in Fig. 10.20 can be developed.

1) **Input shaft:** The directions of the x, y and z are shown in Fig. 10.20. The vibrations considered include the translational vibration along y and the lateral vibrations along x and z, as well as the rotational vibration around its axis. The center line intersects the input shaft axis at point P. The input shaft is represented by an equivalent mass m_3 and a moment of inertia J_3 at point P. The main parameters in the model are defined here:

   θ, $\dot{\theta}$ and $\ddot{\theta}$: the rotation angle, angular velocity and angular acceleration of the input end of the shaft. Without consideration of the driver, $\dot{\theta}$ is a constant and $\ddot{\theta}=0$.

   k_{x3} and k_{z3}: the bending stiffness of the input shaft at point P in x and z directions, respectively.

   k_{y3}: the equivalent stiffness along the y direction.

   x_3, y_3 and z_3: the displacements of point P in the x, y and z directions, respectively.

   θ_3: the actual rotation angle of point P.

   q_3: the elastic rotation angle of point P relative to the input end, which has the following relationship:

$$q_3=\theta_3-\theta. \tag{10.96}$$

2) **Output shaft:** The rotary table are modeled as a moment of inertia J_1 and a mass m_1, while the indexing cam is represented by moment of inertia J_2 and a mass m_2. Then the output shaft is treated as a beam having two lumped masses. The main parameters are described as follows:

   k_{11}: the bending stiffness influence coefficient of the rotary table.

   k_{12} and k_{21}: the stiffness influence coefficients. The subscripts 1 and 2 indicate the rotary table and the indexing wheel, respectively.

   k_{22}: the bending stiffness influence coefficient of the indexing wheel.

   $k_{\theta 1}$: the torsional stiffness of the output shaft.

   k_{z2}: the supporting stiffness of the output shaft in z direction.

   τ, $\dot{\tau}$ and $\ddot{\tau}$: the theoretical output rotation angle, angular velocity and acceleration, respectively.

   x_1 and y_1: the displacement of the rotary table in x and y directions, respectively.

   x_2, y_2 and z_2: the displacement of the indexing wheel in x, y and z directions, respectively.

   θ_1 and θ_2: the rotation angle of the rotary table and the indexing wheel.

   q_1 and q_2: the elastic rotation angle of the rotary table and the indexing wheel; the following relationships exist:

$$\begin{cases} q_1 = \theta_1 - \tau \\ q_2 = \theta_2 - \tau. \end{cases} \tag{10.97}$$

3) **Interface between rollers and cam:** On the engaging interface between the roller and the cam groove, there are basically three types of deformation, including 1) the roller axle's bending deflection, 2) surface contact deformation, and 3) the bearing deformation. Although the deformation is hard to accurately calculate, this book use two stiffness coefficients to model it. One is the bending stiffness of the roller axle; the other is the equivalent rotational stiffness of the indexing wheel, designated as $k_{e\theta 2}$. The latter changes with the engaging position; thus, for make a distinction, we use $k_{e\theta 2i}$ to represent this stiffness when the i-th roller is in engagement.

10.6.1.2 Governing equation

According to the model in Fig. 10.20, the governing equation of motion can be derived using Lagrange's equation method. The torsional deformation of the input shaft is a nonlinear term. For convenience of solution, it is linearized as:

$$\tau(\theta_3) = \tau(\theta) + \theta'(\theta_3 - \theta), \tag{10.98}$$

where

$$\tau' = \frac{\mathrm{d}\tau}{\mathrm{d}\theta}. \tag{10.99}$$

Neglecting the detailed derivation, we directly give the governing equation of motion as:

$$\boldsymbol{M}\ddot{\boldsymbol{X}} + \boldsymbol{C}\dot{\boldsymbol{X}} + \boldsymbol{K}\boldsymbol{X} = \boldsymbol{P}, \tag{10.100}$$

where $\boldsymbol{X}$ is the generalized coordinate vector in the following form:

$$\boldsymbol{X} = \begin{bmatrix} q_1 & q_2 & q_3 & x_1 & y_1 & x_2 & y_2 & z_2 & x_3 & y_3 & z_3 \end{bmatrix}^{\mathrm{T}}. \tag{10.101}$$

The positive directions of x_i, y_i and z_i are consistent with the corresponding frames, and the positive direction of θ_i are shown in Fig. 10.20.

$\boldsymbol{M}$ is the mass matrix of the form:

$$\boldsymbol{M} = \mathrm{diag}\left\{ J_1 \quad J_2 \quad J_3 \quad m_1 \quad m_1 \quad m_2 \quad m_2 \quad m_{e2} \quad m_3 \quad m_3 \quad m_3 \right\}. \tag{10.102}$$

$\boldsymbol{C}$ is the damping matrix, which is treated as modal damping.

$\boldsymbol{P}$ is the inertial load vector as follows:

$$\boldsymbol{P} = \begin{bmatrix} -J_1\ddot{\tau} & -J_2\ddot{\tau} & J_3\ddot{\theta} & 0 & 0 & 0 & 0 & 0 & 0 & 0 & 0 \end{bmatrix}^{\mathrm{T}}. \tag{10.103}$$

Given that, in such a system, the inertial loads are dominant, only inertial loads are included in the matrix $\boldsymbol{P}$.

$\boldsymbol{K}$ is the stiffness matrix in the following form:

$$\boldsymbol{K} = \begin{bmatrix} k_{\theta 1} & -k_{\theta 1} & 0 & 0 & 0 & 0 & 0 & 0 & 0 & 0 & 0 \\ & D + k_{\theta 1} & -D\tau' & 0 & 0 & A & B & C & -A & -B & -C \\ & & D\tau'^2 + k_{\theta 3} & 0 & 0 & -A\tau' & -B\tau' & -C\tau' & A\tau' & B\tau' & C\tau' \\ & & & k_{11} & 0 & k_{12} & 0 & 0 & 0 & 0 & 0 \\ & & & & k_{11} & k_{12} & 0 & 0 & 0 & 0 & \\ & & & & & 2E + k_{22} & F & G & -2E & -F & -G \\ & Sym. & & & & & 2L + k_{22} & M & -F & -2L & -M \\ & & & & & & & 2N + k_{z2} & -G & -M & -2N \\ & & & & & & & & 2E + k_{x3} & F & G \\ & & & & & & & & & 2L + k_{y3} & M \\ & & & & & & & & & & 2N + k_{z3} \end{bmatrix}. \tag{10.104}$$

Some entries in $\boldsymbol{K}$ are listed here:

$$\begin{cases} A = \dfrac{1}{R}\sum_{i=1}^{3} k_{e\theta 2i}\sin\gamma_i \\ B = \dfrac{1}{R}\sum_{i=1}^{3} k_{e\theta 2i}\cos\gamma_i \\ C = \dfrac{1}{R}\sum_{i=1}^{3} k_{e\theta 2i}\tan\phi_i \\ D = \sum_{i=1}^{3} k_{e\theta 2i} \\ E = \dfrac{1}{R^2}\sum_{i=1}^{3} k_{e\theta 2i}\sin^2\gamma_i \\ F = \dfrac{1}{R^2}\sum_{i=1}^{3} k_{e\theta 2i}\cos\gamma_i\sin\gamma_i \\ G = \dfrac{1}{R^2}\sum_{i=1}^{3} k_{e\theta 2i}\tan\phi_i\sin\gamma_i \\ L = \dfrac{1}{R^2}\sum_{i=1}^{3} k_{e\theta 2i}\cos^2\gamma_i \\ M = \dfrac{1}{R^2}\sum_{i=1}^{3} k_{e\theta 2i}\tan\phi_i\cos\gamma_i \\ N = \dfrac{1}{R^2}\sum_{i=1}^{3} k_{e\theta 2i}\tan^2\phi_i\tau' = \dfrac{\mathrm{d}\tau}{\mathrm{d}\theta} \end{cases} \tag{10.105}$$

γ_i and ϕ_i $(i = 1, 2, 3)$ are calculated as:

$$\gamma_i = \tau - (i-1)\frac{2\pi}{n} \tag{10.106}$$

$$\tan\phi_i = \left(\frac{R}{a - R\cos\tau_i}\right)\frac{\mathrm{d}\tau}{\mathrm{d}\theta}. \tag{10.107}$$

R is the radius of the pitch circle of the indexing wheel.

10.6.1.3 Simulation and analysis

The parameters of Eq. (10.100) depend on the position of the mechanism. In other words, the stiffness matrix $\boldsymbol{K}$ changes with mechanism position; thus, the vibration modes changes as well.

This equation set can be solved numerically by either direct integration or modal superposition.

In this book, the modal superposition method is used. To deal with the time-varying stiffness, the time is discretized into multiple short time steps. In each time step, the stiffness matrix is approximately treated as constant; thus, modal superposition can be applied. A closed-form algorithm is used to obtain the solution. The detail of the closed-form algorithm will be introduced in Chapter 11.

The parameters used in simulation are given in Table 10.5. Some other

Table 10.5 Parameters of the indexing cam.

Parameter	Value	Parameter	Value
a	0.07 m	$k_{\theta 1}$	4.151×10^4 N·m/rad
r	0.034 m	$k_{\theta 3}$	1.981×10^4 N·m/rad
d	0.02 m	$k_{e\theta 2}$	8.465×10^5 N·m/rad
J_1	0.01855 kg·m^2	k_{x11}, k_{y11}	1.418×10^7 N/m
J_2	0.0005112 kg·m^2	k_{x12}, k_{y12}	1.65×10^6 N/m
J_3	0.001876 kg·m^2	k_{x21}, k_{y21}	1.65×10^6 N/m
m_1	4.270 kg	k_{x22}, k_{ez2}	2.016×10^8 N/m
m_2	0.945 kg	k_{y3}, k_{y22}	1.172×10^8 N/m
m_3	2.240 kg	k_{z3}, k_{x3}	9.606×10^6 N/m

a is the center distance; r is the radius of the pitch circle of the indexing wheel; d is the diameter of the roller on the indexing wheel.

parameters include: 1) The input speed is 960 rpm; 2) the vibration modes up to the fourth order are retained; 3) the damping ratio is 0.03; 4) one revolution is divided into 720 time steps; 5) modified sinusoidal acceleration and modified constant velocity motions are used for the indexing period.

The simulation results are presented in Fig. 10.21, in which the tangential acceleration of the rotary table at the radius of 0.08 m is shown. One important observation from the figure is that the response at the start and the end of the indexing period has an abrupt jump. This is caused by the impacts of the meshing-in and meshing-out of the rollers. Similar to gear mesh, the rollers abruptly enter or exit mesh with the cam at the moment, thus, creating meshing-in and meshing-out impacts, respectively. The impacts can be controlled by refining the assembly process to make the engagement smoother.

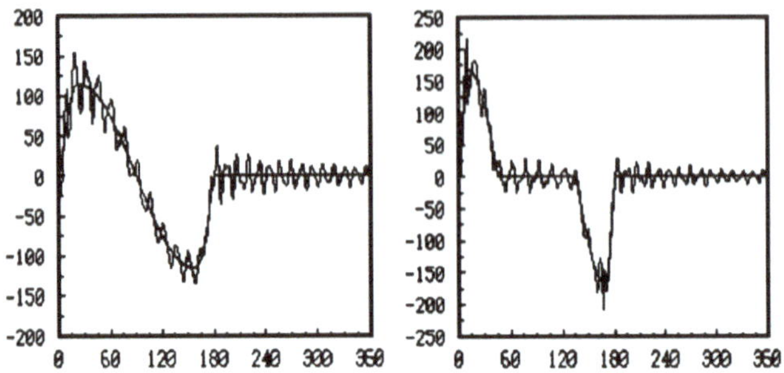

(a) Modified sinusoidal acceleration (b) Modified constant velocity

Figure 10.21 Dynamic response.

10.6.2 Flexible chain system driven by a parallel cam

A typical application of the flexible chain is in die-cutting machines as shown in Fig. 10.22. A parallel indexing cam, numbered as 7, rotates con-

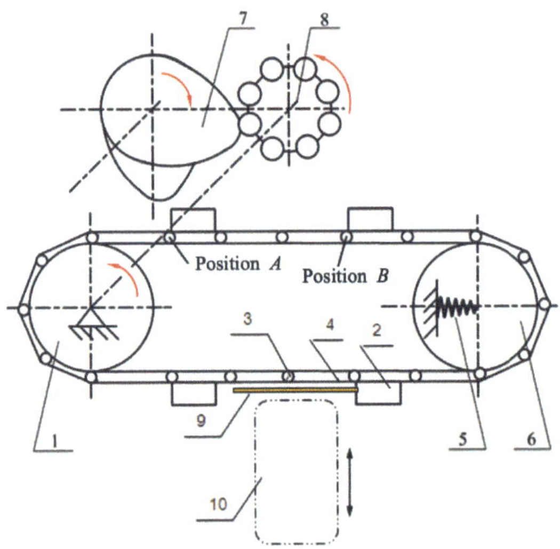

1:driving sprocket; 2:block; 3:chain; 4:track; 5:tensioning spring; 6:driven sprocket; 7:parallel cam; 8:indexing wheel; 9:cupboard to be cut; 10:die

Figure 10.22 Die-cutting mechanism.

tinuously at a constant speed and drives the indexing wheel 8 to rotate intermittently. The wheel 8 is connected to the sprocket 1, which drives the chain to move intermittently so that a series of required operations can be implemented. Due to the frequent starta, accelerationa, decelerationa, and stopa, the dynamics of the system needs a thorough investigation.

10.6.2.1 Dynamic model

Given that the chain is tensioned and supported by the guidance track vertically, the lateral vibration is not considered. The following assumptions are made for the development of the dynamic model:

- Due to the chain being much more flexible than the cam, the flexibility of the chain is considered, while the cam is regarded as rigid.
- Blocks 2 are treated as lumped masses (M_3) located at nodes p and q.
- The two sprockets are modeled as two lumped masses M_1 and M_2.
- The mass, m, of each chain link plate is split to two masses $m/2$ at each of the roller pins at the two ends.
- Each link plate is modeled as a linear spring of k.

The chain system is then represented by Fig. 10.23. The lumped masses of

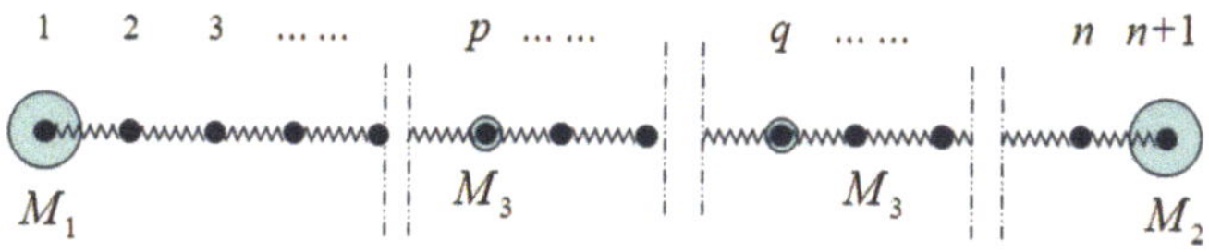

Figure 10.23 Dynamic model of the chain system.

the two sprockets are calculated so:

$$\begin{cases} M_1 = \dfrac{J_1}{R_1^2} \\ M_2 = \dfrac{J_2}{R_2^2}, \end{cases} \tag{10.108}$$

where J and R are the moment of inertia and the radius of the pitch circle. The subscripts 1 and 2 indicate the driving and driven, respectively.

The stiffness of the link plate can be determined through tensile tests. If the applied force is Q, the elongation is δ and the length of the chain under test is of n pitches, the stiffness k is calculated by

$$k = \frac{nQ}{\delta}. \tag{10.109}$$

10.6.2.2 Governing equation of motion

The displacement of each lumped mass in Fig. 10.23 is designated as u_i. The generalized coordinate vector in this problem is

$$U=\left\{u_1 \quad u_2 \quad \cdots, u_p \quad \cdots, \quad u_q \quad \cdots \quad u_n \quad u_{n+1}\right\}^{\mathrm{T}}. \tag{10.110}$$

Then the governing equation of the system can be represented as

$$M\ddot{U}+C\dot{U}+KU=-M\ddot{U}_{\mathrm{r}}. \tag{10.111}$$

The matrices in this equation are explained and given next.

1) M is the mass matrix of $n+1$ dimension that has the following form:

$$\begin{aligned} M = \operatorname{diag}\{M_1+m/2 \quad m \quad \cdots \quad m \quad M_3+m \quad m \quad \cdots \quad m \\ M_3+m \quad m \quad \cdots \quad m \quad M_2+m/2\}. \end{aligned} \tag{10.112}$$

It is noted that the block masses, M_3, are added to the p-th and q-th lumped masses.

2) K is the stiffness matrix. It has the nonzero elements located on a narrow diagonal belt as:

$$K=\begin{bmatrix} k & -k & & & & & \\ -k & 2k & -k & & & & \\ & -k & 2k & -k & & & \\ & & \ddots & \ddots & \ddots & & \\ & & & -k & 2k & -k & \\ & & & & -k & 2k & -k \\ & & & & & -k & k \end{bmatrix}. \tag{10.113}$$

3) C is the damping matrix. It is taken as modal damping.
4) $-M\ddot{U}_{\mathrm{r}}$ is the inertial load vector caused by the rigid motion. $\ddot{U}_{\mathrm{r}}$ is the rigid acceleration vector. Given that the rigid motion of the whole chain is the same, all elements in $\ddot{U}_{\mathrm{r}}$ are the same, which is the tangential acceleration of the pitch circle on the driving sprocket. It is given in this form:

$$a_t=R_1\ddot{\theta}_1, \tag{10.114}$$

where R_1 is the radius of the pitch circle on the driving sprocket, and $\ddot{\theta}_1$ is the angular acceleration of the indexing wheel.

10.6.3 Simulation and analysis

Eq. (10.111) is solved numerically. The driving sprocket is assumed to rotate at an average speed of 300 rpm. Modified sinusoidal acceleration and modified constant velocity motions are used. The parameters in the simulation are given in Table 10.6.

Table 10.6 Simulation parameters of the flexible chain.

Parameter	l (mm)	M_1, M_2 (kg)	M_3 (kg)	R_1, R_2 (mm)
Value	15.875	1.133	1.00	101.17

The acceleration response of the fortieth link is reported in Fig. 10.24.

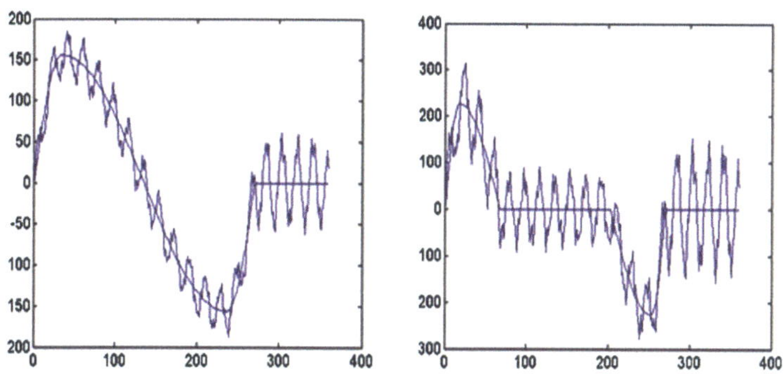

(a) Modified sinusoidal acceleration (b) Modified constant velocity

Figure 10.24 Acceleration response of the fortieth link.

Two important observations can be made from the simulation.

- The further the distance to the driving sprocket, the bigger the acceleration response is. This is also true for the residual vibration response.
- The residual vibration of the modified sinusoidal acceleration motion (Fig. 10.24(a)) is obviously less than that of the modified constant velocity (Fig. 10.24(b)). This is different from the results in the rotary table case.

CHAPTER 11

Elasto-dynamics of linkage

11.1 Introduction

11.1.1 Brief historical review

After World War II, the trend toward faster and lighter became more pronounced in the machine industry.

Inertia loads are proportional to the square of the input angular velocity. Higher speed and more flexible components lead to larger deformation of the links, thus larger motion error. An example of four-bar linkage is shown in Fig. 11.1. Point P on the coupler traces a closed path in one cycle. The solid line represents the theoretical path (or expected) obtained from kinematic analysis under the assumption of rigid links and joints. Due to the elasticity of the links and joints, however, the actual path generated by the point deviates from the theoretical one as shown by the dashed line. The difference of the two motions is referred to as *motion error*. With the increase of operation speed and link flexibility, the motion error may become large enough to significantly affect the machine function and performance. Erdman and Sandor (1972) made the following statement regarding a high-speed gripper mechanism:

"A gripper mechanism in a high-speed printing press may be synthesized kinematically with all links regarded as rigid. In the actual performance, however, such a linkage may be able to accomplish its task only up to one-half its design speed; above that speed, the gripper may miss the target."

Analysis and experiments has shown that link flexibility significantly affects the accuracy and stability of high-speed manipulators. In a paper by Sunada and Dubowsky (1983), the following statement was made:

"The requirements for higher speeds and better system performance make it a necessity to consider the dynamic effects of distributed link flexibility and mass in the design of such manipulators."

The increase of speed not only makes the inertia load more dominant, but also ramps up the excitation frequency. On the other hand, more flexible components mean a lower natural frequency. These two opposing factors increase the risk of resonance and generally cause more severe vibration.

Machinery Dynamics
https://doi.org/10.1016/B978-0-12-815785-5.00022-4

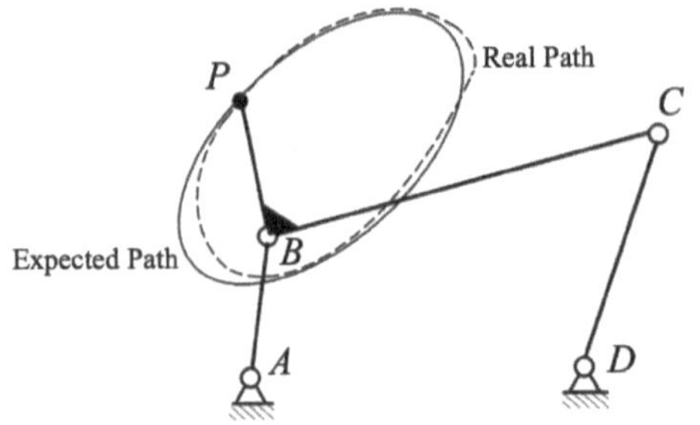

Figure 11.1 Path error of an elastic linkage.

Conventionally, mechanism analysis and synthesis are based on the assumption of rigid links and rigid joints. This practice becomes obsolete with the continuous increase of speed and component flexibility. In this case, a new branch of mechanism dynamics, *elasto-dynamics*, has been developed since the 1970s, starting with linkages.

The elasto-dynamics of linkages studies the real motion and forces (*elasto-dynamic analysis*) and design method (*elasto-dynamic synthesis*), while considering the flexibility of all or some of the links.

Elasto-dynamics is much more complicated than conventional mechanism analysis and synthesis problems. In general, it requires intensive computation; thus, modern computers and numerical computation are indispensable. Erdman et al. (1972) introduced the finite element method into elastic linkage analysis. Since then, many papers have been published on analysis, synthesis, and vibration suppression of elastic linkage mechanisms in various journals and conference proceedings. In 1997, the first monograph on elasto-dynamics of linkages was published in China (Zhang et al., 1997).

Although the majority of study focused on theoretical developments, applications have been reported in the design of industrial sewing machines, packaging machines, textile machines, and high-speed robots.

11.1.2 Review on elasto-dynamic analysis

The motion from kinematic analysis based on the rigid-body assumption is referred to as the *nominal motion*. Generally the elastic deformation of links is very small compared to the nominal motion. Theoretically, the nominal motion and the elastic deformation are coupled. However, common practice is to neglect this coupling in actual elasto-dynamic analysis. This practice is acceptable as long as the link flexibility and the operation speed are not too high.

This practice basically assumes the actual motion of a linkage as the superposition of the nominal motion and the elastic deformation. The former is obtained by kinematic analysis, while the latter is calculated through elasto-dynamic analysis.

The links of linkage are literally continuous elastic bodies. For continuous vibration problems, however, analytical solutions are only available for very simple geometries. This greatly limits the application of the continuous vibration model. In elasto-dynamic analysis, the continuous body is almost always discretized into a discrete system. Two methods of discretization are the most common:

1) **Lumped-parameter method**
   In this method, an elastic body is approximated by a certain number of concentrated masses or concentrated moments of inertia connected by massless springs. The basic principle and procedure of this method were previously introduced in Chapter 6.
2) **Finite element method**
   In this method, an elastic body is divided into a certain number of elements. Each element has a certain number of nodes. Each element has an element mass and stiffness matrices depending on the element type and the shape function chosen. Based on the element mass and stiffness, the global mass and stiffness matrices are generated. This method is presented in Chapter 7.

Both methods transform the partial differential equation (PDE) in continuous system to the ordinary differential equation (ODE). Relatively speaking, the finite element method has higher accuracy than the lumped-parameter method.

Elasto-dynamic analysis is much more complicated than the analysis of rigid-body. In this chapter, a method based on the concept of *instantaneous structure* is introduced. In this method, the status of the linkage, including the mechanism shape and the loads applied on it, is frozen in a specific position; thus, the mechanism in this position can be treated as an instantaneous structure. If the motion cycle is divided into multiple instantaneous moments, at each moment the linkage is frozen into an instantaneous structure. Then the well-established methods in structural dynamics can be directly applied to the elasto-dynamic analysis of high-speed linkages. The basic steps of the method can be summarized as follows:

1) Divide the links into beam elements, making points of interest coincident with element nodes;
2) Choose proper displacement shapes (the shape function), and the generalized coordinates of the element; develop the governing equation of motion of element using the Lagrange's formulation;
3) Assemble the global equation of motion from that of the elements;
4) Solve the equation with proper numerical methods, and analyze the results.

In this chapter, we deal only with the relatively simple planar four-bar linkage. In addition, all moving members are assumed having regular shapes that can be represented by straight beam elements of uniform cross section. The chapter is organized as follows:

- Section 11.2 introduces the development of equations of elements.
- The assemblage of global equations from the counterparts of elements is presented in Section 11.3.
- In Section 11.4, numerical solution of the equation is introduced. In particular a numerical algorithm solving ODEs having periodically-changing parameters is introduced based on the modal superposition.
- Section 11.5 briefly outlines the elasto-dynamic synthesis of linkages and techniques for the suppression of vibration.

11.2 Equation of elements

11.2.1 Generalized coordinates

A straight beam element is shown in Fig. 11.2. Rather than in the beam element introduced in Chapter 7, where only the bending deformations are considered, both bending and axial displacements are considered here. The lateral deflection shape is assumed as a quintic polynomial, while the axial displacements is assumed as linear. The reason for this treatment will

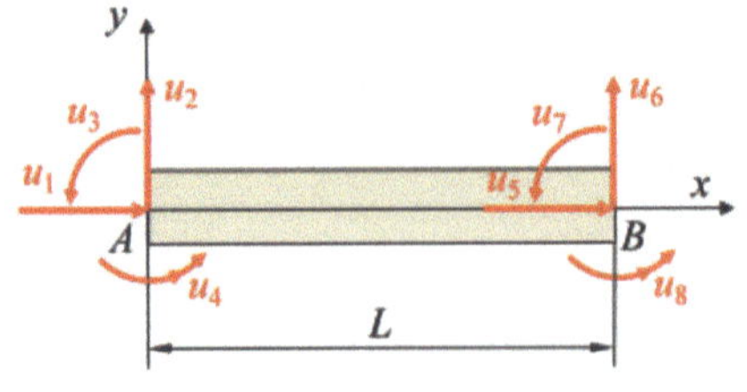

Figure 11.2 Beam element.

be explained in Section 11.4.

$$W(x,t) = a_0 + a_1x + a_2x^2 + a_3x^3 + a_4x^4 + a_5x^5 \quad (11.1)$$

$$V(x,t) = b_0 + b_1x, \quad (11.2)$$

where a_i and b_i are constants. W and V are the lateral and axial displacements, respectively. The notation of $W(x, t)$ and $V(x, t)$ mean that they are functions of x and time t.

The generalized coordinates of an element, u_1 to u_8, are shown in Fig. 11.2 and are explained as follows:

$u_1(t)$, $u_5(t)$ are the axial displacements of nodes A and B, respectively;
$u_2(t)$, $u_6(t)$ are the lateral displacements of nodes A and B, respectively;
$u_3(t)$, $u_7(t)$ are the elastic rotation angles of nodes A and B, respectively;
$u_4(t)$, $u_8(t)$ are the curvatures at nodes A and B, respectively.

The generalized coordinate vector of an element is

$$\boldsymbol{u} = \begin{bmatrix} u_1 & u_2 & u_3 & u_4 & u_5 & u_6 & u_7 & u_8 \end{bmatrix}^{\mathrm{T}}. \quad (11.3)$$

The constants a_i and b_i are to be determined by boundary conditions as follows:

$$\begin{cases} V(0,t) = u_1(t) \\ W(0,t) = u_2(t) \\ W'(0,t) = u_3(t) \\ W''(0,t) = u_4(t) \end{cases} \qquad \begin{cases} V(L,t) = u_5(t) \\ W(L,t) = u_6(t) \\ W'(L,t) = u_7(t) \\ W''(L,t) = u_8(t), \end{cases} \quad (11.4)$$

where W' and W'' mean the first- and second-order derivatives of W with respect to x.

After the eight constants are determined, W and V can be expressed as:

$$\begin{cases} W(x,t) = \sum_i \phi_i(x)u_i(t) \quad (i = 2, 3, 4, 6, 7, 8) \\ V(x,t) = \sum_j \phi_j(x)u_j(t) \quad (j = 1, 5). \end{cases} \quad (11.5)$$

ϕ_i are the shape functions in the following forms:

$$\phi_1(x) = 1 - e \quad (11.6)$$

$$\phi_2(x) = 1 - 10e^3 + 15e^4 - 6e^5 \quad (11.7)$$

$$\phi_3(x) = L(e - 6e^3 + 8e^4 - 3e^5) \quad (11.8)$$

$$\phi_4(x) = L^2(e^2 - 3e^3 + 3e^4 - e^5)/2 \tag{11.9}$$

$$\phi_5(x) = e \tag{11.10}$$

$$\phi_6(x) = 10e^3 - 15e^4 + 6e^5 \tag{11.11}$$

$$\phi_7(x) = L(-4e^3 + 7e^4 - 3e^5) \tag{11.12}$$

$$\phi_8(x) = L^2(e^3 - 2e^4 - e^5)/2, \tag{11.13}$$

where e is defined as the ratio x/L.

11.2.2 Kinematic relationships

The motion period (cycle) needs to be discretized into a series of small time steps. At each moment, the linkage is frozen as an instantaneous structure.

To develop the kinematic relationship within an element before and after the elastic deformation happens, three Cartesian coordinate frames are established as in Fig. 11.3. OXY is the *absolute coordinate system*; Axy is the *moving element coordinate system;* and OX_eY_e is the *rotating coordinate system*, the two axes of which are parallel to the x and y in the moving-element coordinate system at each position of the linkage.

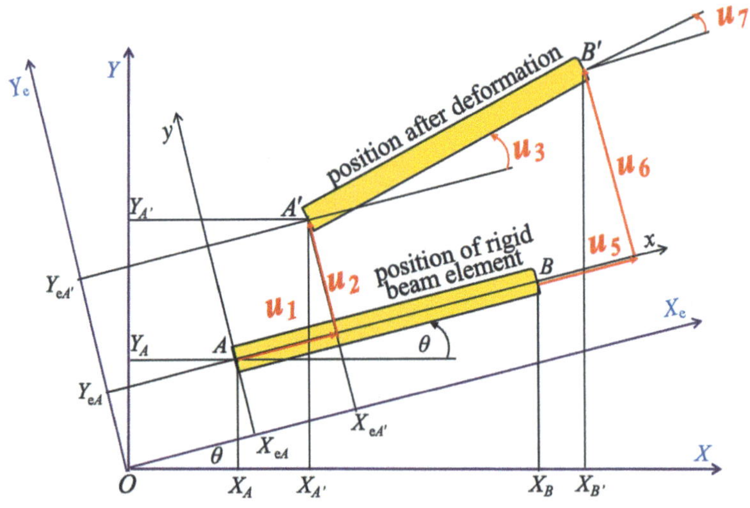

Figure 11.3 Kinematic relation.

AB represents the position of the element before elastic deformation, while $A'B'$ is the position after the deformation. We can easily write the following relationships based on this figure:

$$X_{A'} = X_A + u_1 C - u_2 S \tag{11.14}$$

$$Y_{A'} = Y_A + u_2 C + u_1 S \tag{11.15}$$

$$\theta_{A'} = \theta + u_3 \tag{11.16}$$

$$m_{A'} = u_4, \tag{11.17}$$

where X and Y are the absolute coordinates of the nodes; θ is the angle between the axis of the deformed element and the coordinate axis X; and m is the curvature of the deformed element. The subscripts A and A' represent the nodes shown. S and C represent $\sin\theta$ and $\cos\theta$, respectively.

It is straightforward to write the following transformation between the rotation frame Axy and absolute frame OXY:

$$\begin{cases} \begin{bmatrix} \dot{X}_{A'} \\ \dot{Y}_{A'} \end{bmatrix} = \begin{bmatrix} C & -S \\ S & C \end{bmatrix} \begin{bmatrix} \dot{x}_{A'} \\ \dot{y}_{A'} \end{bmatrix} \\ \begin{bmatrix} \dot{X}_{B'} \\ \dot{Y}_{B'} \end{bmatrix} = \begin{bmatrix} C & -S \\ S & C \end{bmatrix} \begin{bmatrix} \dot{x}_{B'} \\ \dot{y}_{B'} \end{bmatrix} \end{cases}. \tag{11.18}$$

We can represent the absolute velocity as

$$\dot{\boldsymbol{u}}_a = \dot{\boldsymbol{u}}_r + \dot{\boldsymbol{u}} + \dot{\boldsymbol{u}}_{re}, \tag{11.19}$$

where $\dot{\boldsymbol{u}}_a$, $\dot{\boldsymbol{u}}_r$, $\dot{\boldsymbol{u}}$, and $\dot{\boldsymbol{u}}_{re}$ are called the *absolute velocity vector*, the *rigid-body velocity vector*, the *elastic deformation velocity vector*, and the *relative velocity vector*, respectively. They are defined as:

$$\begin{cases} \dot{\boldsymbol{u}}_a = \begin{bmatrix} \dot{X}^e_{A'} & \dot{Y}^e_{A'} & \dot{\theta}_{A'} & \dot{m}_{A'} & \dot{X}^e_{B'} & \dot{Y}^e_{B'} & \dot{\theta}_{B'} & \dot{m}_{B'} \end{bmatrix}^T \\ \dot{\boldsymbol{u}}_r = \begin{bmatrix} \dot{X}^e_A & \dot{Y}^e_A & \dot{\theta} & 0 & \dot{X}^e_B & \dot{Y}^e_B & \dot{\theta} & 0 \end{bmatrix}^T \\ \dot{\boldsymbol{u}} = \begin{bmatrix} \dot{u}_1 & \dot{u}_2 & \dot{u}_3 & \dot{u}_4 & \dot{u}_5 & \dot{u}_6 & \dot{u}_7 & \dot{u}_8 \end{bmatrix}^T \\ \dot{\boldsymbol{u}}_{re} = \dot{\theta} \begin{bmatrix} -u_2 & u_1 & 0 & 0 & -u_6 & u_5 & 0 & 0 \end{bmatrix}^T \end{cases}. \tag{11.20}$$

Similar to the transformation of Eq. (11.18), we have the transformation of acceleration as

$$\begin{cases} \begin{bmatrix} \ddot{X}_{A'} \\ \ddot{Y}_{A'} \end{bmatrix} = \begin{bmatrix} C & -S \\ S & C \end{bmatrix} \begin{bmatrix} \ddot{x}_{A'} \\ \ddot{y}_{A'} \end{bmatrix} \\ \begin{bmatrix} \ddot{X}_{B'} \\ \ddot{Y}_{B'} \end{bmatrix} = \begin{bmatrix} C & -S \\ S & C \end{bmatrix} \begin{bmatrix} \ddot{x}_{B'} \\ \ddot{y}_{B'} \end{bmatrix} \end{cases}. \tag{11.21}$$

Taking advantage of this transformation, we have the following relationship:

$$\ddot{\boldsymbol{u}}_{\rm a}=\ddot{\boldsymbol{u}}_{\rm r}+\ddot{\boldsymbol{u}}+\ddot{\boldsymbol{u}}_{\rm n}+\ddot{\boldsymbol{u}}_{\rm t}+\ddot{\boldsymbol{u}}_{\rm c},\tag{11.22}$$

where

$$\begin{cases}\ddot{\boldsymbol{u}}_{\rm a}=\begin{bmatrix}\ddot{X}^{\rm e}_{\rm A'} & \ddot{Y}^{\rm e}_{\rm A'} & \ddot{\theta}_{\rm A'} & \ddot{m}_{\rm A'} & \ddot{X}^{\rm e}_{\rm B'} & \ddot{Y}^{\rm e}_{\rm B'} & \ddot{\theta}_{\rm B'} & \ddot{m}_{\rm B'}\end{bmatrix}^{\rm T}\\ \ddot{\boldsymbol{u}}_{\rm r}=\begin{bmatrix}\ddot{X}^{\rm e}_{\rm A} & \ddot{Y}^{\rm e}_{\rm A} & \ddot{\theta} & 0 & \ddot{X}^{\rm e}_{\rm B} & \ddot{Y}^{\rm e}_{\rm B} & \ddot{\theta} & 0\end{bmatrix}^{\rm T}\\ \ddot{\boldsymbol{u}}=\begin{bmatrix}\ddot{u}_1 & \ddot{u}_2 & \ddot{u}_3 & \ddot{u}_4 & \ddot{u}_5 & \ddot{u}_6 & \ddot{u}_7 & \ddot{u}_8\end{bmatrix}^{\rm T}\\ \ddot{\boldsymbol{u}}_{\rm n}=\dot{\theta}^2\begin{bmatrix}-u_1 & -u_2 & 0 & 0 & -u_5 & -u_6 & 0 & 0\end{bmatrix}^{\rm T}\\ \ddot{\boldsymbol{u}}_{\rm t}=\ddot{\theta}\begin{bmatrix}-u_2 & u_1 & 0 & 0 & -u_6 & u_5 & 0 & 0\end{bmatrix}^{\rm T}\\ \ddot{\boldsymbol{u}}_{\rm c}=2\dot{\theta}\begin{bmatrix}-\dot{u}_2 & \dot{u}_1 & 0 & 0 & -\dot{u}_6 & \dot{u}_5 & 0 & 0\end{bmatrix}^{\rm T}.\end{cases}\tag{11.23}$$

They are referred to as the *absolute acceleration*, the *rigid-body acceleration*, the *elastic acceleration*, the *normal relative acceleration*, the *tangential relative acceleration*, and the *Coriolis acceleration*, respectively.

$\ddot{\boldsymbol{u}}_{\rm n}$, $\ddot{\boldsymbol{u}}_{\rm t}$, and $\ddot{\boldsymbol{u}}_{\rm c}$ are the coupling terms between the rigid motion parameters ($\dot{\theta}$ and $\ddot{\theta}$) and the elastic deformation (u_i and $\dot{u}_i$). These coupling terms would greatly complicate the computation. As stated at the beginning of this chapter, they can be neglected as long as the link flexibility and operation speed are not too high. With this simplification, Eqs. (11.19) and (11.22) are shortened to

$$\dot{\boldsymbol{u}}_{\rm a}=\dot{\boldsymbol{u}}_{\rm r}+\dot{\boldsymbol{u}}\tag{11.24}$$

$$\ddot{\boldsymbol{u}}_{\rm a}=\ddot{\boldsymbol{u}}_{\rm r}+\ddot{\boldsymbol{u}}.\tag{11.25}$$

These two expressions are the simplified kinematic relation. It simply states that *the actual motion can be treated as the superposition of rigid-body motion and elastic deformation.*

11.2.3 Equation of elements

In the following, we will derive the equations of element using Lagrange's equation.

11.2.3.1 Kinetic energy

Assuming the center of mass of the element is located on the axis, so that we can neglect the kinetic energy of rotation, we have:

$$\begin{cases} E_{\mathrm{k}} = E_{\mathrm{W}} + E_{\mathrm{V}} \\ E_{\mathrm{W}} = \dfrac{1}{2}\int_0^L \rho A\left[\dot{W}_{\mathrm{a}}(x,t)\right]^2 \mathrm{d}x \\ E_{\mathrm{V}} = \dfrac{1}{2}\int_0^L \rho A\left[\dot{V}_{\mathrm{a}}(x,t)\right]^2 \mathrm{d}x, \end{cases} \tag{11.26}$$

where E_{k}, E_{W}, and E_{V} are the total kinetic energy, the kinetic energy in lateral motion, and the kinetic energy in axial motion, respectively. A is the cross-sectional area, ρ is the density of material, and $\dot{W}_{\mathrm{a}}(x,t)$ and $\dot{V}_{\mathrm{a}}(x,t)$ are the absolute velocities in lateral and axial directions, respectively.

Now, let's look at how to obtain $\dot{W}_{\mathrm{a}}(x,t)$. According to Eq. (11.24), we have

$$\dot{W}_{\mathrm{a}}(x,t) = \dot{W}_{\mathrm{r}}(x,t) + \dot{W}(x,t), \tag{11.27}$$

where $\dot{W}_{\mathrm{r}}(x,t)$ and $\dot{W}(x,t)$ are the rigid-body velocity and elastic velocity, respectively.

Representing the elastic velocity with the shape functions from Eq. (11.5), we have

$$\dot{W}(x,t) = \sum_i \phi_i(x)\dot{u}_i(t) \quad (i=2,3,4,6,7,8). \tag{11.28}$$

In rigid-body motion, a beam element remains straight. Thus, the rigid velocity in y direction is linear. The shape function in the lateral direction is a quintic polynomial. Thus, it can be used to represent a linear function. For this reason, the rigid velocity can also be represented in the following form:

$$\dot{W}_{\mathrm{r}}(x,t) = \sum_i \phi_i(x)\dot{u}_{\mathrm{r}i}(t) \quad (i=2,3,4,6,7,8), \tag{11.29}$$

where $\dot{u}_{\mathrm{r}i}(t)$ is the element of $\dot{\boldsymbol{u}}_{\mathrm{r}i}$.

Combining Eqs. (11.28) and (11.29) gives

$$\dot{W}_{\mathrm{a}}(x,t) = \sum_i \phi_i\dot{u}_{\mathrm{a}i}(t) \quad (i=2,3,4,6,7,8). \tag{11.30}$$

With exactly the same procedure, we get a representation of the $\dot{V}_a(x,t)$ as follows:

$$\dot{V}_a(x,t) = \sum_k \phi_k \dot{u}_{ak}(t) \qquad (k = 1, 5). \tag{11.31}$$

Substituting Eqs. (11.30) and (11.31) to Eq. (11.26), we obtain

$$E_W = \frac{1}{2}\int_0^L \sum_i \sum_j \rho A \phi_i(x)\phi_j(x)\dot{u}_{ai}\dot{u}_{aj}\,\mathrm{d}x \quad (i,j = 2,3,4,6,7,8) \tag{11.32}$$

$$E_V = \frac{1}{2}\int_0^L \sum_k \sum_l \rho A \phi_k(x)\phi_l(x)\dot{u}_{ak}\dot{u}_{al}\,\mathrm{d}x \quad (k,l = 1,5). \tag{11.33}$$

Changing the order of the integration and summation, the following two equations are obtained:

$$E_W = \frac{1}{2}\sum_i \sum_j \dot{u}_{ai}\bar{m}_{ij}\dot{u}_{aj}\,\mathrm{d}x \quad (i,j = 2,3,4,6,7,8) \tag{11.34}$$

$$E_V = \frac{1}{2}\sum_k \sum_l \dot{u}_{ak}\bar{m}_{kl}\dot{u}_{al}\,\mathrm{d}x \quad (k,l = 1,5), \tag{11.35}$$

where

$$\bar{m}_{ij} = \int_0^L \rho A \phi_i(x)\phi_j(x)\,\mathrm{d}x \quad (i,j = 2,3,4,6,7,8) \tag{11.36}$$

$$\bar{m}_{kl} = \int_0^L \rho A \phi_k(x)\phi_l(x)\,\mathrm{d}x \quad (k,l = 1,5). \tag{11.37}$$

Combining Eqs. (11.34) and (11.35) leads to

$$E_k = \frac{1}{2}\sum_{i=1}^{8}\sum_{j=1}^{8} \dot{u}_{ai}\bar{m}_{ij}\dot{u}_{aj} = \frac{1}{2}\dot{\boldsymbol{u}}_a^{\mathrm{T}}\bar{\boldsymbol{m}}\dot{\boldsymbol{u}}_a, \tag{11.38}$$

where $\bar{\boldsymbol{m}}$ is a 8 × 8 matrix as follows:

$$\bar{\boldsymbol{m}} = \begin{bmatrix} \frac{1}{3} & 0 & 0 & 0 & \frac{1}{6} & 0 & 0 & 0 \\ & \frac{181}{462} & \frac{311L}{4620} & \frac{281L^2}{55440} & 0 & \frac{25}{231} & \frac{-151L}{4620} & \frac{181L^2}{55440} \\ & & \frac{52L^2}{3465} & \frac{23L^3}{18480} & 0 & \frac{151L}{4620} & \frac{-19L^2}{1980} & \frac{13L^3}{13860} \\ & & & \frac{L^4}{9240} & 0 & \frac{181L^2}{55440} & \frac{-13L^3}{13860} & \frac{L^4}{11088} \\ & & & & \frac{1}{3} & 0 & 0 & 0 \\ & & Sym. & & & \frac{181}{462} & \frac{-311L}{4620} & \frac{281L^2}{55440} \\ & & & & & & \frac{52L^2}{3465} & \frac{-23L^3}{18480} \\ & & & & & & & \frac{L^4}{9240} \end{bmatrix}. \tag{11.39}$$

11.2.3.2 Potential energy

Neglecting gravitational potential energy, the only form of potential energy is caused by elastic deformations. Thus, the potential energy of an element can be represented as

$$E_\mathrm{p} = \frac{1}{2}\int_0^L EI\left[W''(x,t)\right]^2 \mathrm{d}x + \frac{1}{2}\int_0^L EA\left[V'(x,t)\right]^2 \mathrm{d}x, \tag{11.40}$$

where E and I are the modulus of elasticity of the material and the moment of inertia of the cross section, $W''(x, t)$ and $V'(x, t)$ are the second-order derivative of W and the first-order derivative of V with respective to x, respectively. From Eq. (11.5), we have

$$\begin{cases} \left[W''(x,t)\right]^2 = \sum\limits_i \sum\limits_j \phi_i''(x)\phi_j''(x)u_i(t)u_j(t) & (i,j = 2,3,4,6,7,8) \\ \left[V'(x,t)\right]^2 = \sum\limits_k \sum\limits_l \phi_k'(x)\phi_l'(x)u_k(t)u_l(t) & (k,l = 1,5). \end{cases} \tag{11.41}$$

Substituting these two equations into Eq. (11.40) and reversing the order of integration and summation, we have

$$E_\mathrm{p} = \frac{1}{2}\boldsymbol{u}^\mathrm{T}\bar{\boldsymbol{k}}\boldsymbol{u}. \tag{11.42}$$

$\bar{\boldsymbol{k}}$ is the element stiffness matrix of 8 × 8. Its elements are calculated by

$$\begin{cases} \bar{k}_{ij} = \int_0^L EI\phi_i''(x)\phi_j''(x)\,\mathrm{d}x & (i,j = 2,3,4,6,7,8) \\ \bar{k}_{kl} = \int_0^L EA\phi_k'(x)\phi_l'(x)\,\mathrm{d}x & (k,l = 1,5). \end{cases} \tag{11.43}$$

Substituting the shape function into the previous form and completing the integration give the form of $\bar{\boldsymbol{k}}$ as

$$\bar{\boldsymbol{k}}=\begin{bmatrix} \frac{A}{IL} & 0 & 0 & 0 & \frac{-A}{IL} & 0 & 0 & 0 \\ & \frac{120}{7L^3} & \frac{60}{7L^2} & \frac{3}{7L} & 0 & \frac{-120}{7L^3} & \frac{60}{7L^2} & \frac{-3}{7L} \\ & & \frac{192}{35L} & \frac{11}{35} & 0 & \frac{-60}{7L^2} & \frac{108}{35L} & \frac{-4}{35} \\ & & & \frac{3L}{35} & 0 & \frac{-3}{7L} & \frac{4}{35} & \frac{L}{70} \\ & & & & \frac{A}{IL} & 0 & 0 & 0 \\ & & Sym. & & & \frac{120}{7L^3} & \frac{-60}{7L^2} & \frac{3}{7L} \\ & & & & & & \frac{192}{35L} & \frac{-11}{35} \\ & & & & & & & \frac{3L}{35} \end{bmatrix}. \tag{11.44}$$

11.2.3.3 Equation of motion in rotating coordinate

Lagrange's equation for a beam element can be written as

$$\frac{\mathrm{d}}{\mathrm{d}t}\left(\frac{\partial E_{\mathrm{k}}}{\partial \dot{\boldsymbol{u}}}\right)-\frac{\partial E_{\mathrm{k}}}{\partial \boldsymbol{u}}+\frac{\partial E_{\mathrm{p}}}{\partial \boldsymbol{u}}=\bar{\boldsymbol{f}}, \tag{11.45}$$

where $\boldsymbol{u}$ is the generalized coordinate vector of the element and $\bar{\boldsymbol{f}}$ is the generalized force vector.

From Eq. (11.24), we have

$$\dot{u}_{ai}=\dot{u}_{ri}+\dot{u}_i. \tag{11.46}$$

From Eqs. (11.38) and (11.42), we have

$$\frac{\partial E_{\mathrm{k}}}{\partial u_i}=\sum_{j=1}^{8}\bar{m}_{ij}\ddot{u}_{aj} \tag{11.47}$$

$$\frac{\partial E_{\mathrm{p}}}{\partial u_i}=\sum_{j=1}^{8}\bar{k}_{ij}u_j \tag{11.48}$$

$$\frac{\partial E_{\mathrm{k}}}{\partial u_i}=0. \tag{11.49}$$

Substituting Eqs. (11.47) through (11.49) into Eq. (11.45) yields

$$\sum_{j=1}^{8}\bar{m}_{ij}\ddot{u}_{aj}+\sum_{j=1}^{8}\bar{k}_{ij}u_j=\bar{f}_i \quad (i=1\text{–}8). \tag{11.50}$$

It can be written in matrix form as

$$\bar{m}\ddot{u}_a + \bar{k}u = \bar{f}. \tag{11.51}$$

Taking advantage of Eq. (11.25), this is transformed into

$$\bar{m}\ddot{u} + \bar{k}u = \bar{f} - \bar{m}\ddot{u}_r. \tag{11.52}$$

This is the equation of motion for an element in the rotating coordinate frame. u is defined in the element moving coordinate frame, while $\ddot{u}_r$ is the rigid body acceleration defined in the rotating coordinate frame.

11.2.3.4 Equation of motion in absolute coordinate

To study the whole linkage system, it is necessary to transform the equation into the absolute coordinate frame. To this end, we define the element generalized coordinate in the absolute frames as

$$U^e = \begin{bmatrix} U_1 & U_2 & U_3 & U_4 & U_5 & U_6 & U_7 & U_8 \end{bmatrix}^T. \tag{11.53}$$

The superscript e here indicates this refers to an element. It is noted that only the linear elements in u and U^e need to be transformed, while the rotational ones remain the same. The transformation between the two coordinate frames shown in Fig. 11.4 are as follows:

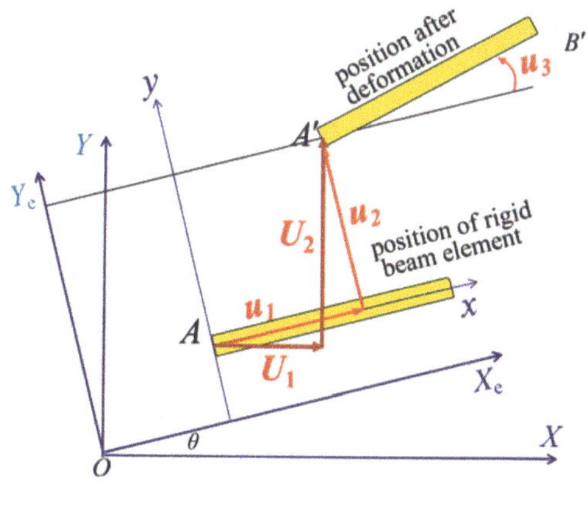

Figure 11.4 Coordinate transformation.

$$\begin{cases} u_1 = U_1C + U_2S \\ u_2 = -U_1S + U_2C \\ u_3 = U_3 \\ u_4 = U_4 \\ u_5 = U_5C + U_6S \\ u_6 = -U_5S + U_6C \\ u_7 = U_7 \\ u_8 = U_8, \end{cases} \tag{11.54}$$

where $C = \cos\theta$ and $S = \sin\theta$.

Eq. (11.54) can be put in the more compact matrix form:

$$\boldsymbol{u} = \boldsymbol{R}\boldsymbol{U}^{\mathrm{e}}. \tag{11.55}$$

$\boldsymbol{R}$, called the *coordinate transformation matrix*, is

$$\boldsymbol{R} = \begin{bmatrix} C & S & 0 & 0 & 0 & 0 & 0 & 0 \\ -S & C & 0 & 0 & 0 & 0 & 0 & 0 \\ 0 & 0 & 1 & 0 & 0 & 0 & 0 & 0 \\ 0 & 0 & 0 & 1 & 0 & 0 & 0 & 0 \\ 0 & 0 & 0 & 0 & C & S & 0 & 0 \\ 0 & 0 & 0 & 0 & -S & C & 0 & 0 \\ 0 & 0 & 0 & 0 & 0 & 0 & 1 & 0 \\ 0 & 0 & 0 & 0 & 0 & 0 & 0 & 1 \end{bmatrix}. \tag{11.56}$$

The rigid-body acceleration $\ddot{\boldsymbol{U}}_{\mathrm{r}}^{\mathrm{e}}$ and the acceleration in the element moving frame $\ddot{\boldsymbol{u}}$ can also be transformed in the same way:

$$\ddot{\boldsymbol{U}}_{\mathrm{r}}^{\mathrm{e}} = \begin{bmatrix} \ddot{X}_{\mathrm{A}} & \ddot{Y}_{\mathrm{A}} & \ddot{\theta} & 0 & \ddot{X}_{\mathrm{B}} & \ddot{Y}_{\mathrm{B}} & \ddot{\theta} & 0 \end{bmatrix}^{\mathrm{T}} \tag{11.57}$$

$$\ddot{\boldsymbol{u}}_{\mathrm{r}} = \boldsymbol{R}\ddot{\boldsymbol{U}}_{\mathrm{r}}^{\mathrm{e}}. \tag{11.58}$$

The transformation matrix $\boldsymbol{R}$ is constant at each position because, at that specific mechanism position θ, the linkage is frozen as an instantaneous structure.

From Eq. (11.55) we have

$$\ddot{\boldsymbol{u}} = \boldsymbol{R}\ddot{\boldsymbol{U}}^{\mathrm{e}}. \tag{11.59}$$

Substituting Eqs. (11.55), (11.58), (11.59) into Eq. (11.52) yields

$$\boldsymbol{m}\ddot{\boldsymbol{U}}^{\mathrm{e}} + \boldsymbol{k}\boldsymbol{U}^{\mathrm{e}} = \boldsymbol{f} - \boldsymbol{m}\ddot{\boldsymbol{U}}_{\mathrm{r}}^{\mathrm{e}}, \tag{11.60}$$

where

$$\boldsymbol{m} = \boldsymbol{R}^{\mathrm{T}}\bar{\boldsymbol{m}}\boldsymbol{R} \tag{11.61}$$

$$\boldsymbol{k} = \boldsymbol{R}^{\mathrm{T}}\bar{\boldsymbol{k}}\boldsymbol{R} \tag{11.62}$$

are the *equivalent element mass matrix* and the *equivalent element stiffness matrix*, respectively.

The generalized force vector is obtained as

$$\boldsymbol{f} = \boldsymbol{R}^{\mathrm{T}}\bar{\boldsymbol{f}}. \tag{11.63}$$

Eq. (11.60) is the equation of motion of an element in the absolute coordinate system.

11.3 Global equation of motion

11.3.1 Generalized coordinates

Without a loss of generality, a four-bar linkage shown in Fig. 11.5 is taken as an example to illustrate the procedures. The rocker and the coupler are divided into two elements each, while the crank is treated as one element. In the figure nodes are directly numbered as integers $(1, 2\cdots,)$, while elements are numbered by circled integers (①, ②, $\cdots$). The link number is put in parenthesis as $(1), (2), \cdots$. As stated in the previous section, four displacements of nodes, including the elastic displacements in X and Y directions, the elastic rotation angle and curvature of the beam, are included in the generalized coordinates. Nodes 3 and 5 have all the four displacements. Nodes 2 and 4 are coincident with the center of the revolute joints and belong to two adjacent links. Thus, they have two different rotation angles, but have no curvature. Node 6 connects the rocker and the frame. If the frame is assumed rigid, its elastic displacements in both X and Y directions are zero (constrained). In this case, it has only a rotation.

The rigid-body motion leads to a singular stiffness matrix. To solve this problem, the crank is assumed as a cantilever beam having node 1 fixed. Due to the elasticity of the crank being considered, there is a curvature at node 1. The treatment of cantilever beam is in agreement with the instantaneous-structure assumption. It is reasonable if the crankshaft has a

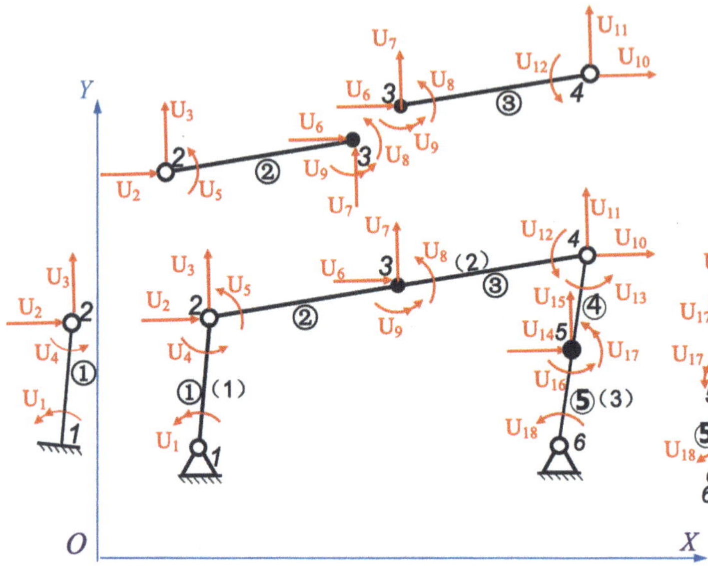

Figure 11.5 Finite element model of a four-bar linkage.

large moment of inertia, such as a flywheel installed or a very large motor rotor.

With the above analysis, 18 generalized coordinates, $U_1, U_2, \cdots, U_{18}$, are needed for this four-bar linkage. In the general case, we define the global generalized coordinate vector as

$$\boldsymbol{U} = \begin{bmatrix} U_1 & U_2 & \cdots & U_{\mathrm{N_u}} \end{bmatrix}^{\mathrm{T}}, \tag{11.64}$$

where $\mathrm{N_u}$ is the total number of DOFs. For the specific case of this four-bar mechanism, $\mathrm{N_u} = 18$.

To map the element generalized coordinate vector U^e to the global one U, a *mapping matrix* $\boldsymbol{I}_{\mathrm{u}}$ is used, in which the number of rows equals the number of elements, and the number of columns equals the number of DOF of the element generalized coordinate. For the specific case of four-bar linkage being considered, $\boldsymbol{I}_{\mathrm{u}}$ is this matrix of 5×8 dimension:

$$\boldsymbol{I}_{\mathrm{u}} = \begin{bmatrix} 0 & 0 & 0 & 1 & 2 & 3 & 4 & 0 \\ 2 & 3 & 5 & 0 & 6 & 7 & 8 & 9 \\ 6 & 7 & 8 & 9 & 10 & 11 & 12 & 0 \\ 14 & 15 & 16 & 17 & 10 & 11 & 13 & 0 \\ 0 & 0 & 18 & 0 & 14 & 15 & 16 & 17 \end{bmatrix}. \tag{11.65}$$

The row indicates the element number, while the column indicates which is the generalized coordinate of the element. The entries give the position where the element generalized coordinate is mapped to. For example, the eight generalized coordinates of element ② are mapped to the global generalized coordinates $\{U_2, U_3, U_5, 0, U_6, U_7, U_8, U_9\}$. This can be represented as

$$\boldsymbol{U}_2^{\mathrm{e}} = \begin{bmatrix} U_2 & U_3 & U_5 & 0 & U_6 & U_7 & U_8 & U_9 \end{bmatrix}^{\mathrm{T}}. \tag{11.66}$$

Note that, for the elements ④ and ⑤ in Fig. 11.5, the nodes 5 and 6 correspond to the point A in Fig. 11.4 because the rotation angle of the rocker θ_3 is measured at node 6 as shown in Fig. 11.6.

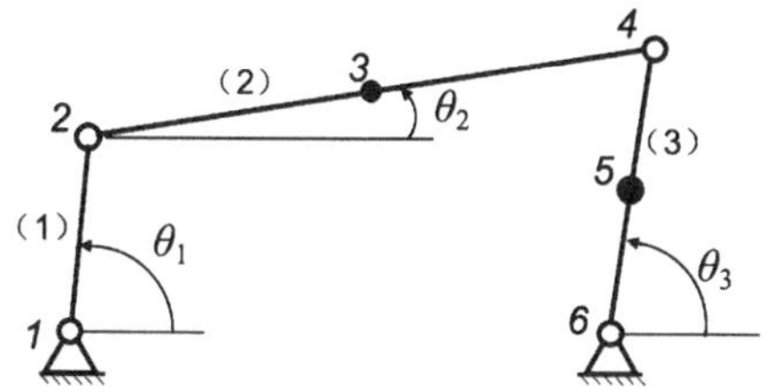

Figure 11.6 Rotation angle of links.

11.3.2 Formation of global equation of motion

The equation of motion of an element Eq. (11.60) is based on the element's generalized coordinates. For an arbitrary element ⓘ, the equation is

$$\boldsymbol{m}\ddot{\boldsymbol{U}}_i^{\mathrm{e}} + \boldsymbol{k}_i\boldsymbol{U}_i^{\mathrm{e}} = \boldsymbol{f}_i - \boldsymbol{m}_i\ddot{\boldsymbol{U}}_{\mathrm{r}i}^{\mathrm{e}}. \tag{11.67}$$

This is the same with Eq. (11.60) except that a subscript i is added to indicate the element number.

We define a global rigid-body acceleration vector $\ddot{\boldsymbol{U}}_{\mathrm{r}}$ as

$$\ddot{\boldsymbol{U}}_{\mathrm{r}} = \begin{bmatrix} \ddot{\boldsymbol{U}}_{\mathrm{r}1}^{\mathrm{T}} & \ddot{\boldsymbol{U}}_{\mathrm{r}2}^{\mathrm{T}} & \cdots & \ddot{\boldsymbol{U}}_{\mathrm{rN_u}}^{\mathrm{T}} \end{bmatrix}^{\mathrm{T}}. \tag{11.68}$$

$\ddot{\boldsymbol{U}}_{\mathrm{r}}$ is of the same dimension as $\ddot{\boldsymbol{U}}$, and their entries are mapped well. For example, if U_i is a linear displacement, $\ddot{U}_{\mathrm{r}i}$ will be linear acceleration, whereas, if U_i is a rotation angle, $\ddot{U}_{\mathrm{r}i}$ will be angular acceleration. For the case in Fig. 11.5, $\ddot{\boldsymbol{U}}_{\mathrm{r}}$ is as follows:

$$\ddot{\boldsymbol{U}}_{\mathrm{r}} = \begin{bmatrix} 0 \ \ddot{X}_2 \ \ddot{Y}_2 \ \ddot{\theta}_1 \ \ddot{\theta}_2 \ \ddot{X}_3 \ \ddot{Y}_3 \ \ddot{\theta}_2 \ 0 \ \ddot{X}_4 \ \ddot{Y}_4 \ \ddot{\theta}_2 \ \ddot{\theta}_3 \ \ddot{X}_5 \ \ddot{Y}_5 \ \ddot{\theta}_3 \ 0 \ \ddot{\theta}_3 \end{bmatrix}^{\mathrm{T}}, \tag{11.69}$$

where the subscript of the linear accelerations indicates the node number, and the subscript of the angular accelerations means the link number (see Fig. 11.6).

$\boldsymbol{U}_i^{\mathrm{e}}$ has an 8×1 dimension, and $\boldsymbol{U}$ is $N_u \times 1$. To map $\boldsymbol{U}_i^{\mathrm{e}}$ into $\boldsymbol{U}$, we use a *coordination matrix*, $\boldsymbol{B}_i$, to represent the relationship as

$$\boldsymbol{U}_i^{\mathrm{e}} = \boldsymbol{B}_i \boldsymbol{U}. \tag{11.70}$$

$\boldsymbol{B}_i$ is $8 \times N_u$. Its entries are either 0 or 1. For the linkage in Fig. 11.5, the coordination matrix for element ② is

$$\boldsymbol{B}_2 = \begin{bmatrix} 0 & 1 & 0 & 0 & 0 & & & & & \\ 0 & 0 & 1 & 0 & 0 & & & \boldsymbol{0}_{(4\times13)} & & \\ 0 & 0 & 0 & 0 & 1 & & & & & \\ 0 & 0 & 0 & 0 & 0 & & & & & \\ & & & & 0 & 1 & 0 & 0 & 0 & \\ & \boldsymbol{0}_{(4\times4)} & & & 0 & 0 & 1 & 0 & 0 & \boldsymbol{0}_{(4\times9)} \\ & & & & 0 & 0 & 0 & 1 & 0 & \\ & & & & 0 & 0 & 0 & 0 & 1 & \end{bmatrix}. \tag{11.71}$$

Similarly, we have

$$\ddot{\boldsymbol{U}}_{\mathrm{r}i}^{\mathrm{e}} = \boldsymbol{B}_i \ddot{\boldsymbol{U}}_{\mathrm{r}}. \tag{11.72}$$

Substituting Eqs. (11.70) and (11.72) into Eq. (11.67) and left-multiplying both sides by $\boldsymbol{B}_i^{\mathrm{T}}$ yield

$$\boldsymbol{M}_i^{\mathrm{e}} \ddot{\boldsymbol{U}} + \boldsymbol{K}_i^{\mathrm{e}} \boldsymbol{U} = \boldsymbol{F}_i^{\mathrm{e}} - \boldsymbol{M}_i^{\mathrm{e}} \ddot{\boldsymbol{U}}_{\mathrm{r}}, \tag{11.73}$$

where

$$\begin{cases} \boldsymbol{M}_i^{\mathrm{e}} = \boldsymbol{B}_i^{\mathrm{T}} \boldsymbol{m}_i \boldsymbol{B}_i \\ \boldsymbol{K}_i^{\mathrm{e}} = \boldsymbol{B}_i^{\mathrm{T}} \boldsymbol{k}_i \boldsymbol{B}_i \\ \boldsymbol{F}_i^{\mathrm{e}} = \boldsymbol{B}_i^{\mathrm{T}} \boldsymbol{f}_i. \end{cases} \tag{11.74}$$

Eq. (11.73) is the equation of motion of element i mapped into the global generalized coordinates. Assembling equations of all elements gives the global equation of motion

$$\boldsymbol{M}\ddot{\boldsymbol{U}} + \boldsymbol{K}\boldsymbol{U} = \boldsymbol{F} - \boldsymbol{M}\ddot{\boldsymbol{U}}_{\mathrm{r}}, \tag{11.75}$$

where $\boldsymbol{M}$, $\boldsymbol{K}$, and $\boldsymbol{F}$ are the global mass, global stiffness, and global generalized force matrices, respectively. They are

$$\begin{cases} \boldsymbol{M} = \sum_{i=1}^{N_e} \boldsymbol{M}_i^e \\ \boldsymbol{K} = \sum_{i=1}^{N_e} \boldsymbol{K}_i^e \\ \boldsymbol{F} = \sum_{i=1}^{N_e} \boldsymbol{F}_i^e. \end{cases} \tag{11.76}$$

N_e is the total number of elements.

As indicated in Chapter 7, the generalized forces include both the external forces and the internal forces between elements. When superimposing the element generalized forces into the global generalized force vector, however, the internal forces between adjacent elements cancel out because they are action and interaction forces. Thus, the global generalized force vector contains only external loads.

In the process of calculating the masses in this example, only the distributed masses of elements are considered. If there are concentrated masses in the mechanism, such as the bearing mass at joint 2 and the disk installed on the rocker output shaft (joint 4) in Fig. 11.7, they should be superimposed on the global mass matrix as concentrated parameters. The process is explained subsequently.

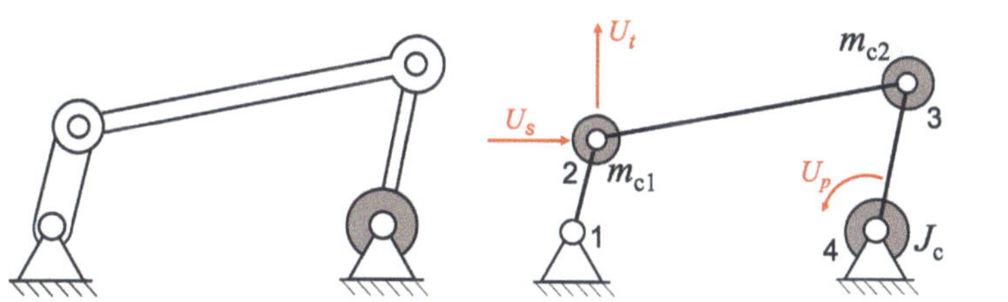

(a) The linkage (b) Concentrated masses and moment of inertia

Figure 11.7 Concentrated mass and moment of inertia.

If a concentrated mass m_c is located at a node, which has linear displacements U_s and U_t in the global generalized coordinate vector, the mass m_c is added on the s-th and t-th diagonal elements of the global mass matrix. If there is a concentrated moment of inertia J_c at a node, which has a rotation angle U_p in the global generalized coordinate vector, J_c is added to the p-th diagonal element of the global mass matrix.

Eq. (11.75) does not include damping. For simplicity, we add a modal damping to the equation and it becomes

$$\boldsymbol{M}\ddot{\boldsymbol{U}} + \boldsymbol{C}\dot{\boldsymbol{U}} + \boldsymbol{K}\boldsymbol{U} = \boldsymbol{F} - \boldsymbol{M}\ddot{\boldsymbol{U}}_{\mathrm{r}}, \tag{11.77}$$

where $\boldsymbol{C}$ is the modal damping matrix.

11.4 Solution of equation and analysis

11.4.1 Solution of equation

Eq. (11.77) looks similar to Eq. (6.23) (see p. 150). However, $\boldsymbol{M}$ and $\boldsymbol{K}$ in this equation are not constants: they vary with the mechanism position. This can be clearly seen from the element stiffness matrix, Eqs. (11.56), (11.61) and (11.62). This type of vibration problem having time varying parameters is referred to as *parametric vibration* (Nayfeh and Mook, 2008).

It is almost impossible to solve Eq. (11.77) in closed form. Numerical solution is, thus, the only option. The general-purpose numerical algorithms for solving ODE can be found in well-written texts, such as those by Sauer (2012), and by Chapra and Canale (2011). There are many methods available in this group, such as Euler's method and the widely used Runge–Kutta method. This type of method gives both the transit and steady-state responses.

If we are more interested in the steady-state response, there is a method, called the *closed form-algorithm*, available. This method was first proposed by Midha et al. (1979); it combines the modal superposition principle and the periodicity of the motion. The basic steps of the method are summarized as follows:

1) The period of motion is first discretized into short time steps;
2) The time-varying $\boldsymbol{M}$ and $\boldsymbol{K}$ are treated as constants as long as the time step is short enough. Then the natural frequencies and vibration modes in each time step are obtained by solving the eigenvalue problem;
3) The steady-state response in each time step is represented in closed form by the modal superposition method;
4) Taking advantage of the periodicity of the motion, an algebraic equation set in terms of the response in each time step is formed by combining the equations in step 3);
5) Solving the algebraic equation gives all the coefficients to form the steady-state response.

In the following, this method is introduced.

If the motion period T is divided into n time steps (intervals) of equal length Δt, the length of time step is

$$\Delta t=\frac{T}{n}. \tag{11.78}$$

The time moments (nodes) at the ends of time steps are a series of linkage positions to be analyzed. The relationship between the time moments and the time steps is as follows:

$$t_0 \xrightarrow{\text{step 1}} t_1 \xrightarrow{\text{step 2}} t_2 \cdots t_{i-1} \xrightarrow{\text{step i}} t_i \cdots t_{n-1} \xrightarrow{\text{step n}} t_n.$$

Within the i-th time step $t_{i-1} \leq t \leq t_i$, the time-varying $\boldsymbol{M}$ and $\boldsymbol{K}$ are treated as constants because Δt is very short. For example, we take the values of $\boldsymbol{M}$ and $\boldsymbol{K}$ at the starting moment t_{i-1}. Then an eigenvalue problem within the time step can be formed:

$$\left[\boldsymbol{K}^{(i)}-\left(\omega^{(i)}\right)^2 \boldsymbol{M}^{(i)}\right]\boldsymbol{\varphi}^{(i)}=\mathbf{0}. \tag{11.79}$$

Then, natural frequencies $\omega_r^{(i)}$ and the corresponding vibration mode $\boldsymbol{\varphi}_r^{(i)}$ can be obtained by solving the eigenvalue problem. Retaining the vibration mode up to n_r order ($r=1,2,\cdots,n_r$) and neglecting the contribution from higher orders, the principal vibration of each retained mode can be represented as

$$\ddot{\eta}_r(t)+2\zeta_r\dot{\eta}_r(t)+(\omega_r^{(i)})^2\eta_r(t)=N_r(t) \quad (r=1,2,\cdots,n_r), \tag{11.80}$$

where $\eta(t)_r$, $\dot{\eta}_r(t)$, and $\ddot{\eta}_r(t)$ are the r-th modal coordinate and its first- and second-order derivatives, respectively. $N_r(t)$ is the excision of the r-th mode. ζ_r is the modal damping ratio.

According to Duhamel's integral (Eq. (5.70)), the response at the end moment of the step can, if the responses at the starting moment $\eta_r(t_{i-1})$, $\dot{\eta}_r(t_{i-1})$ are known, be represented as:

$$\begin{aligned}\eta_r(t_i)=&\left(\frac{1}{\omega_{\mathrm{d}r}^{(i)}}\right)\int_{t_{i-1}}^{t_i} N_r(\tau)e^{-\zeta_r\omega_r^{(i)}(t_i-\tau)}\sin\omega_{\mathrm{d}r}^{(i)}(t_i-\tau)\,\mathrm{d}\tau+\\ &\frac{\eta_r(t_{i-1})}{\sqrt{1-\zeta_r^2}}e^{-\zeta_r\omega_r^{(i)}\Delta t}\cos\left(\omega_{\mathrm{d}r}^{(i)}\Delta t-\Psi_r\right)+\frac{\dot{\eta}_r(t_{i-1})}{\omega_{\mathrm{d}r}^{(i)}}e^{-\zeta_r\omega_r^{(i)}\Delta t}\sin\omega_{\mathrm{d}r}^{(i)}\Delta t\end{aligned} \tag{11.81}$$

$$\dot{\eta}_r(t_i)=\left(\frac{1}{\omega_{\mathrm{d}r}^{(i)}}\right)\int_{t_{i-1}}^{t_i} N_r(\tau)\left[-\zeta_r\omega_r^{(i)}e^{-\zeta_r\omega_r^{(i)}(t_i-\tau)}\sin\omega_{\mathrm{d}r}^{(i)}(t_i-\tau)\right]\mathrm{d}\tau+$$

$$\left(\frac{1}{\omega_{\text{dr}}^{(i)}}\right)\int_{t_{i-1}}^{t_i} N_r(\tau)\left[\omega_{\text{dr}}^{(i)} e^{-\zeta_r\omega_r^{(i)}(t_i-\tau)}\cos\omega_{\text{dr}}^{(i)}(t_i-\tau)\right]\mathrm{d}\tau +$$
$$-\eta_r(t_{i-1})\frac{\omega_{\text{dr}}^{(i)}}{\sqrt{1-\zeta_r^2}}e^{-\zeta_r\omega_r^{(i)}\Delta t}\sin\left(\omega_{\text{dr}}^{(i)}\Delta t-\Psi_r\right)-$$
$$\eta_r(t_{i-1})\frac{\zeta_r\omega_r^{(i)}}{\sqrt{1-\zeta_r^2}}e^{-\zeta_r\omega_r^{(i)}\Delta t}\cos\left(\omega_{\text{dr}}^{(i)}\Delta t-\Psi_r\right)-$$
$$\dot{\eta}_r(t_{i-1})e^{-\zeta_r\omega_r^{(i)}\Delta t}\left[\frac{\zeta_r\omega_r^{(i)}}{\omega_{\text{dr}}^{(i)}}\sin\omega_{\text{dr}}^{(i)}\Delta t-\cos\omega_{\text{dr}}^{(i)}\Delta t\right], \tag{11.82}$$

where

$$\Psi_r = \arctan\left(\frac{\zeta_r}{\sqrt{1-\zeta_r^2}}\right). \tag{11.83}$$

Eqs. (11.81) and (11.82) can be shortened to

$$\begin{cases} a_1^{(i)}\eta(t_{i-1}) + b_1^{(i)}\dot{\eta}(t_{i-1}) + c_1^{(i)}\eta(t_i) + d_1^{(i)}\dot{\eta}(t_i) = e_1^{(i)} \\ a_2^{(i)}\eta(t_{i-1}) + b_2^{(i)}\dot{\eta}(t_{i-1}) + c_2^{(i)}\eta(t_i) + d_2^{(i)}\dot{\eta}(t_i) = e_2^{(i)}. \end{cases} \tag{11.84}$$

The coefficients in these equations are summarized here:

$$\begin{cases} a_1^{(i)} = \dfrac{-e^{-\zeta_r\omega_r^{(i)}\Delta t}}{\sqrt{1-\zeta_r^2}}\cos\left(\omega_{\text{dr}}^{(i)}\Delta t-\Psi_r\right) \\ b_1^{(i)} = \dfrac{-e^{-\zeta_r\omega_r^{(i)}\Delta t}}{\omega_{\text{dr}}^{i}}\sin\omega_{\text{dr}}^{(i)}\Delta t \\ c_1^{(i)} = 1 \\ d_1^{(i)} = 0 \\ a_2^{(i)} = \omega_r^{(i)}e^{-\zeta_r\omega_r^{(i)}\Delta t}\left\{\sin\left(\omega_{\text{dr}}^{(i)}\Delta t-\Psi_r\right)+\dfrac{\zeta_r}{\sqrt{1-\zeta_r^2}}\cos\left(\omega_{\text{dr}}^{(i)}\Delta t-\Psi_r\right)\right\} \\ b_2^{(i)} = e^{-\zeta_r\omega_r^{(i)}\Delta t}\left\{\dfrac{\zeta_r}{\sqrt{1-\zeta_r^2}}\sin\omega_{\text{dr}}^{(i)}\Delta t-\cos\omega_{\text{dr}}^{(i)}\Delta t\right\} \\ c_2^{(i)} = 1 \\ d_2^{(i)} = 0 \\ e_1^{(i)} = \dfrac{1}{\omega_{\text{dr}}^{(i)}}\int_{t_{i-1}}^{t_i} N_r(\tau)e^{-\zeta_r\omega_r^{(i)}(t_i-\tau)}\sin\omega_{\text{dr}}^{(i)}(t_i-\tau)\,\mathrm{d}\tau \\ e_2^{(i)} = \int_{t_{i-1}}^{t_i} N_r(\tau)e^{-\zeta_r\omega_r^{(i)}(t_i-\tau)}\left\{\dfrac{-\zeta_r}{\sqrt{1-\zeta_r^2}}\sin\omega_{\text{dr}}^{(i)}(t_i-\tau)+\cos\omega_{\text{dr}}^{(i)}(t_i-\tau)\right\}\mathrm{d}\tau. \end{cases} \tag{11.85}$$

For each time step, an algebraic equation set Eq. (11.84) is formed. Notice that the motion is periodic, and we have

$$\begin{cases} \eta_r(t_n) = \eta_r(t_0) \\ \dot{\eta}_r(t_n) = \dot{\eta}_r(t_0). \end{cases} \tag{11.86}$$

Combining Eq. (11.84) of each time step and taking advantage of Eq. (11.86), the following equation in matrix form can be formed:

$$\boldsymbol{A}\eta = \boldsymbol{E}, \tag{11.87}$$

where

$$\boldsymbol{A} = \begin{bmatrix} a_1^{(1)} & b_1^{(1)} & c_1^{(1)} & d_1^{(1)} & 0 & & \cdots & & 0 \\ a_2^{(1)} & b_2^{(1)} & c_2^{(1)} & d_2^{(1)} & 0 & & \cdots & & 0 \\ 0 & 0 & a_1^{(2)} & b_1^{(2)} & c_1^{(2)} & d_1^{(2)} & 0 & \cdots & 0 \\ 0 & 0 & a_2^{(2)} & b_2^{(2)} & c_2^{(2)} & d_2^{(2)} & 0 & \cdots & 0 \\ \vdots & \vdots & \vdots & \vdots & \vdots & \vdots & \vdots & \vdots & \vdots \\ 0 & & \cdots & & 0 & a_1^{(n-1)} & b_1^{(n-1)} & c_1^{(n-1)} & d_1^{(n-1)} \\ 0 & & \cdots & & 0 & a_2^{(n-1)} & b_2^{(n-1)} & c_2^{(n-1)} & d_2^{(n-1)} \\ c_1^{(n)} & d_1^{(n)} & 0 & & \cdots & & 0 & a_1^{(n)} & b_1^{(n)} \\ c_2^{(n)} & d_2^{(n)} & 0 & & \cdots & & 0 & a_2^{(n)} & b_2^{(n)} \end{bmatrix}$$

$$\eta = \begin{bmatrix} \eta_r(t_0) \\ \dot{\eta}_r(t_0) \\ \eta_r(t_1) \\ \dot{\eta}_r(t_1) \\ \vdots \\ \eta_r(t_{n-2}) \\ \dot{\eta}_r(t_{n-2}) \\ \eta_r(t_{n-1}) \\ \dot{\eta}_r(t_{n-1}) \end{bmatrix}; \qquad \boldsymbol{E} = \begin{bmatrix} e_1^{(1)} \\ e_2^{(1)} \\ e_1^{(2)} \\ e_2^{(2)} \\ \vdots \\ e_1^{(n-1)} \\ e_2^{(n-1)} \\ e_1^{(n)} \\ e_2^{(n)} \end{bmatrix}.$$

$\eta_r(t_0), \eta_r(t_1), \cdots, \eta_r(t_{n-1})$ can be easily obtained by solving Eq. (11.87). Then, the physical response can be calculated by the modal superposition principle:

$$U(t) = \sum_{r=1}^{n_r} \eta_r(t)\varphi_r. \tag{11.88}$$

The modal force $N_r(\tau)$ in Eqs. (11.81) and (11.82) varies with time. In a small time step, it can be approximated by a linear function as

$$N_r(\tau) = N_r(t_{i-1}) + \frac{N_r(t_i) - N_r(t_{i-1})}{\Delta t}(\tau - t_{i-1}). \tag{11.89}$$

To use the modal superposition method, both the mass matrix and the stiffness matrix are required to be symmetric. This condition is met in the linkage dynamic equation.

11.4.2 Result analysis

The dynamic response can be used for many technical analyse, such as the analysis of 1) the error of motion and 2) the dynamic stress.

11.4.2.1 Error of motion

The actual motion of a mechanism is the nominal motion superimposed with the elastic deformation. The nominal motion is obtained through kinematic analysis. The elastic deformation $\boldsymbol{U}(t)$ is obtained through solving Eq. (11.77). $\dot{\boldsymbol{U}}(t)$ and $\ddot{\boldsymbol{U}}(t)$ can be solved by numerical differentiation methods, such as the finite difference method.

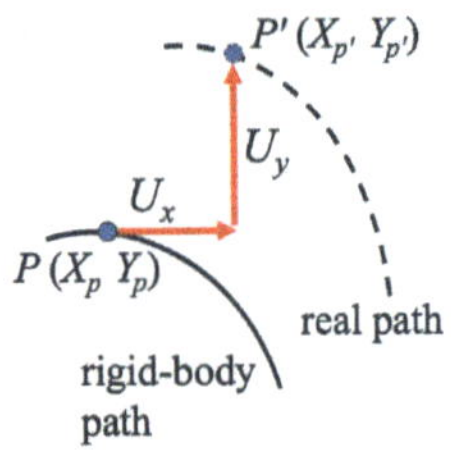

Figure 11.8 Actual and theoretical paths.

For a path-generating linkage, for example, it is required to determine the deviation of the path traced by a point P in the coupler. The nominal (theoretical) path and the real path are shown in Fig. 11.8. Point P on the theoretical path moves to the point P' on the real path. The relationship between the two are as follows:

$$X_{P'}(t) = X_P(t) + U_x(t) \tag{11.90}$$

$$Y_{P'}(t) = Y_P(t) + U_x(t). \tag{11.91}$$

$X_P(t)$ and $Y_P(t)$ are given by the kinematic analysis, while $U_x(t)$ and $U_y(t)$ are obtained by the elasto-dynamic analysis.

11.4.2.2 Dynamic stress

With the elasto-dynamic analysis result, the dynamic stress within a link can be analyzed. For a beam element, the dynamic stress is a function of both time t and the point location. It can be represented as

$$\sigma(x, t) = Eh|W''(x, t)| + E|V'(x, t)|. \tag{11.92}$$

The first and the second terms on the right-hand side represent the stresses caused by bending and tensile/compression, respectively. E is the modulus of elasticity, and h is the distance from the neutral axis to the outermost edge of the section. Using the shape function defined in Eq. (11.5), it becomes

$$\sigma(x, t) = Eh|\varphi_i''(x)u_i(t)| + \frac{E}{L}|u_5(t) - u_1(t)| \quad (i = 2, 3, 4, 6, 7, 8). \tag{11.93}$$

At a specific time moment t, the bending stress depends on the position x. To find the maximum stress, we have

$$\frac{\partial \sigma_b}{\partial x} = Eh\left|\sum_i \varphi_i'''(x)u_i(t)\right| = 0 \quad (i = 2, 3, 4, 6, 7, 8). \tag{11.94}$$

Since the shape functions are quintic polynomials of x, Eq. (11.94) can be reduced to the quadratic equation

$$Ae^2 + Be + C = 0, \tag{11.95}$$

where $e = x/L$, and A, B and C are:

$$\begin{cases} A = 120(u_6 - u_2) - 60L(u_3 + u_7) + 10L^2(u_8 - u_4) \\ B = 120(u_2 - u_6) + 8L(8u_3 + 7u_7) + 4L^2(3u_4 - 2u_8) \\ C = 20(u_6 - u_2) - 4L(3u_3 + 2u_7) + L^2(u_8 - 3u_4). \end{cases} \tag{11.96}$$

These procedures to find the maximum dynamic stress can be summarized as:

1) By solving Eq. (11.77), $\boldsymbol{U}(t)$ is obtained. Using Eq. (11.70) gives $\boldsymbol{U}_i^e(t)$, which is substituted into Eq. (11.55) to find $\boldsymbol{u}_i(t)$;
2) Solving Eq. (11.95) yields the e at which the maximum stress happens. Substituting e into Eq. (11.93) gives the maximum stress at this specific moment. If $e \leq 0$ or $e \geq 1$, it simply means the maximum stress appears at the end of the element. The maximum stress can be calculated by substituting $e = 0$ or $e = 1$ into Eq. (11.93);

3) Repeating the process for all positions, the maximum stress in the i-th element can be obtained. By comparing the stress of elements on the same link, the maximum stress in the link can be obtained.

Now, we see the point for using the quintic polynomial shape function more clearly. In this case, we have a quadratic equation to obtain the maximum stress. However, if cubic polynomial is used for the shape function, the bending stress in the element will be a constant.

11.4.2.3 Analysis in different levels

The dynamic analysis can be conducted in two different levels. Analysis based on Eq. (11.77) is referred to as *kineto-elastodynamic analysis* (KED) in the literature. This analysis needs to solve the time-varying ODE sets. It is relatively time consuming and requires higher computation power.

If the inertia and the damping terms on the left-hand side are neglected, Eq. (11.77) is reduced to

$$\boldsymbol{KU} = \boldsymbol{F} - \boldsymbol{M}\ddot{\boldsymbol{U}}_{\mathrm{r}}. \tag{11.97}$$

Analysis based on this equation is referred to as *kineto-elastostatic analysis* (KES) or *quasi-static analysis*.

The definition of the analysis and the relevant physical implications are not always consistent in the literature. In view of this, we make the following clarification and explanation on the two types of analysis.

- KED treats a linkage as a moving elastic system and studies the vibration of the linkage under the excitation of external forces and rigid-body inertia forces.
- KES treats a linkage as a moving elastic system as well, but studies the static deformation of the linkage under the action of external forces and the rigid-body inertia forces.
- *Elasto-dynamic analysis* is a general term including both KED and KES.

In general, KES analysis is much simpler than KED analysis, requiring much less computation time.

The results of the two analyses may be very close to each other in some cases, or differ from each other very much in some other cases. A comparison of the two types of analytical results for several cases is shown in Fig. 11.9, in which the solid lines are the KED, while the dashed lines represent the KES. The three linkages have completely identical lengths of links; however, the cross sections, thus the rigidity, of the links are different. The rigidities decrease from Fig. 11.9 (a) to (c). Fig. 11.9(a) has the largest

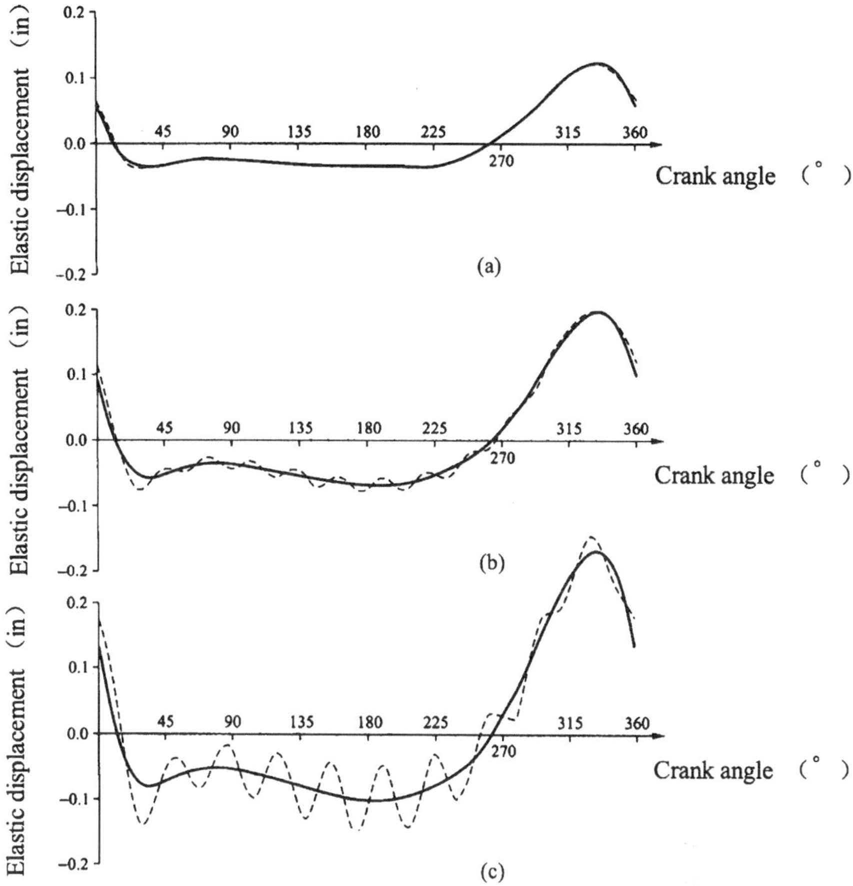

Figure 11.9 KES and KED analysis.

stiffness; the KED and KES results are almost the same. Fig. 11.9(c) has the smallest stiffness; the difference between the KES and KED is obvious. While Fig. 11.9(b) has the stiffness in the middle, difference between KED and KES is also between (a) and (c).

In general, KED is required if the links have relatively large flexibility. While if the mechanism is relatively stiff, KES may be adequate.

11.5 Elasto-dynamic synthesis and suppression of vibration

The elasto-dynamic response has a series of detrimental effects on the linkage, such as causing motion error and vibration, radiating noise and quickening fatigue, etc. In the following, mechanism synthesis with an objective of improving motion accuracy and vibration suppression is briefly discussed.

11.5.1 Elasto-dynamic synthesis

KED synthesis is very complicated and time-consuming due to the fact that the synthesis generally involves many cycles of KED analysis.

I. Imam et al. (1979) formulated the design of elastic linkage as a nonlinear programming problem. They took the cross-sectional dimensions of links as the design variables and the weight of mechanism as the objective function. The motion error was treated as a design constraint. As long as the motion error is within a certain range, it is regarded as acceptable. This method involves many cycles of KED analysis in the iterative process, thus, it is very time-consuming. W. Thornton et al. (1979) introduced the optimality criterion technique from the structural design field into linkage design. The basic idea is that the minimum mechanism weight can be obtained by fully stressing each linked. In this method, the number of iterations required is greatly reduced. A new elasto-dynamic optimum synthesis method was formulated by Zhang and Grandin (1982, 1983), who introduced the *kinematic refinement technique* in combination with the optimality criterion technique. In this method, the cross-sectional parameters are taken as the first group of design variables, and a fully-stressed linkage is obtained by using the optimality criterion technique. Geometric parameters are used as the second group of design variables, and a linkage with an improved criterion of kinematic performance is obtained by means of the kinematic refinement technique.

In the following section, the kinematic refinement technique is introduced to demonstrate the basic idea of reducing dynamic motion error through elasto-dynamic synthesis.

The principle of kinematic refinement technique is to adjust the geometric parameters of the linkage, such as the link length and the position of the fixed joints, so that the real motion of linkage having elastic deformation approaches the expected motion.

Fig. 11.10 shows a path-generating four-bar linkage. The expected path, described as the design requirement and represented by T^{E}, is given by a series of coordinates X_i^{E}, and Y_i^{E} $(i = 1, 2, \cdots, n)$. n is the number of given points on the path. The path traced by the rigid-body linkage with the kinematic synthesis is referred to as the rigid-body path. The difference

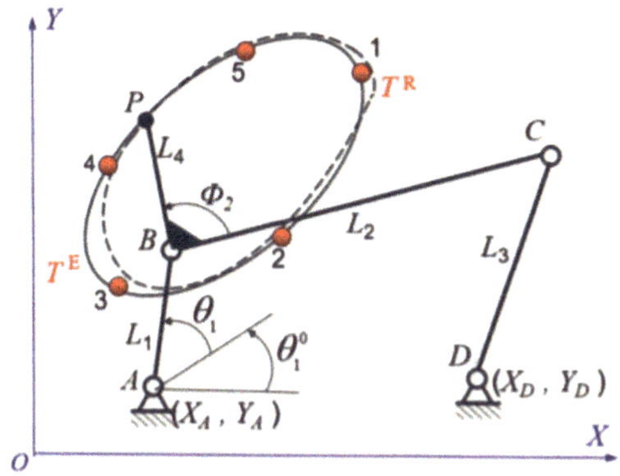

Figure 11.10 Kinematic refinement technique.

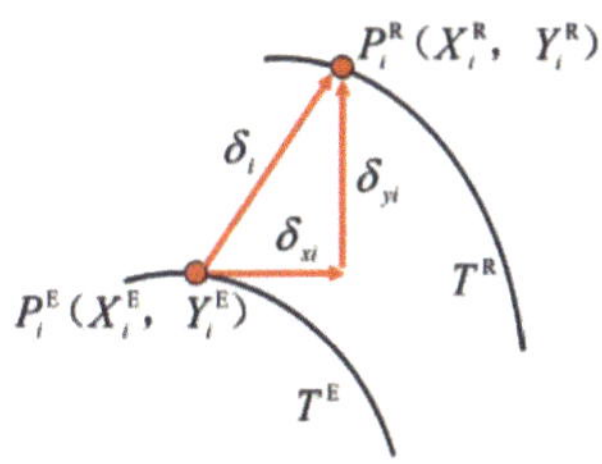

Figure 11.11 Dynamic path error.

between the expected path and the rigid-body path is called the structural error, which is very small in practice. The path traced by the elastic linkage is called the real path, represented by T^R, which is given by the a series of coordinates X_i^R, and Y_i^R $(i = 1, 2, \cdots, n)$. T^E and T^R are shown in Fig. 11.10.

The dynamic-path error is defined as the difference between the expected path and the real path in X and Y directions as shown in Fig. 11.11:

$$\delta_{xi} = X_i^R - X_i^E \tag{11.98}$$

$$\delta_{yi} = Y_i^R - Y_i^E. \tag{11.99}$$

The path deviation is defined as

$$\delta_i = \sqrt{(\delta_{xi})^2 + (\delta_{yi})^2}. \tag{11.100}$$

The dynamic-path error and deviation can both be obtained through elasto-dynamic analysis.

In the linkage of Fig. 11.10, ten design variables are chosen as follows:

$$V = \begin{bmatrix} L_1 & L_2 & L_3 & L_4 & \phi_2 & \theta_1^0 & X_A & Y_A & X_D & Y_D \end{bmatrix}^{\mathrm{T}}. \tag{11.101}$$

The meaning of the variables is shown in Fig. 11.10. For example, θ_1^0 is the angle between the crank and X at the initial point.

The kinematic refinement technique is based on the assumption that a design parameter affects the rigid-body path and the real path in the same way.

There are two criteria accepted in the control of error in mechanism study, namely, the maximum error or the root-mean-square error. In the following, we only use the maximum error. We choose five points on the path to evaluate the closeness of the two paths.

If a slight change ΔV_j $(j = 1, 2, \cdots, 10)$ is made in a design variable, the change of the rigid-body path coordinates at the i-th path point is

$$\begin{cases} \Delta X_i^{\mathrm{R}} = \sum\limits_{j=1}^{10} \dfrac{\partial X_i^{\mathrm{R}}}{\partial V_j} \Delta V_j \\ \Delta Y_i^{\mathrm{R}} = \sum\limits_{j=1}^{10} \dfrac{\partial Y_i^{\mathrm{R}}}{\partial V_j} \Delta V_j. \end{cases} \tag{11.102}$$

If this change of rigid-body path is a part of the dynamic path error, it would be reasonable to assume that the real path of the new linkage will be closer to the rigid-body path. Let

$$\begin{cases} \Delta X_i^{\mathrm{R}} = -C\delta_{xi} \\ \Delta Y_i^{\mathrm{R}} = -C\delta_{yi}. \end{cases} \tag{11.103}$$

C is called the *relaxation coefficient* ranging between 0 and 1. Its physical meaning will be explained later.

From this, the following linear relationship can be obtained:

$$\boldsymbol{J}\Delta \boldsymbol{V} = \boldsymbol{B}, \tag{11.104}$$

where $\Delta \boldsymbol{V}$ is the parameter adjustment vector:

$$\Delta \boldsymbol{V} = \begin{bmatrix} \Delta V_1 & \Delta V_2 & \cdots & \Delta V_{10} \end{bmatrix}^{\mathrm{T}}. \tag{11.105}$$

$\boldsymbol{B}$ is the displacement correction vector:

$$\boldsymbol{B} = -C\begin{bmatrix} \delta_{x1} & \delta_{y1} & \delta_{x2} & \delta_{y2} & \cdots \delta_{x5} & \delta_{y5} \end{bmatrix}^{\mathrm{T}}. \tag{11.106}$$

$\boldsymbol{J}$ is the Jacobin matrix or the sensitivity matrix. Its element is calculated as follows:

$$J_{\mathrm{kl}} = \begin{cases} \dfrac{\partial X_i^{\mathrm{R}}}{\partial V_i} & \text{if } k = 2i-1, l = j \\ \dfrac{\partial Y_i^{\mathrm{R}}}{\partial V_i} & \text{if } k = 2i, l = j. \end{cases} \tag{11.107}$$

The sensitivity matrix generally has to be calculated numerically. For this purpose, we conduct the kinematic analysis 11 times:

$$\begin{cases} \boldsymbol{V}^{(0)} = \begin{bmatrix} V_1 & V_2 & \cdots & V_{10} \end{bmatrix}^{\mathrm{T}} \\ \boldsymbol{V}^{(1)} = \begin{bmatrix} V_1 + \epsilon_1 & V_2 & \cdots & V_{10} \end{bmatrix}^{\mathrm{T}} \\ \boldsymbol{V}^{(2)} = \begin{bmatrix} V_1 & V_2 + \epsilon_2 & \cdots & V_{10} \end{bmatrix}^{\mathrm{T}} \\ \cdots \\ \boldsymbol{V}^{(10)} = \begin{bmatrix} V_1 & V_2 & \cdots & V_{10} + \epsilon_{10} \end{bmatrix}^{\mathrm{T}}. \end{cases} \tag{11.108}$$

ϵ_j is chosen based on the rule:

$$\epsilon_j = \begin{cases} \dfrac{\delta_{\max}}{10} & \text{for } (j = 1, 2, 3, 4, 7, 8, 9, 10) \\ \dfrac{\delta_{\max}}{10L_1} & \text{for } (j = 5, 6). \end{cases} \tag{11.109}$$

$\delta_{\max}$ is the maximum deviation in one period.

If the coordinates of the rigid-body path at the i-th refinement point obtained from the kinematic analysis of the j-th linkage are $\left(X_i^{\mathrm{R}}\right)_j$ and $\left(Y_i^{\mathrm{R}}\right)_j$, then the elements of sensitivity matrix can be calculated as:

$$J_{\mathrm{kl}} = \begin{cases} \dfrac{\partial X_i^{\mathrm{R}}}{\partial V_i} \approx \dfrac{\left(X_i^{\mathrm{R}}\right)_j - \left(X_i^{\mathrm{R}}\right)_0}{\epsilon_j} & \text{if } k = 2i-1, l = j \\ \dfrac{\partial Y_i^{\mathrm{R}}}{\partial V_i} \approx \dfrac{\left(Y_i^{\mathrm{R}}\right)_j - \left(Y_i^{\mathrm{R}}\right)_0}{\epsilon_j} & \text{if } k = 2i, l = j. \end{cases} \tag{11.110}$$

$\Delta \boldsymbol{V}$, the parameter adjustment vector, can be obtained by solving Eq. (11.104). Then the new parameter of the linkage is

$$V_j' = V_j + \Delta V_j \quad (j = 1, 2, \cdots, 10), \tag{11.111}$$

where V_j is the parameter of the original linkage.

Up to this point, a round of parameter adjustment and kinematic refinement have been completed. In general, multiple iterations are required to complete the process. Each iteration needs elasto-dynamic analysis and the calculation of δ_{xi} and δ_{yi} according to Eqs. (11.98) and (11.99). X_i^R and Y_i^R come from the rigid-body path.

Calculation shows that this refinement process converges rather fast. After three–five iterations, the path deviation may be reduced by 50% to 70%, or even 80%. The essence of this refinement is to compensate the path error by elastic deformation. However, this deviation could not be eliminated completely. After several improving iterations, the error may increase again as shown in Fig. 11.12. The reason for this may be that some points improve, but some other points are worsened.

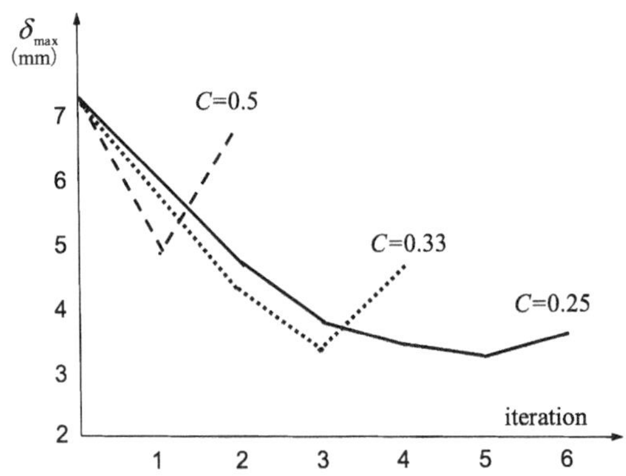

Figure 11.12 Influence of relaxation coefficient.

The introduction of relaxation coefficient C is to prevent excessive correction in an iteration. Excessive correction may break the basic assumptions of kinematic refinement technique (the change of parameters affects the rigid-body path and the real path in the same way). The selection of C value significantly affects the effectiveness of improvement and the number of iterations required. As a rule of thumb, C is in the range 0.2–0.3.

11.5.2 Suppression of elasto-dynamic response

Proper design may have less vibration response in the mechanism. In addition, the response can also be controlled through the measures described next.

11.5.2.1 Composite materials

Thompson et al. (1983) studied the use of composite materials to make the links of high-speed linkages. Composite materials generally have high stiffness-to-weight ratios; mechanisms made from composite materials have higher natural frequency. In addition, composite materials have relatively large damping and can restrain the vibration. Generally, high-speed linkages work with relatively low external loads. The high strength-to-weight ratio, good fatigue resistance, and easiness of forming make composite materials very suitable for construction of elastic linkages.

11.5.2.2 Active control

The suppression of elasto-dynamic response by active control is the application of vibration control in mechanism dynamics. Elastic linkages are complex time-varying and nonlinear systems with multivariables. In general, two types of active control strategies are use:

1) **Smart Mechanism**
   Some researchers explored the active control of elastic linkages by using piezoelectric ceramic materials. The piezoelectric material is used as either sensors to pick up vibration response, or actuators to create control torque required in the feedback control. More details can be found in the paper by (Song et al., 2001), in which artificial neural networks are applied to active vibration control of a flexible linkage.
2) **Controllable Mechanism**
   Oliver et al. (1985) proposed the idea of retrofitting a four-bar linkage with an elastic coupler. The fixed joint at the rocker was replaced by a slider so that the linkage has two rigid-body DOFs. The crank was driven by a motor and the exciter controlled by a microprocessor acted

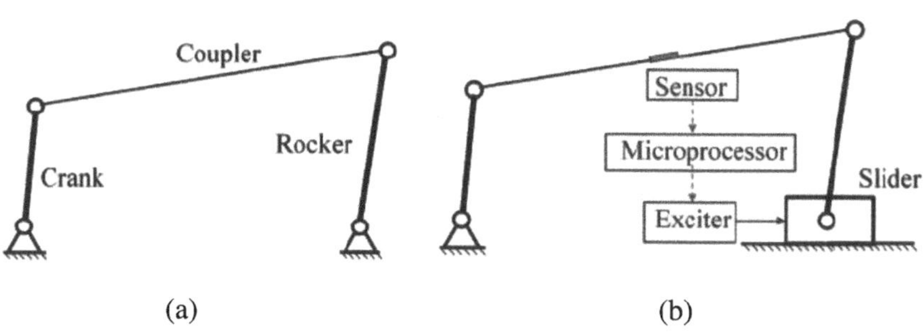

Figure 11.13 Retrofitment design of a four-bar linkage.

on the slider to provide the control input. The control quantity of the microprocessor is related to the dynamic response measured on the elastic coupler, which constitutes a closed-loop feedback control system as shown in Fig. 11.13 and improves the quasi-static response of the linkage.

CHAPTER 12

Elasto-dynamics of gear trains

12.1 Introduction

12.1.1 Historical review

Humans started using primitive gears more than 2000 years ago. During the 13th century, cycloidal gears were invented and first used in the watch and clock industry in Switzerland. Involute gears appeared during the First Industrial Revolution. The application of involute gears greatly improved the quality of mechanical transmission, being critical to the nonstop growth of operation speed and power of various machines (Zhang and Yang, 2020).

Gears are mainly used in mechanical transmissions, which are integral to almost all types of machines, such as machine tools, automobile, aircraft, and ships, among many others.

Study of gear dynamics was directly motivated by the practical need to estimate the dynamic load (Özgüven and Houser, 1988). The concept of dynamic load factor, which is defined as the ratio of dynamic load to static load, was originally suggested in 1868 by H. Walker. In 1925 a team led by E. Buckingham conducted a series of experimental study on the dynamic load of gears. The report was published in 1931 (Buckingham, 1931) as the first authoritative document on gear dynamics. It described the procedures to determine the tooth load variation caused by mesh dynamics and gear-tooth error. It was shown that the tooth load variation depends largely upon the effective masses, the effective error, and the speed of the gears. Buckingham's method (Buckingham, 1949) was based on the impact theory, yet without taking into account vibration theory.

The first spring–mass model of gears was introduced by Tuplin (1950). It treated a gear system as an elastic system for the first time, signaling a new era in gear dynamics. Afterwards, various refined and more complicated dynamic models have been proposed. In these models, the time-varying meshing stiffness, gear-tooth backlash, gear-tooth profile machining error, etc., are considered.

12.1.2 Features of gear dynamics

Gear dynamics generally deals with the vibration of gears, shaft, and bearings without inclusion of the supporting structures, such as the gear box

Machinery Dynamics
https://doi.org/10.1016/B978-0-12-815785-5.00023-6

housing. If the supporting structure is incorporated into the dynamic model, it is referred to as *gear system dynamics*. In this chapter, only gear dynamics is presented, without involving the ambient structures. Gear dynamics has a series of unique features that differentiate it from common vibratory systems. These features are briefly summarized as follows:

1. **Excitation:** In gear dynamics, there are both *internal excitation* and *external excitation*. The focus is placed on the response to internal excitation.
2. **Dynamic model:** Depending on the research objectives, models of various levels of complexity are used. These models can be roughly grouped as: 1) pure rotational models, 2) translational–rotational models, and 3) gear rotor system model, etc.
3. **Solution method:** The governing equation of motion in gear dynamics has the form of an ODE with time-varying coefficients and sometimes clearance-type nonlinearity. Due to the complexity, the main methods of solution are various numerical methods. Some analytical methods, such as the harmonic balance method and the perturbation method, also find application with proper simplification to the model.

The chapter is organized as follows:

- Section 12.2 analyzed the different types of excitation in gear dynamics.
- In Sections 12.3.1 and 12.4, pure rotational model and translational–rotational models are presented, respectively.
- Section 12.5 concludes the chapter with a short review of gear dynamics.

12.2 Excitation in gear dynamics

The excitation sources in a gear system are very complicated and are broadly divided into two categories: *external excitation* and *internal excitation*.

External excitation refers to the driving torque and working load. Internal excitation refers to the excitation directly produced by the meshing action, including: 1) *time-varying stiffness excitation*, 2) *tooth profile error excitation*, and 3) *meshing impact excitation*.

The concept of internal excitation was first described by Harris (1958). Today, it has become one of the central concepts in gear dynamics.

12.2.1 Stiffness excitation

Stiffness excitation is caused by the change of number of tooth pairs in contact during the meshing process. For example, if a gear pair has a *contact ratio* between 1 and 2, it means that one pair of teeth is in contact for some time, while two pairs of teeth are in mesh in the other time. If one tooth pair in contact is treated as a spring, two pairs of teeth in contact means two springs in parallel. In another words, the gear meshing stiffness changes alternatively between one spring and two springs. This time-varying stiffness is referred to as *parametric excitation* (Nayfeh and Mook, 2008). In the following, the calculation of meshing stiffness is presented.

12.2.1.1 Gear-tooth deformation

Gear teeth undergo elastic deformation under the meshing load. The deformation is evaluated conventionally along the *line of action*. The deformation consists of three parts: namely, the bending deformation, the shearing deformation and the surface contact deformation. These deformations can be calculated by relatively simple mechanics of materials, or more accurate finite element method. Some of the literature also give more accurate estimation techniques based on the size of the gear teeth (Smith, 2003). A method proposed by Lin et al. (1988) is given in Appendix C.

12.2.1.2 Meshing stiffness

Once the tooth deformation along the line of action is obtained, the stiffness of a single tooth at this point can be calculated as

$$k_k = \frac{F_k}{\delta_k}, \tag{12.1}$$

where δ_k is the deformation of the tooth at an arbitrary contact point k, F_k is the force applied on the gear tooth along the line of action, and k_k is the tooth stiffness of point k. It should be noted that the tooth deformation, thus the tooth stiffness, varies at different points.

A gear pair is equivalent to two springs connected in series. Thus, the stiffness of one gear pair is calculated as

$$k_{vi} = \frac{k_1 k_2}{k_1 + k_2}. \tag{12.2}$$

Generally, the *contact ratio* ϵ for a pair of spur gears is between 1 and 2. When one pair of gear teeth in contact, the meshing stiffness is the stiffness of one pair spring. When two gear pairs are in contact, this is equivalent to two pairs of springs being in parallel. Thus, the overall meshing stiffness is calculated by

$$k_v = k_{v1} + k_{v2}. \tag{12.3}$$

k_{v1} and k_{v2} are the stiffness of individual tooth-pair springs. Assuming the gears rotate at a constant angular velocity, the change of k_v, k_{v1}, and k_{v2} is schematically shown in Fig. 12.1, in which the stiffness of the two individual tooth-pair springs are drawn as dashed line, while the meshing stiffness is a solid line.

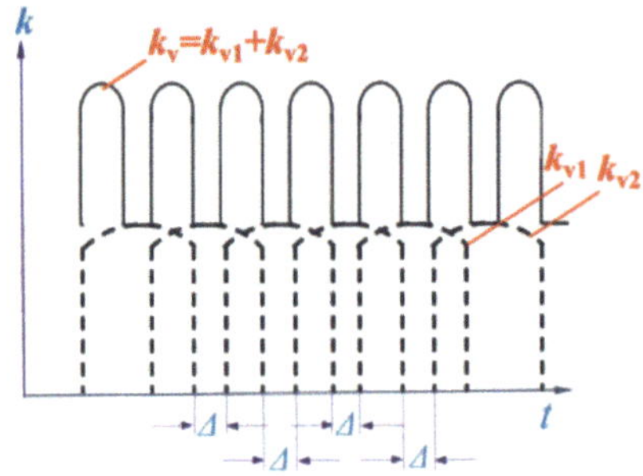

Figure 12.1 Meshing stiffness.

Obviously, the meshing stiffness changes abruptly when the meshing status changes from one pair to two pair, or vice versa. However, the variation is small during a one-pair or two-pair mesh. Thus, it is common practice to simplify the stiffness as an alternating rectangular wave in practice.

12.2.2 Error excitation

There are inevitably many types of errors in gear machining and assembly processes. All these errors will influence the dynamic response. For example, some types of machining errors excite the system as displacements.

Gear errors, which are roughly categorized into three groups: *motion accuracy*, *smoothness* and *contact accuracy,* are very complicated .

The effect of gear errors on dynamics is traditionally illustrated with the concept of *transmission error*. For a pair of ideal rigid gears, when the driving gear rotates at a constant speed, the driven gear is expected to rotate at a constant speed as well. However, the actual motion of the driven gear always deviates slightly from the expected one because of various errors and material elasticity. The difference between the expected motion and the real motion of the driven gear is defined as the transmission error (Smith, 2003). One example of transmission error measured is given in Fig. 12.2. It is seen that the transmission error consists of a relatively long wave and high-frequency short waves. Also, the long wave has a roughly sinusoidal shape,

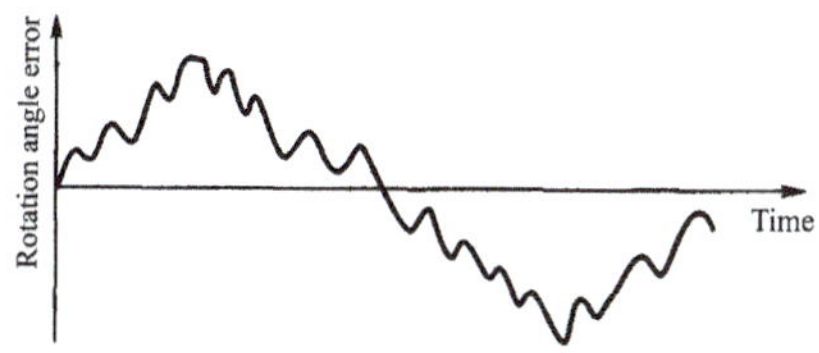

Figure 12.2 Transmission error.

mainly caused by the eccentricity of the gears. Thus, it has the rotation frequency. This long wave mainly affects the kinematic accuracy, but has little impact on the dynamics. The short wave, however, is mainly caused by two types of machining errors, namely, the tooth profile error and the tooth spacing error. Its frequency is the *meshing frequency* f_{m}, which is much higher than that of the long wave:

$$f_{\mathrm{m}} = \frac{n_1 z_1}{60} = \frac{n_2 z_2}{60}, \tag{12.4}$$

where n_1, n_2 are the rotational speeds and z_1, z_2 are the gear-tooth numbers. Subscripts 1 and 2 means the driving and driven gears.

The short-wave error excites the gear dynamics and significantly affects the smoothness of transmission. The two components in the short-wave error, however, have a different excitation mechanism. The tooth profile error is a displacement excitation between the meshing gears as shown in Fig. 12.3. The tooth spacing error is more complicated, exciting the system as meshing impacts which will be discussed later.

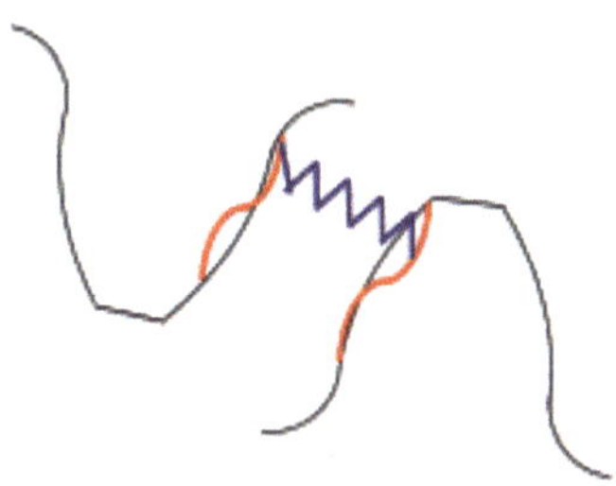

Figure 12.3 Tooth profile error.

The tooth profile error, designated as $e(t)$, can be represented by a Fourier series with the meshing frequency as the fundamental wave:

$$e(t) = \sum_j e_j \sin(j\omega_{\mathrm{m}} t + \Psi_j), \tag{12.5}$$

where ω_{m} is the circular meshing frequency. e_j and Ψ_j are the amplitude and phase angle of the j-th order component, respectively.

$$\omega_{\mathrm{m}} = 2\pi f_{\mathrm{m}}. \tag{12.6}$$

The harmonics in Eq. (12.5) have the potential to excite resonance. Thus, the tooth profile error should be well-controlled for the smoothness of transmission.

12.2.3 Meshing impact

Meshing impact is caused by the combined action of tooth deformation and tooth spacing error. Both has the effect of changing the base pitches of the driving and the driven gears; consequently, the gear pair gets in and out of mesh at points deviating from the smooth involute profile. Impacts happen at the meshing-in and meshing-out points. This can be clarified by viewing Fig. 12.4.

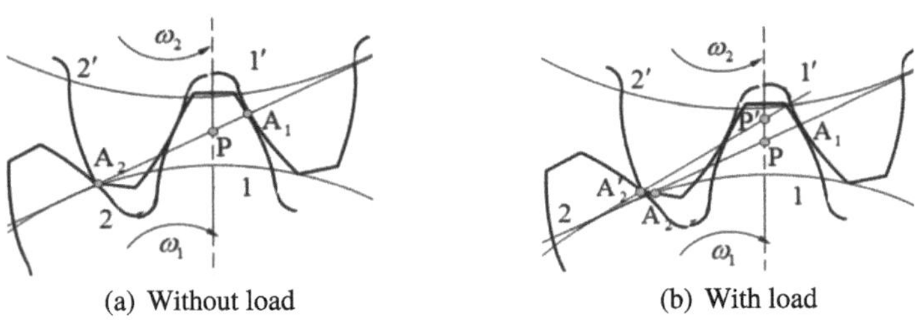

Figure 12.4 Mesh-in impact.

In the absence of a load, the meshing process of a gear pair is shown in Fig. 12.4(a). When the tooth pair $1-1'$ engages at point A_1, the tooth pair $2-2'$ enters meshing at point A_2. The distance between A_1 and A_2 is referred to as the *base pitch*. For involute profile gears, all the meshing points are on the line of action, and the transmission is perfectly smooth.

When the gears transmit loads, however, the gear teeth in contact will experience some deformation as shown in Fig. 12.4(b). Due to the deformation, the pair $2-2'$ comes into contact at point A_2', which is slightly earlier than the expected point A_2. At point A_2' the normal line $A_2'P'$ does not pass the theoretical pitch point P, causing a *meshing-in impact* at this point. Similarly, *meshing-out impact* happens when the gear teeth $1-1'$ get out of contact. Tooth spacing error also causes meshing impacts.

Tooth profile error has been widely accepted in publications on gear dynamics. The effect of meshing impact, however, is much less studied. In this chapter, we limit our discussion to the tooth profile error.

12.3 Pure rotational models for spur gear trains

If the supporting bearings and the shaft are stiff enough, the translational motions can be neglected. In this case, pure rotational models can be used to investigate gear dynamics. Two types of rotational models are often used.

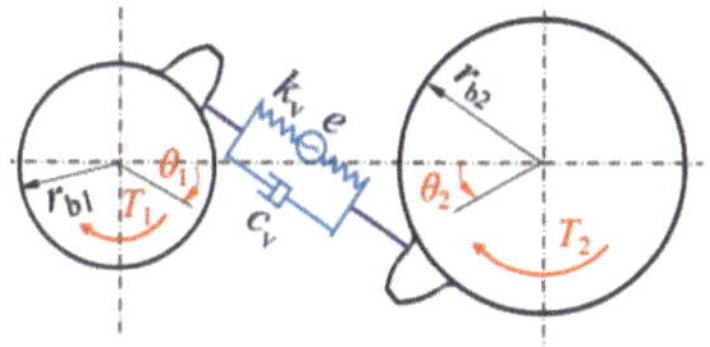

Figure 12.5 Rotational model.

12.3.1 Rotational model for gear pairs

This is the simplest dynamic model in gear dynamic. In this model only the rotation of the gears are considered. The gear mesh is modeled as elastic, while the supporting structure, including the housing, the bearings and the shafts, are considered as rigid. The model is shown in Fig. 12.5. The two gear bodies are treated as two disks, and the gear teeth in contact are modeled as a spring k_v, along with a damper c_v. The tooth profile error $e(t)$ is inserted between the meshing gear teeth as excitation. r_b, T, J, and θ are the radius of the base circle, torque, moment of inertia, and elastic rotation angle. The subscripts 1 and 2 represent the driving and driven gears, respectively. The positive directions of variables are as shown.

The elastic deformation along the line of action is:

$$\delta_i = \theta_1 r_{b1} - \theta_2 r_{b2} - e_i \tag{12.7}$$

$$e_i = e_{i1} + e_{i2}, \tag{12.8}$$

where e_{i1} and e_{i2} are the profile errors of the driving and driven gear, respectively. Then the dynamic meshing force is calculated as:

$$\begin{aligned} F &= \sum_i F_i = \sum_i \left(k_{vi}\delta_i + c_{vi}\dot{\delta}_i\right) \\ &= \sum_i \left[k_{vi}\left(\theta_1 r_{b1} - \theta_2 r_{b2} - e_i\right) + c_{vi}\left(\dot{\theta}_1 r_{b1} - \dot{\theta}_2 r_{b2} - \dot{e}_i\right)\right], \end{aligned} \tag{12.9}$$

where i is the number index of tooth pairs in mesh. For the case with a contact ratio ϵ between 1 and 2, $i = 1, 2$.

The equations of motion in rotation of the two gears are

$$\begin{cases} J_1\ddot{\theta}_1 = T_1 - r_{b1}F \\ J_2\ddot{\theta}_2 = -T_2 + r_{b2}F. \end{cases} \tag{12.10}$$

Substituting Eq. (12.9) into Eq. (12.10) yields

$$\begin{cases} J_1\ddot{\theta}_1 + r_{b1}\sum_i \left[k_{vi}(\theta_1 r_{b1} - \theta_2 r_{b2} - e_i) + c_{vi}\left(\dot{\theta}_1 r_{b1} - \dot{\theta}_2 r_{b2} - \dot{e}_i\right)\right] = T_1 \\ J_2\ddot{\theta}_2 - r_{b2}\sum_i \left[k_{vi}(\theta_1 r_{b1} - \theta_2 r_{b2} - e_i) + c_{vi}\left(\dot{\theta}_1 r_{b1} - \dot{\theta}_2 r_{b2} - \dot{e}_i\right)\right] = -T_2. \end{cases} \tag{12.11}$$

It can be seen that even when the load on gears is stable, the time-varying meshing stiffness still leads to vibration of the gears (see Fig. 12.6).

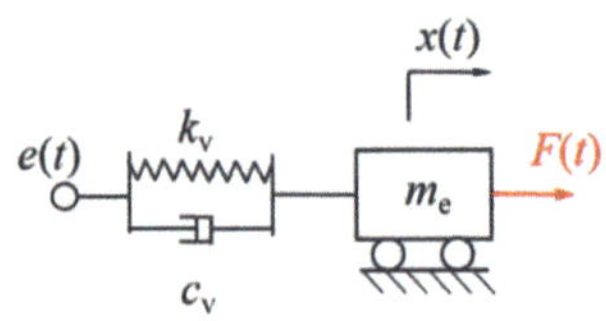

Figure 12.6 Equivalent single-DOF model.

The gears have rigid motion; thus, Eq. (12.11) has singular stiffness. This issue can be resolved by defining a new generalized coordinate, x, which is defined as the relative linear displacement along the line of action as follows:

$$x = \theta_1 r_{b1} - \theta_2 r_{b2}. \tag{12.12}$$

From static equilibrium, we have

$$F_0 = \frac{T_1}{r_{b1}} = \frac{T_2}{r_{b2}}. \tag{12.13}$$

Multiplying the first equation by r_{b1} and the second by r_{b2} of Eq. (12.11), and taking advantage of Eqs. (12.12) and (12.13), Eq. (12.11) is transformed into a single-DOF equation as follows:

$$m_e\ddot{x} + c_v\dot{x} + k_v x = F_d, \tag{12.14}$$

where

$$m_e = \frac{J_1 J_2}{J_2 r_{b1}^2 + J_1 r_{b2}^2} \tag{12.15}$$

is the equivalent mass of the system. The excitation F_d is in the following form:

$$F_d = F_0 + \sum_i c_{vi} \dot{e}_i + \sum_i k_{vi} e_i. \tag{12.16}$$

Eq. (12.14) is the vibration equation of a pair of gears. It is noted that the meshing stiffness, the meshing damping, and the tooth profile error are all functions of time. According to the vibration theory in Chapter 5, the natural frequency is found to be

$$f = \frac{1}{2\pi}\sqrt{\frac{k_v}{m_e}} = \frac{1}{2\pi}\sqrt{\frac{\sum k_{vi}}{m_e}}. \tag{12.17}$$

Due to k_v being time-varying, the natural frequency f is also time-varying.

To use the pure rotational model, the torsional stiffness of shafts is expected to be much larger than the meshing stiffness. As a rule of thumb, the pure rotational model gives satisfactory results if the natural frequency of shaft torsional vibration is at least 10 times of the natural frequency of the gear model.

Based on Eq. (12.14), various analyses can be conducted. Fig. 12.7 shows the rotational acceleration of a gear pair and the stress at the root of gear teeth calculated at the following parameter: tooth number

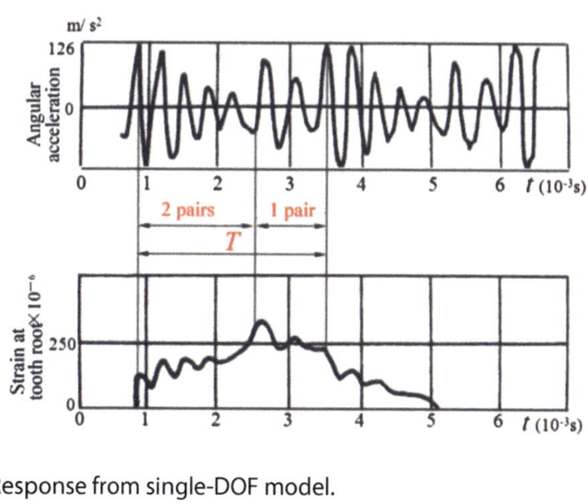

Figure 12.7 Response from single-DOF model.

$z_1 = z_2 = 56$; module $m = 3$ mm, face width $b = 8$ mm, rotation speed $n = 400$ rpm, and torque $T = 39.2$ N·m.

12.3.2 Rotational model of gear–rotor system

The dynamic model of a gear pair shown in Fig. 12.5, and Eq. (12.14) considers only the flexibility of gear teeth. When the torsional flexibility of shaft is nonnegligible, the flexibility of both the shaft and the gear teeth have to be included. The model of this type is called the rotational model of a gear–rotor system. In the following, the procedures to develop such a model is presented based on the work by Lin et al. (1988).

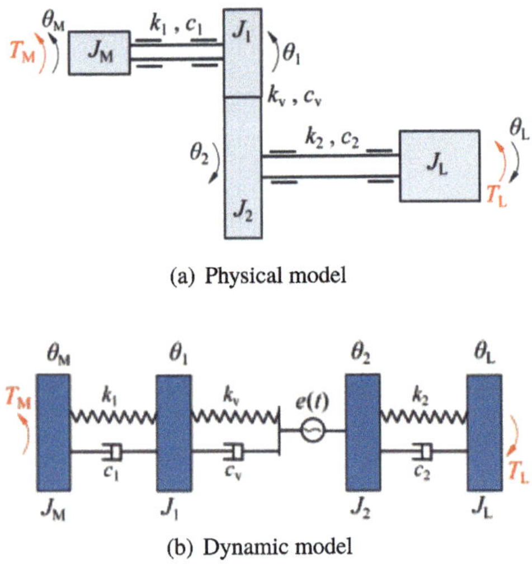

Figure 12.8 Torsional model of gear rotor system.

A gear rotor system is shown in Fig. 12.8(a), in which the driving motor and the external load are included. The parameters in the figure are explained as follows:

J_M, J_L, J_1, and J_2 are the moments of inertia of the driving motor, the external load, gear 1 and gear 2, respectively;

θ_M, θ_L, θ_1, and θ_2 are the rotational displacement of the driving motor, the external load, gear 1 and gear 2, respectively;

T_M and T_L are the motor driving torque and the external torque;

k_1 and k_2 are the torsional stiffness of the driving and driven shafts, respectively;

c_1 and c_2 are the damping coefficients of the driving and the driven shafts, respectively;

k_v and c_v are the meshing stiffness and the meshing damping in the gear pair;

$e(t)$ is the gear-tooth profile error.

Taking θ_M, θ_L, θ_1, and θ_2 as the generalized coordinate variables, a dynamic model can be developed as shown Fig. 12.8(b). The equation of motion can be easily obtained by Newton's second law as follows:

$$\begin{cases} J_M\ddot{\theta}_M + c_1\left(\dot{\theta}_M - \dot{\theta}_1\right) + k_1\left(\theta_M - \theta_1\right) = T_M \\ J_1\ddot{\theta}_1 + c_1\left(\dot{\theta}_1 - \dot{\theta}_M\right) + k_1\left(\theta_1 - \theta_M\right) + r_{b1}F = 0 \\ J_2\ddot{\theta}_2 + c_2\left(\dot{\theta}_2 - \dot{\theta}_L\right) + k_2\left(\theta_2 - \theta_L\right) - r_{b2}F = 0 \\ J_L\ddot{\theta}_L + c_2\left(\dot{\theta}_L - \dot{\theta}_2\right) + k_2\left(\theta_L - \theta_2\right) = -T_L, \end{cases} \tag{12.18}$$

where F is the meshing force in the following form:

$$F = c_v\left(r_{b1}\dot{\theta}_1 - r_{b2}\dot{\theta}_2 - \dot{e}\right) + k_v\left(r_{b1}\theta_1 - r_{b2}\theta_2 - e\right). \tag{12.19}$$

Similar to the pure rotation model in the previous section, there is a rigid motion in the equation. This is resolved by defining the following three new variables:

$$\begin{cases} \varphi_1 = \theta_1 - \theta_M \\ x = r_{b1}\theta_1 - r_{b2}\theta_2 \\ \varphi_2 = \theta_L - \theta_2, \end{cases} \tag{12.20}$$

where x is the relative linear displacement of the two gears along the line of action and index φ_1 and φ_2 are the relative rotational angles of the driving and driven gears, respectively. Substituting Eq. (12.20) into Eq. (12.18), the four equations are reduced to three after some manipulation:

$$\begin{cases} J_1\ddot{\varphi}_1 + c_1\dot{\varphi}_1 + k_1\varphi + r_{b1}F = 0 \\ J_2J_1\ddot{x} + J_1c_2r_{b2}\dot{\varphi}_2 + J_1k_2r_{b2}\varphi_2 + J_1r_{b2}^2F + J_2c_1r_{b1}\dot{\varphi}_1 + J_2k_1r_{b1}\varphi_1 + J_2r_{b1}^2F = 0 \\ J_LJ_2\ddot{\varphi}_2 + \left(J_L + J_2\right)c_2\dot{\varphi}_2 + \left(J_L + J_2\right)k_2\varphi_2 + J_Lr_{b2}F = -J_2T_L. \end{cases} \tag{12.21}$$

Recasting these in matrix form, we have

$$\boldsymbol{M}\ddot{\boldsymbol{q}} + \boldsymbol{C}\dot{\boldsymbol{q}} + \boldsymbol{K}\boldsymbol{q} = \boldsymbol{P}, \tag{12.22}$$

where

$$\boldsymbol{q} = \begin{bmatrix} \varphi_1 & x & \varphi_2 \end{bmatrix}^T \tag{12.23}$$

is the generalized coordinate vector. $\boldsymbol{M}$, $\boldsymbol{C}$, $\boldsymbol{K}$, and $\boldsymbol{P}$ are the mass matrix, damping matrix, stiffness matrix, and excitation vector, respectively. They are

$$\boldsymbol{M}=\begin{bmatrix} J_1 & 0 & 0 \\ 0 & J_2/r_{b1} & 0 \\ 0 & 0 & J_L \end{bmatrix} \tag{12.24}$$

$$\boldsymbol{K}=\begin{bmatrix} k_1 & r_{b1}k_v & 0 \\ \tau_1 k_1 & (r_{b2}u+\tau_1 r_{b1})k_v & uk_2 \\ 0 & \tau_2 r_{b2}k_v & (\tau_2+1)k_2 \end{bmatrix} \tag{12.25}$$

$$\boldsymbol{C}=\begin{bmatrix} c_1 & r_{b1}c_v & 0 \\ \tau_1 c_1 & (r_{b2}u+\tau_1 r_{b1})c_v & uc_2 \\ 0 & \tau_2 r_{b2}c_v & (\tau_2+1)c_2 \end{bmatrix} \tag{12.26}$$

$$\boldsymbol{P}=\begin{bmatrix} r_{b1}c_v\dot{e}+r_{b1}k_v e \\ (r_{b2}u+\tau_1 r_{b1})(c_v\dot{e}+k_v e) \\ \tau_2 r_{b2}(c_v\dot{e}+k_v e)-T_L \end{bmatrix}. \tag{12.27}$$

In these equations, $u=r_{b2}/r_{b1}$, $\tau_1=J_2/J_1$, and $\tau_2=J_L/J_2$.

12.4 Translational–rotational model

In the pure rotation models, the translational motion is neglected. If the translation motion is nonnegligible, more refined translational–rotational models are often used. Since the 1980s, many such type of models have been proposed by various researchers. In the next section, the model is introduced based on the work by Sun et al. (2000).

12.4.1 Dynamic model

12.4.1.1 Basic assumption

Before the development of the dynamic model, the following assumptions are made:

1) Only cylindrical spur gears are discussed;
2) Only the in-plane motion of gears are considered, while out-of-plane vibration is neglected;
3) Viscous damping is assumed;
4) Gravity and friction between gear teeth are ignored.

12.4.1.2 Geometric and kinematic relationship

To develop the geometric and kinematic relation needed in the dynamic model, we first establish the static model in Fig. 12.9. In this model, we established two groups of coordinate frames:

- Gear center frame: $G_i x_i y_i z_i$ $(i = 1, 2)$ has its origin coincident with the theoretical center of the gears.
- Bearing center frame: $B_j x_{bj} y_{bj} z_{bj}$ $(j = 1–4)$ has its origin at the center of the bearings.

In the coordinate frames, the y coordinate, including y_i and y_{bj}, are in the direction of the line of action before deflection. z is in line with the axis of the shafts. These can be seen clearly in Fig. 12.9.

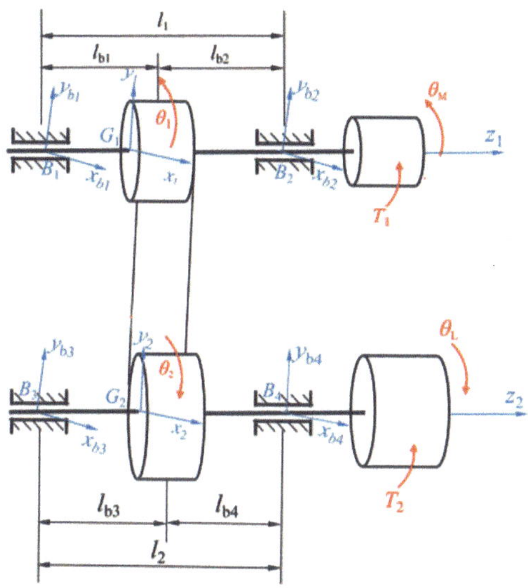

Figure 12.9 Static model of a gear transmission system.

The dynamic model is shown in Fig. 12.10. Each shaft is discretized into three lumped masses located at the centers of the two bearings and the gear as in Fig. 12.10(a). Some parameters in this figure are explained as follows:

m_{bj} $(j = 1–4)$: The lumped masses discretized from the shaft at the bearing centers;

m_i $(i = 1, 2)$: the lumped masses discretized from the shaft at the gear centers;

J_i $(i = 1, 2)$: the moments of inertia of the gears;

J_M: the moment of inertia of the driving motor;

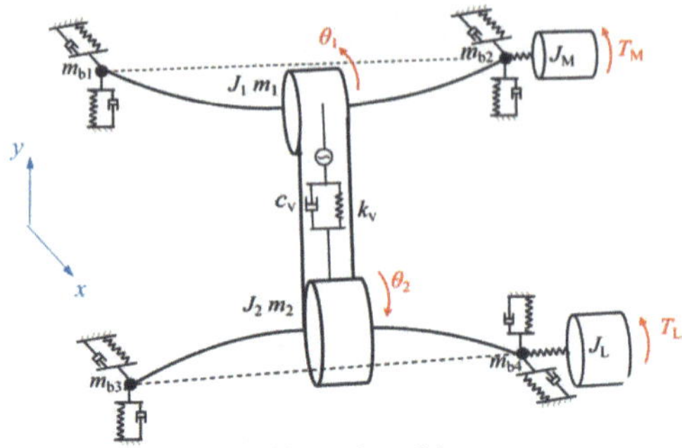

(a) Physical model

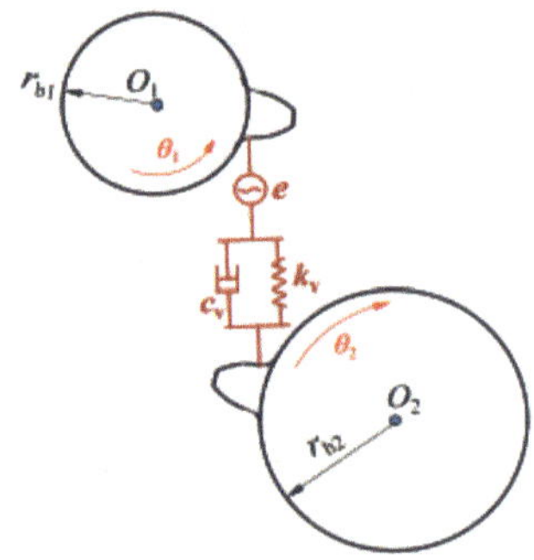

(b) Dynamic model

Figure 12.10 Dynamic model of gear transmission system.

J_L: the moment of inertia of the external load;

θ_i $(i = 1, 2)$: the rotational displacements of the gears.

The shafts deflect under the gear meshing force. Taking the driving shaft as an example illustrated in Fig. 12.11, it is originally a straight line $B_1G_1B_2$. After the deflection, its position changes to the curve $B_1''O_1B_2''$. Decomposing the displacement into x and y components, and considering both the rigid motion and the elastic deformation, we have:

$$\begin{cases} \delta_{1x} = x_1 - \xi_1 x_{b2} - \xi_2 x_{b1} \\ \delta_{1y} = y_1 - \xi_1 y_{b2} - \xi_2 y_{b1}, \end{cases} \tag{12.28}$$

where δ_{1x} and δ_{1y} are the elastic deformations in x and y directions, respectively, of driving shaft at the gear 1 center. x_{bj} and y_{bj} $(j = 1, 2)$ are the displacements along x and y directions, respectively, of the j-th bearing center. $\xi_j = l_{bj}/l_1$ $(j = 1, 2)$.

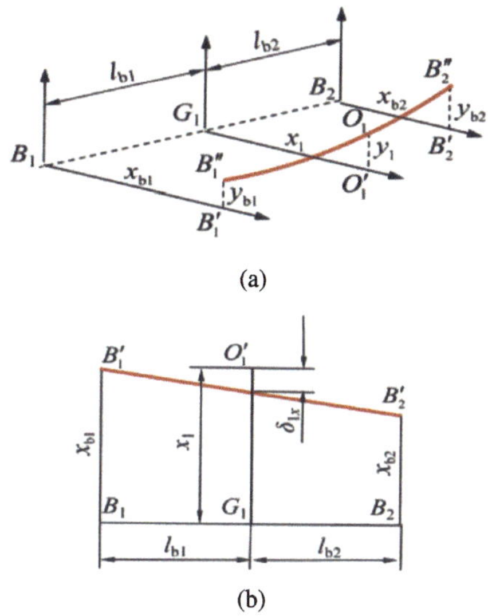

Figure 12.11 Bending deflection.

By the same procedure, we have the similar relationship for the driven shaft:

$$\begin{cases} \delta_{2x} = x_2 - \xi_3 x_{b4} - \xi_4 x_{b3} \\ \delta_{2y} = y_2 - \xi_3 y_{b4} - \xi_4 y_{b3}, \end{cases} \tag{12.29}$$

where δ_{2x} and δ_{2y} are the elastic deformations in x and y directions, respectively, of driven shaft at the gear 2 center. x_{bj} and y_{bj} $(j = 3, 4)$ are the displacements along x and y directions, respectively, of the j-th bearing center. $\xi_j = l_{bj}/l_2$ $(j = 3, 4)$.

12.4.1.3 Force analysis

For the i-th engaging tooth pair, the meshing force is

$$F_i = k_{vi}\delta_i + c_{vi}\dot{\delta}_i, \tag{12.30}$$

where k_{vi} and δ_i are the meshing stiffness and the relative displacement along the line of action. c_{vi} is the damping coefficient in the mesh. If the contact ratio ϵ is between 1 and 2, $i = 1, 2$, we have

$$\delta_i = y_1 - y_2 + \theta_1 r_{b1} - \theta_2 r_{b2} - e_i, \tag{12.31}$$

where e_i is the profile error of the i-th pair of teeth. r_{b1} and r_{b2} are the radii of the base circle of the driving and driven gears, respectively.

Summation of the two tooth pairs in contact gives the total meshing force as

$$F=\sum_i F_i=\sum_i\Big[k_{vi}\left(y_1-y_2+\theta_1 r_{b1}-\theta_2 r_{b2}-e_i\right)$$
$$+c_{vi}\left(\dot{y}_1-\dot{y}_2+\dot{\theta}_1 r_{b1}-\dot{\theta}_2 r_{b2}-\dot{e}_i\right)\Big]. \tag{12.32}$$

Backlash (clearance) exists between gear teeth for lubrication and assembly consideration. If the clearance is considered, three cases exist in the calculation of the meshing force:

1) If $\delta_i \geq 0$, using Eq. (12.32);
2) If $-c \leq \delta_i < 0$, $F_i = 0$;
3) If $\delta_i < -c$, the opposite side gets in contact and the meshing force will flip over. The i-th meshing force is calculated by

$$F_i=-k_{vi}\left(-y_1+y_2-\theta_1 r_{b1}+\theta_2 r_{b2}-e_i-c\right)$$
$$+c_{vi}\left(\dot{y}_1-\dot{y}_2+\dot{\theta}_1 r_{b1}-\dot{\theta}_2 r_{b2}-\dot{e}_i\right), \tag{12.33}$$

where c is the clearance.

12.4.1.4 Equation of motion

If the driving angular velocity of the motor, $\dot{\theta}$, is constant, the rotational angle of the input shaft at the gear relative to the input end is

$$\phi_1=\theta_1-\theta. \tag{12.34}$$

The elastic rotational angle of the driven gear relative to its theoretical position is

$$\phi_2=\theta_2-\frac{1}{u}\theta, \tag{12.35}$$

where u is the gear ratio.

Assuming the output end has a rotational angle θ_L, its elastic rotational angle relative to the driven gear is

$$\phi_L=\theta_L-\theta_2. \tag{12.36}$$

Choosing the generalized coordinate vector as

$$\boldsymbol{q}=\left[y_1\ x_1\ \phi_1\ y_2\ x_2\ \phi_2\ y_{b1}\ x_{b1}\ y_{b2}\ x_{b2}\ y_{b3}\ x_{b3}\ y_{b4}\ x_{b4}\ \phi_L\right]^{\mathrm{T}}. \tag{12.37}$$

Using the Lagrange's equation as below

$$\frac{\mathrm{d}}{\mathrm{d}t}\left(\frac{\partial E_{\mathrm{k}}}{\partial \dot{\boldsymbol{q}}}\right)-\frac{\partial E_{\mathrm{k}}}{\partial \boldsymbol{q}}+\frac{\partial E_{\mathrm{p}}}{\partial \boldsymbol{q}}+\frac{\partial E_{\mathrm{d}}}{\partial \dot{\boldsymbol{q}}}=\mathbf{Q}. \tag{12.38}$$

The quantities involved in this equation are explained in the following.

1) Generalized force vector Q

$$\begin{cases} Q_1=-\sum_i F_i \qquad Q_3=r_{\mathrm{b}1}\sum_i F_i \qquad Q_4=\sum_i F_i \\ Q_6=r_{\mathrm{b}2}\sum_i F_i \qquad Q_{15}=-T_{\mathrm{L}}. \end{cases} \tag{12.39}$$

Other entries are all zero.

2) Kinetic energy E_{k}

$$E_{\mathrm{k}}=E_{\mathrm{k}1}+E_{\mathrm{k}2}, \tag{12.40}$$

where $E_{\mathrm{k}1}$ and $E_{\mathrm{k}2}$ are the kinetic energies of the driving and driven gears, respectively, are:

$$\begin{cases} E_{\mathrm{k}1}= \frac{1}{2}m_1\left(\dot{x}_{01}^2+\dot{y}_{01}^2\right)+\frac{1}{2}J_1\dot{\theta}_1^2+\frac{1}{2}m_{\mathrm{b}1}\left(\dot{x}_{\mathrm{b}1}^2+\dot{y}_{\mathrm{b}1}^2\right) \\ \qquad +\frac{1}{2}m_{\mathrm{b}2}\left(\dot{x}_{\mathrm{b}2}^2+\dot{y}_{\mathrm{b}2}^2\right)+\frac{1}{2}J_{\mathrm{M}}\dot{\theta}^2 \\ E_{\mathrm{k}2}= \frac{1}{2}m_2\left(\dot{x}_{02}^2+\dot{y}_{02}^2\right)+\frac{1}{2}J_2\dot{\theta}_2^2+\frac{1}{2}m_{\mathrm{b}3}\left(\dot{x}_{\mathrm{b}3}^2+\dot{y}_{\mathrm{b}3}^2\right) \\ \qquad +\frac{1}{2}m_{\mathrm{b}4}\left(\dot{x}_{\mathrm{b}4}^2+\dot{y}_{\mathrm{b}4}^2\right)+\frac{1}{2}J_{\mathrm{L}}\dot{\theta}_{\mathrm{L}}^2. \end{cases} \tag{12.41}$$

3) Potential energy E_{p}

$$E_{\mathrm{p}}=E_{\mathrm{p}1}+E_{\mathrm{p}2}, \tag{12.42}$$

where $E_{\mathrm{p}1}$ and $E_{\mathrm{p}2}$ are the elastic potential energies of the driving and driven gears, respectively. They are:

$$\begin{cases} E_{\mathrm{p}1}= \frac{1}{2}k_{\mathrm{S}1}\left(\delta_{1\mathrm{x}}^2+\delta_{1\mathrm{y}}^2\right)+\frac{1}{2}k_{\mathrm{T}1}\phi_1^2+\frac{1}{2}k_{\mathrm{bx}1}x_{\mathrm{b}1}^2+\frac{1}{2}k_{\mathrm{by}1}y_{\mathrm{b}1}^2 \\ \qquad +\frac{1}{2}k_{\mathrm{bx}2}x_{\mathrm{b}2}^2+\frac{1}{2}k_{\mathrm{by}2}y_{\mathrm{b}2}^2 \\ E_{\mathrm{p}2}= \frac{1}{2}k_{\mathrm{S}2}\left(\delta_{2\mathrm{x}}^2+\delta_{2\mathrm{y}}^2\right)+\frac{1}{2}k_{\mathrm{T}2}\phi_{\mathrm{L}}^2+\frac{1}{2}k_{\mathrm{bx}3}x_{\mathrm{b}3}^2+\frac{1}{2}k_{\mathrm{by}3}y_{\mathrm{b}3}^2 \\ \qquad +\frac{1}{2}k_{\mathrm{bx}4}x_{\mathrm{b}4}^2+\frac{1}{2}k_{\mathrm{by}4}y_{\mathrm{b}4}^2, \end{cases} \tag{12.43}$$

where $k_{\mathrm{S}1}$ and $k_{\mathrm{S}2}$ are the bending stiffness of the driving and driven shafts at the position where gear is installed. $k_{\mathrm{T}1}$ and $k_{\mathrm{T}2}$ are the torsional stiffness of the driving and driven shafts. k_{bxj} and k_{byj} are the bearing stiffness ($j = 1$–4) in x and y directions, respectively.

4) **Dissipation function E_d**

$$E_d = E_{d1} + E_{d2}, \tag{12.44}$$

where E_{d1} and E_{d2} are the dissipation function of the driving and driven gear, respectively. They are

$$\begin{cases} E_{d1} = & \frac{1}{2}c_{S1}\left(\dot{\delta}_{1x}^2 + \dot{\delta}_{1y}^2\right) + \frac{1}{2}c_{T1}\dot{\phi}_1^2 + \frac{1}{2}c_{bx1}\dot{x}_{b1}^2 + \frac{1}{2}c_{by1}\dot{y}_{b1}^2 \\ & + \frac{1}{2}c_{bx2}\dot{x}_{b2}^2 + \frac{1}{2}c_{by2}\dot{y}_{b2}^2 \\ E_{d2} = & \frac{1}{2}c_{S2}\left(\dot{\delta}_{2x}^2 + \dot{\delta}_{2y}^2\right) + \frac{1}{2}c_{T2}\dot{\phi}_L^2 + \frac{1}{2}c_{bx3}\dot{x}_{b3}^2 + \frac{1}{2}c_{by3}\dot{y}_{b3}^2 \\ & + \frac{1}{2}c_{bx4}\dot{x}_{b4}^2 + \frac{1}{2}c_{by4}\dot{y}_{b4}^2, \end{cases} \tag{12.45}$$

where c_{Si} and c_{Ti} ($i = 1, 2$) are the damping coefficients of the driving and driven shafts, respectively.

Substituting Eqs. (12.39) through (12.45) into Eq. (12.38), and taking advantage of Eqs. (12.28) and (12.29), the governing equation of motion as below is obtained as:

$$\boldsymbol{M}\ddot{q} + \left(\boldsymbol{C} + \boldsymbol{C}'\right)\dot{q} + \left(\boldsymbol{K} + \boldsymbol{K}'\right)q = \boldsymbol{T} + \boldsymbol{E} + \boldsymbol{F}, \tag{12.46}$$

where $\boldsymbol{M}$, $\boldsymbol{C}$, and $\boldsymbol{K}$ are the mass, damping, and stiffness matrices, respectively. $\boldsymbol{K}'$ and $\boldsymbol{C}'$ are the meshing stiffness and meshing damping matrices. $\boldsymbol{E}$, $\boldsymbol{T}$, and $\boldsymbol{F}$ are the profile error excitation, external excitation, and coupled inertial excitation vectors, respectively. Their detailed forms are given in Appendix D.

12.4.2 Case study

A pair of spur gears having a transmission ratio $u = 1$ is driven by an electric motor. The input speed is 500 rpm. The external load at the output end is $T_L = 100$ N·m. Some other key parameters are given in Table 12.1.

12.4.2.1 Natural frequencies

The natural frequencies of the system can be obtained by solving the eigenvalue problem of Eq. (12.46):

$$\left(\boldsymbol{K} + \boldsymbol{K}' - \omega_n^2\right)\boldsymbol{\Phi} = \boldsymbol{0}. \tag{12.47}$$

Table 12.1 Parameters of gear system.

Parameter	Value	Parameter	Value	Parameter	Value
z_1	30	m_1	2.5 kg	J_M	0.004 kg·m^2
z_2	30	m_2	3.0 kg	J_L	0.010 kg·m^2
m	3 mm	m_{b1}	2.5 kg	J_1	0.006 kg·m^2
b	12 mm	m_{b2}	2.5 kg	J_2	0.008 kg·m^2
ξ_1	0.5	m_{b3}	3.0 kg	k_{T1}	1.2×10^7 N·m/rad
ξ_2	0.5	m_{b4}	3.0 kg	k_{T2}	2.0×10^7 N·m/rad
ξ_3	0.5	k_{S1}	2.5×10^8 N/m	k_{S2}	4.4×10^8 N/m
ξ_4	0.5	k_{bxj} (shaft 1)	3.0×10^8 N/m	k_{byj} (shaft 1)	3.0×10^8 N/m
ϵ	1.65	k_{bxj} (shaft 2)	9.0×10^8 N/m	k_{byj} (shaft 2)	9.0×10^8 N/m
ζ_S (ζ_T)	0.07	ζ	0.1	ζ_V	0.1

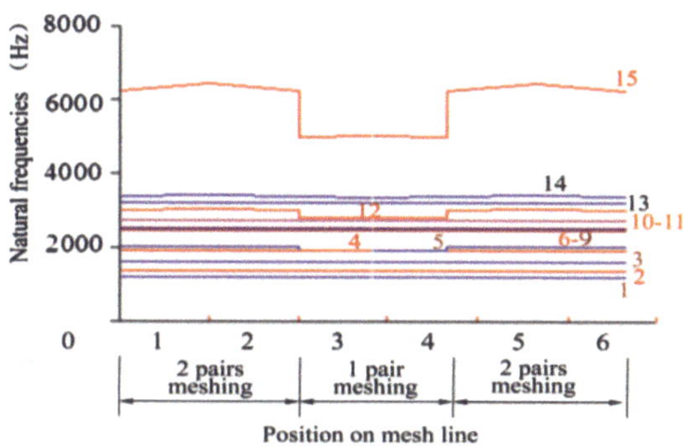

Figure 12.12 Variation of natural frequency.

In Matlab®, this is easily implemented through the embedded eigenvalue solver.

The results of natural frequencies are shown in Fig. 12.12. Clearly the time-varying meshing stiffness causes the natural frequencies to vary with time as well. In this specific case, however, the variation of meshing stiffness affects mainly the higher-order natural frequencies.

12.4.2.2 Influence of meshing stiffness

Numerically solving Eq. (12.46) gives the forced response of the system. The response of angular acceleration of the driven gear is presented in Fig. 12.13. T is the time between entering and exiting the mesh. T_b is the time for a rotation range corresponding to the base pitch. It is seen that the acceleration has abrupt jumps or drops at the moment of meshing-in and meshing-out that are obviously caused by the abrupt changes of the meshing stiffness.

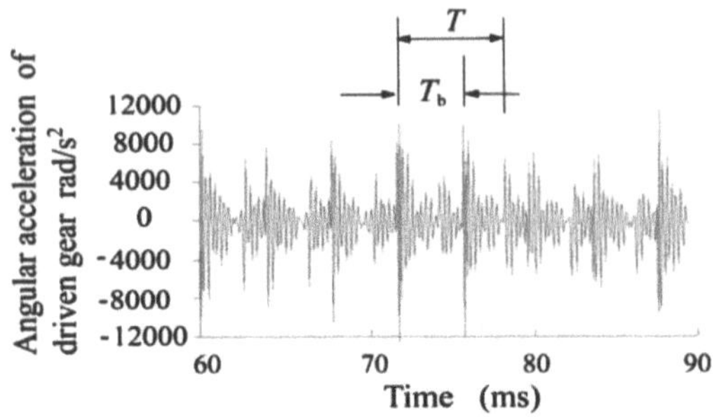

Figure 12.13 Time history of angular acceleration of driving gear.

12.4.2.3 *Dynamic load factor*

With Eq. (12.32) we can calculate the dynamic meshing force, while the static meshing force is given in Eq. (12.13). The dynamic load factor is defined as

$$k_{\mathrm{d}} = \frac{F}{F_0}. \tag{12.48}$$

The dynamic factor is a critical variable in gear strength calculation.

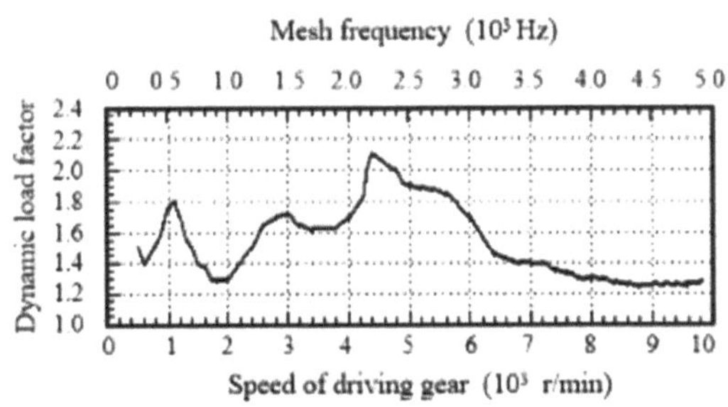

Figure 12.14 Variation of dynamic load factor.

Changing the rotation speed, and calculating the dynamic load factor repeatedly, the relationship between the dynamic load factor and the rotation speed as shown in Fig. 12.14 is obtained. Clearly, k_{d} takes peak values at some speeds, and the reason for these peaks is the resonances. However, these resonances are different from what introduced in Chapter 5. The resonance here is much more complicated. It is caused by the time-varying parameters and referred to as *parametric resonance*. Details can be found in the monograph by Nayfeh and Mook (2008).

12.5 Short review of gear dynamics

Since the 1950s, great progress has been made in gear dynamics, and both linear and nonlinear models have been developed. Since the 1970s, research expanded from deterministic vibration to random vibration. Some researchers already made very comprehensive reviews on this topic (Özgüven and Houser, 1988; Wang et al., 2003). In this section, we do not intend to cover all the details; interested readers can consult the two literature review papers previously cited. Instead, we only briefly outline the main progresses made in recent decades.

12.5.1 Linear gear dynamics

In the early stage, linear models were widely used. In this group of models, the time-varying meshing stiffness is generally considered. Özgüven and Houser (1988) surveyed and commented on almost all the linear models used before the 1980s in gear dynamics.

12.5.2 Nonlinear gear dynamics

After the 1980s, researchers started to consider nonlinear factors. These nonlinear factors include gear-tooth clearance, friction between contact profiles, and gear rattle, among others. The nonlinear response is much richer and complicated than that in linear theory, including multiple solutions, bifurcation, and chaos. Wang et al. (2003) reviewed most of the nonlinear models used before 2003.

12.5.3 Random gear dynamics

There are many uncertain factors in gear systems, such as the manufacturing error which is a main internal excitation to gear systems. One method to deal with uncertainties is to use the random vibration theory. In the 1970s a few researchers started to look at the random dynamics of gear systems, while progress had been very limited due to the inherent complexity of random vibration theory. In recent years, the interest in random gear dynamics resumed, and more papers have been appeared in journals and conference proceedings. Wei et al. (2016) summarized the main progress in random gear dynamics.

12.5.4 Fault diagnosis and condition monitoring

With the knowledge of dynamics, especially the knowledge concerning the frequency domain, fault diagnosis can be conducted in combination with vibration testing. Also, real-time condition monitoring is now applied for very important systems failures of which have severe consequences.

12.5.5 Gear-tooth profile modification

One of the most important application of gear dynamics is vibration reduction through modification of tooth profile as shown in Fig. 12.15. Modification can be made along the profile, such as *tip relief*, or along the flank, such as *crowning* or *end relief*.

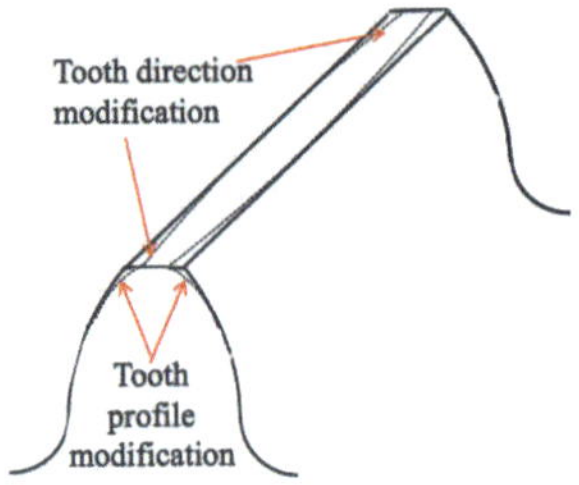

Figure 12.15 Gear-tooth modification.

The basic idea of the modification is to intentionally create an incorrect involute profile so that the meshing-in and meshing-out impacts can be reduced. All these are implemented by removing a small amount of material from the gear tooth. No matter which method of modification, it is only effective when properly implemented. Too much material removal could worsen the vibration and noise.

CHAPTER 13

Dynamics of planetary gear trains

13.1 Introduction

Compared to ordinary fixed-axis gear trains, planetary gear trains have a series of advantages, including compactness, light weight, and high load capacity. Due to these unique superior features, planetary gear trains are widely used in various machines, including helicopters, automobile, and ships, among many others. For some applications, vibration and noise are the main design concerns. For example, it was reported that the noise generated by the planetary gear train alone in a helicopter can exceed 100 dB, being the main source of noise (Schlegel and Mard, 1967). For these applications, high-speed applications in particular, dynamic study is needed.

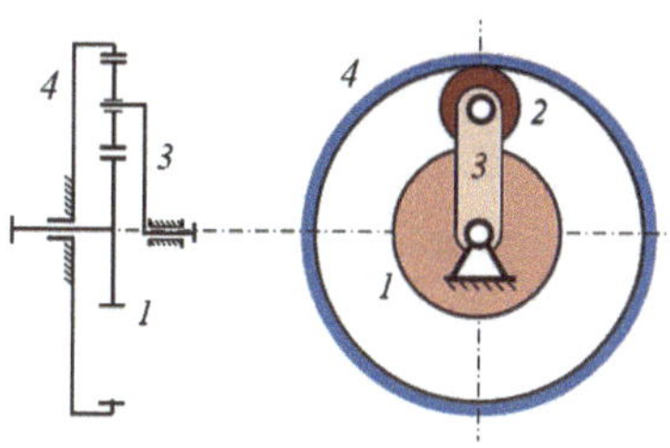

Figure 13.1 Simple planetary gear train.

Planetary gear trains appeared first in Germany around the end of 19th century. However, systematic study of planetary gear dynamics did not start until the 1960s. The main motivation for dynamic study came from the requirement to reduce and control vibration and noise.

The simplest planetary gear train is shown in Fig. 13.1, which visualizes some commonly used terms in planetary gear trains. Gear 1 is the *sun gear*, which is very often the input member. Part 3 is called the *planet carrier*, or simply carrier, which is generally the output member. Gear 4 is an internal *ring gear* or annulus gear. The sun gear, ring gear, and the carrier are often referred to as the *main members* of the planetary gear train. Gear 2 is the *planet gear*. Generally, there are 3–5 planet gears in a planetary gear train.

This chapter is dedicated to the dynamic modeling and analysis of planetary gear trains. It starts with a simple pure rotational model in Section 13.2.

Machinery Dynamics
https://doi.org/10.1016/B978-0-12-815785-5.00024-8

In Section 13.3 the more complicated translational–rotational model is presented. In both sections, the focus is model development and free vibration analysis. In Section 13.4, some other aspects of dynamics, including planet phasing and parameter selection, are discussed.

13.2 Pure rotational model

The pure rotational model considers only the rotation of the parts. Although the model is relatively simple, it offers particular advantages in some cases. For example, 1) the less complexity requires less computation capacity; 2) it is possible to obtain closed-form representation of the natural frequency. For these reasons, it is often used to estimate the dynamic characteristics of the system at an early stage of design when detailed parameters are not determined yet. In addition, when the planetary gear train has relatively large supporting stiffness, the analytical results obtained from the pure rotational model are very close to that with the more complicated translational–rotational model (Kahraman, 1994a).

13.2.1 Dynamic model

The following assumptions are made in the development of the dynamic model:

1) All members are treated as lumped moments of inertia;
2) Only rotational motion is considered;
3) Damping is ignored.

A model is shown in Fig. 13.2, in which each gear mesh is modeled as a spring and each of the main members (carrier, sun gear and ring gear) is connected to the frame through a spring in the tangential direction. The coordinate frame OXY is stationary with the origin located at the center of the sun gear, while a rotating coordinate frame Oxy is attached to the carrier. Its x axis points to the first planet center.

Although this is a rotation model, we choose linear displacement u_i ($u_i = \theta_i r_i$, $i = c, r, s, 1, 2, \cdots, N$) as the generalized coordinate. The subscripts c, r, and s mean the carrier, the ring gear, and the sun gear, respectively. The numerical subscripts i ($i = 1 \cdots N$) represent the i-th planet. N is the total number of the planets.

Some parameters in Fig. 13.2 are explained follows:

k_{rn} is the meshing stiffness of the n-th ring-planet (shorten as $r-p$) mesh. k_{sn} is the meshing stiffness of the n-th sun–planet (shorten as $s-p$)

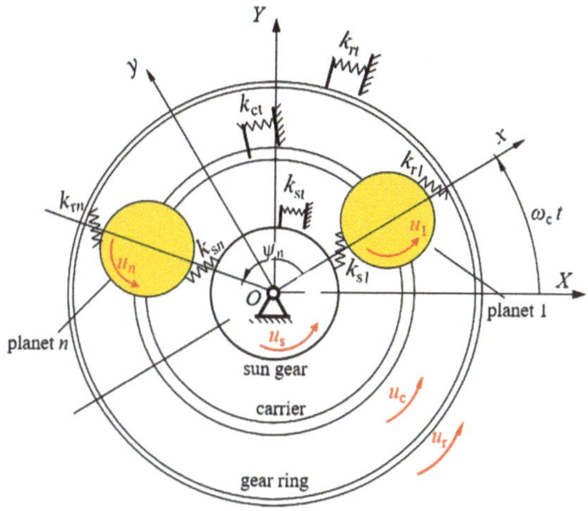

Figure 13.2 Pure rotational model.

mesh. k_{it} (i = c, r, s) is the supporting stiffness in the tangential direction of member.

The gyroscopic effect is neglected in the model. In other words, the rotating frame Oxy is frozen as an instantaneously fixed coordinate frame in each position.

According to Newton's second law of motion, the equation of motion of each member is given as follows:

$$\begin{cases} \dfrac{J_c + N m_p r_c^2}{r_c^2}\ddot{u}_c + k_{ct}u_c = \sum\limits_{n=1}^{N}\left(k_{sn}\delta_{sn}\cos\alpha_s + k_{rn}\delta_{rn}\cos\alpha_r\right) \\ \dfrac{J_r}{r_r^2}\ddot{u}_r + k_{rt}u_r = \sum\limits_{n=1}^{N}\left(-k_{rn}\delta_{rn}\right) \\ \dfrac{J_s}{r_s^2}\ddot{u}_s + k_{st}u_s = \sum\limits_{n=1}^{N}\left(-k_{sn}\delta_{sn}\right) \\ \dfrac{J_1}{r_1^2}\ddot{u}_1 = \left(k_{r1}\delta_{r1} - k_{s1}\delta_{s1}\right) \\ \cdots \\ \dfrac{J_N}{r_N^2}\ddot{u}_N = \left(k_{rN}\delta_{rN} - k_{sN}\delta_{sN}\right), \end{cases} \tag{13.1}$$

where α_s and α_r are the pressure angles of the $s-p$ and $r-p$ meshes, respectively. r_c is the radius of the circle on the carrier where the planets are installed. r_i $(i = s, r, 1, \cdots, N)$ are the radii of the base circle of gears. J_i $(i = s, c, r, 1, \cdots, N)$ are the moments of inertia of members. m_p is the mass of the planet. δ_{rn} and δ_{sn} are the relative displacement along the line of action of the n-th $r-p$ and $s-p$ meshes, respectively. They are given by:

$$\begin{cases} \delta_{rn} = u_{\rm r} - u_{\rm c}\cos\alpha_{\rm r} - u_n + e_{rn}(t) \\ \delta_{sn} = u_{\rm s} - u_{\rm c}\cos\alpha_{\rm s} + u_n + e_{sn}(t), \end{cases} \tag{13.2}$$

where $e_{rn}(t)$ and $e_{sn}(t)$ are the profile errors of the $r-p$ and $s-p$ meshes, respectively. $\alpha_{\rm r}$ and $\alpha_{\rm s}$ are the pressure angles of the $r-p$ and $s-p$ meshes.

Eq. (13.1) can be cast in more compact matrix form as:

$$\boldsymbol{M}\ddot{\boldsymbol{q}} + (\boldsymbol{K}_{\rm b} + \boldsymbol{K}_{\rm m})\boldsymbol{q} = \boldsymbol{T} + \boldsymbol{F}, \tag{13.3}$$

where $\boldsymbol{q}$ is the generalized coordinate vector. It is given by:

$$\boldsymbol{q} = \begin{bmatrix} u_{\rm c} & u_{\rm r} & u_{\rm s} & u_1 & \cdots & u_N \end{bmatrix}^{\rm T}. \tag{13.4}$$

$\boldsymbol{M}$ is the mass matrix given as

$$\boldsymbol{M} = {\rm diag}\begin{bmatrix} \dfrac{J_{\rm c} + Nm_{\rm p}r_{\rm c}^2}{r_{\rm c}^2} & \dfrac{J_{\rm r}}{r_{\rm r}^2} & \dfrac{J_{\rm s}}{r_{\rm s}^2} & \dfrac{J_1}{r_1^2} & \cdots & \dfrac{J_N}{r_N^2} \end{bmatrix}. \tag{13.5}$$

$\boldsymbol{K}_{\rm b}$ is the supporting stiffness matrix given as follows:

$$\boldsymbol{K}_{\rm b} = {\rm diag}\begin{bmatrix} k_{\rm ct} & k_{\rm rt} & k_{\rm st} & 0 & \cdots & 0 \end{bmatrix}. \tag{13.6}$$

$\boldsymbol{K}_{\rm m}$ is the gear-meshing stiffness matrix given as follows:

$$\boldsymbol{K}_{\rm m} = \begin{bmatrix} \sum\limits_{n=1}^{N}(k_{sn}c_{\rm s}^2 + k_{rn}c_{\rm r}^2) & & & & & \\ -\sum\limits_{n=1}^{N} k_{rn}c_{\rm r} & -\sum\limits_{n=1}^{N} k_{rn} & & symm. & & \\ -\sum\limits_{n=1}^{N} k_{sn}c_{\rm s} & 0 & -\sum\limits_{n=1}^{N} k_{sn} & & & \\ k_{\rm r1}c_{\rm r} - k_{\rm s1}c_{\rm s} & -k_{\rm r1} & k_{\rm s1} & k_{\rm r1} + k_{\rm s1} & & \\ \cdots & \cdots & \cdots & & \cdots & \\ k_{{\rm r}N}c_{\rm r} - k_{{\rm s}N}c_{\rm s} & -k_{{\rm r}N} & k_{{\rm s}N} & 0 & \cdots & k_{{\rm r}N} + k_{{\rm s}N} \end{bmatrix}, \tag{13.7}$$

where $c_{\rm r} = \cos\alpha_{\rm r}$, $c_{\rm s} = \cos\alpha_{\rm s}$.

$\boldsymbol{T}$ is the external excitation given by:

$$\boldsymbol{T} = \begin{bmatrix} -\frac{T_c}{r_c} & 0 & \frac{T_s}{r_s} & 0 & \cdots & 0 \end{bmatrix}^{\mathrm{T}}. \tag{13.8}$$

$\boldsymbol{F}$ is the internal excitation given by:

$$\boldsymbol{F} = \begin{bmatrix} \sum_{n=1}^{N} \left[k_{sn} e_{sn}(t) \cos\alpha_s + k_{rn} e_{rn}(t) \cos\alpha_r \right] \\ -\sum_{n=1}^{N} \left[k_{rn} e_{rn}(t) \right] \\ -\sum_{n=1}^{N} \left[k_{sn} e_{sn}(t) \right] \\ k_{r1} e_{r1}(t) - k_{s1} e_{s1}(t) \\ \cdots \\ k_{rN} e_{rN}(t) - k_{sN} e_{sN}(t) \end{bmatrix}. \tag{13.9}$$

13.2.2 Free-vibration analysis

Dropping the excitation and solving the eigenvalue problem of Eq. (13.3) give the vibration modes and natural frequencies. For the time-varying component of the meshing stiffness, the average is taken. Then the eigenvalue problem is given by:

$$\omega_i^2 \boldsymbol{M} \boldsymbol{\varphi}_i = \left[\boldsymbol{K}_b + \overline{\boldsymbol{K}}_m \right] \boldsymbol{\varphi}_i \quad (i = 1, 2, \cdots, N+3), \tag{13.10}$$

where $\overline{\boldsymbol{K}}_m$ is the average meshing stiffness. ω_i and $\boldsymbol{\varphi}_i$ are the i-th natural frequency, and the i-th vibration mode vector, respectively:

$$\boldsymbol{\varphi}_i = \begin{bmatrix} \varphi_{ic} & \varphi_{ir} & \varphi_{is} & \varphi_{i1} & \cdots & \varphi_{iN} \end{bmatrix}^{\mathrm{T}}. \tag{13.11}$$

Solving Eq. (13.10) numerically gives the natural frequencies and vibration mode.

A case study was conducted with the parameters of the planetary gear train given in Table 13.1, and the calculation results are given in Table 13.2.

It was found that, when the number of planet gear $N \geq 3$, there is a repeated frequency, 6444.0 Hz, with a multiplicity if $N - 1$. In addition, this frequency is independent of the planet number N provided $N \geq 3$. There is also a rigid motion, 0 Hz, regardless of the number of planet gears N.

Table 13.1 Parameters of planetary gear train.

Parameter	Value	Parameter	Value
J_c	4.9 kg·m^2	J_r	1.5 kg·m^2
J_s	0.28 kg·m^2	J_p	5.81 kg·m^2
m_p	0.66 kg	k_{rt}	1×10^9 N/m
k_{st}, k_{ct}	0	$\bar{k}_{sn}$, $\bar{k}_{rn}$	5×10^8 N/m
α_r, α_s	24.6°		

Table 13.2 Natural frequencies.

N	Natural frequency (Hz)				
2	0	2304.2	5456.5	9684.1	6444.0
3	0	2206.2	6123.7	11,193	6444.0[a]
4	0	2120.6	6635.2	12,554	6444.0[a]
5	0	2046.2	7066.6	13,793	6444.0[a]
6	0	1980.7	7448.5	14,933	6444.0[a]
7	0	1922.3	7796.4	15,994	6444.0[a]
8	0	1869.7	8118.7	16,988	6444.0[a]

[a] Repeated frequencies with multiplicity $N-1$.

Table 13.3 Vibration modes ($N = 3$).

Natural frequency (Hz)	2206.2	6123.7	11,193	6444.0	6444.0
Vibration mode	Rotational modes			Planet modes	
	1-st mode	2-nd mode	5-th mode	3-rd mode	4-th mode
φ_{ic}	−0.12496	−0.12947	0.03157	0.0	0.0
φ_{ir}	−0.45035	0.42189	0.02481	0.0	0.0
φ_{is}	0.36871	0.50492	−0.91425	0.0	0.0
φ_{i1}	−0.46390	−0.42829	−0.23277	0.73030	0.11547
φ_{i2}	−0.46390	−0.42829	−0.23277	−0.04892	−0.75774
φ_{i3}	−0.46390	−0.42829	−0.23277	−0.68138	0.64227

The calculation results reveal that the vibration modes are well-structured. In Table 13.3, the modes of $N = 3$ are given, and some features of the vibration modes are summarized as follows:

1. The repeated frequencies, 6444.0 Hz, has a multiplicity of 2. The two vibration modes, including the third and the fourth, involve only the motion of planet gears. Consequently, the other members have zero motion. This is clearly seen from the two right-most columns in Table 13.3. These two modes are thus referred to as *planet modes*.

2. In the planet modes, the summation of all the planet's displacements is zero. For example, in the third mode, $0.73030 - 0.04892 - 0.68138 = 0$, while in the fourth mode, $0.11547 - 0.75774 + 0.64227 = 0$. These three vibration modes are called *rotational modes*.

These two types of vibration modes are illustrated in Fig. 13.3.

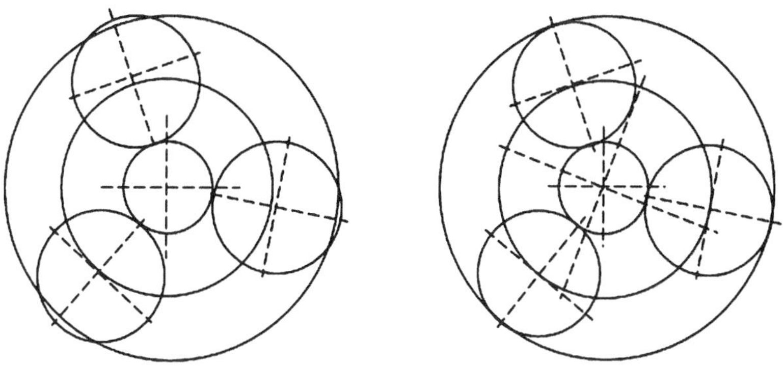

(a) Planet modes (b) Rotational modes

Figure 13.3 Vibration modes.

13.2.3 Analytical analysis of natural frequencies

As analyzed in the previous section, the vibration modes of the planetary gear train are well-structured. We can take advantage of the structure to develop closed-form representation of the natural frequency, which is especially useful in an early design stage to estimate the natural frequencies (Kahraman, 1994a; Wang et al., 2005).

In the following derivation, it is assumed that the pressure angles of the $s-p$ and the $r-p$ meshes are the same, or $\alpha_s = \alpha_r = \alpha$.

13.2.3.1 Planet modes

From the three features of the vibration modes given in the previous subsection, we can assume:

$$\varphi_{ij} = \beta_{ij}\varphi_i \qquad (j = 1, 2, \cdots, N) \tag{13.12}$$

$$\sum_{j=1}^{N} \beta_{ij} = 0, \tag{13.13}$$

where φ_i is a real number. Then the planet modes can be represented as:

$$\boldsymbol{\varphi}_i = \begin{bmatrix} 0 & 0 & 0 & \beta_{1j}\phi_i & \beta_{2j}\phi_i & \cdots & \beta_{Nj}\phi_i \end{bmatrix}^{\mathrm{T}}. \tag{13.14}$$

Substituting Eqs. (13.5) through (13.7) and (13.14) into Eq. (13.10), we have

$$\omega_i = \sqrt{\frac{\overline{k}_{\mathrm{sp}} + \overline{k}_{\mathrm{rp}}}{m_{\mathrm{p}}}}, \tag{13.15}$$

where $\overline{k}_{\mathrm{sp}}$ and $\overline{k}_{\mathrm{rp}}$ are the average meshing stiffness of the $s-p$ and $r-p$ meshes, respectively.

Obviously, the frequency for the planet mode is related to $\overline{k}_{\mathrm{sp}}$, $\overline{k}_{\mathrm{rp}}$, and m_{p}. The number of planet gears N and the supporting stiffness of the three main members (k_{ct}, k_{rt} and k_{st}) have no influence on the frequency.

13.2.3.2 Rotational modes

Similarly, we assume the rotational modes as

$$\varphi_{ij} = \varphi_{i1} \qquad (j = 1, 2, \cdots, N). \tag{13.16}$$

In a planetary gear train, one of the three main members is generally fixed. Thus, the tangential supporting stiffness is taken as infinite. Depending on which member (c, r or s) is fixed, we have the following combinations of the supporting stiffness:

$$\begin{pmatrix} k_{\mathrm{ct}} & k_{\mathrm{rt}} & k_{\mathrm{st}} \end{pmatrix} = \begin{cases} (\infty \quad 0 \quad 0) & \text{if the carrier is fixed} \\ (0 \quad \infty \quad 0) & \text{if the ring gear is fixed} \\ (0 \quad 0 \quad \infty) & \text{if the sun gear is fixed.} \end{cases} \tag{13.17}$$

Substituting Eqs. (13.5) through (13.7), (13.16) and (13.17) into Eq. (13.10), we have

$$\lambda_i \left(A_1 \lambda_i^2 + A_2 \lambda_i + A_3 \right) = 0. \tag{13.18}$$

The coefficients in the equation are:

$$A_1 = \begin{cases} m_{\mathrm{s}} m_{\mathrm{r}} m_{\mathrm{p}} & \text{if carrier is fixed} \\ m_{\mathrm{c}} m_{\mathrm{s}} m_{\mathrm{p}} & \text{if ring gear is fixed} \\ m_{\mathrm{c}} m_{\mathrm{r}} m_{\mathrm{p}} & \text{if sun gear is fixed} \end{cases} \tag{13.19}$$

$$A_2 = \begin{cases} -\left[N\bar{k}_{sp}m_r m_p + N\bar{k}_{rp}m_s m_p + \left(\bar{k}_{sp} + \bar{k}_{rp}\right)m_r m_s\right] & \text{if carrier is fixed} \\ -\Big[N\bar{k}_{sp}m_c m_p + \left(\bar{k}_{sp} + \bar{k}_{rp}\right)m_c m_s + & \\ \quad N\left(\bar{k}_{sp} + \bar{k}_{rp}\right)m_p m_s \cos^2\alpha\Big] & \text{if ring gear is fixed} \\ -\Big[N\bar{k}_{rp}m_c m_p + \left(\bar{k}_{sp} + \bar{k}_{rp}\right)m_c m_r + & \\ \quad N\left(\bar{k}_{sp} + \bar{k}_{rp}\right)m_p m_r \cos^2\alpha\Big] & \text{if sun gear is fixed} \end{cases} \tag{13.20}$$

$$A_3 = \begin{cases} N\bar{k}_{sp}\bar{k}_{rp}\left(m_s + m_r + Nm_p\right) & \text{if carrier is fixed} \\ N\bar{k}_{sp}\bar{k}_{rp}\left(m_c + Nm_p\cos^2\alpha + 4m_s\cos^2\alpha\right) & \text{if ring gear is fixed} \\ N\bar{k}_{sp}\bar{k}_{rp}\left(m_c + Nm_p\cos^2\alpha + 4m_r\cos^2\alpha\right) & \text{if sun gear is fixed.} \end{cases} \tag{13.21}$$

Solving Eq. (13.18) yields:

$$\lambda_i = \frac{-A_2 \pm \sqrt{A_2^2 - 4A_1A_3}}{2A_1} \tag{13.22}$$

$$\omega_i = \sqrt{\lambda_i}. \tag{13.23}$$

For the case of $N = 3$, if all the three main members (c, s and r) are not fixed, there are six DOFs (see Table 13.2). If one of the main members is fixed, however, the number of DOF is reduced to five. There is a zero frequency for rigid motion and two others for the rotational vibration modes. All are unique single roots.

Eqs. (13.15) and (13.23) are the closed-form representations of the natural frequencies. It is very convenient to estimate the free-vibration features with them in the early design stage when the detailed parameters are not determined yet.

Letting

$$\Delta = \sqrt{A_2^2 - 4A_1A_3}, \tag{13.24}$$

we know from Eq. (13.22) that a smaller Δ gives a higher value of the lower rotational natural frequency. Thus, it is expected to improve the dynamic performance. For the fixed ring case, assuming $\bar{k}_{sp} = \bar{k}_{rp}$, Δ will be zero if

$$m_c = Nm_p\cos^2\alpha. \tag{13.25}$$

This is of particular importance in dynamic design. The mass of carrier can be estimated with this relation to have relatively high rotational natural frequency.

13.3 Translational–rotational dynamic model

Sometimes, the pure rotational model may be inadequate to give satisfactory accuracy. More refined translational–rotational models are, thus, needed. This is discussed in the following section (Lin and Parker, 1999).

13.3.1 Dynamic model

The dynamic model is developed with reference to Fig. 13.4, in which both the rotational and translation motions of the members are taken into account.

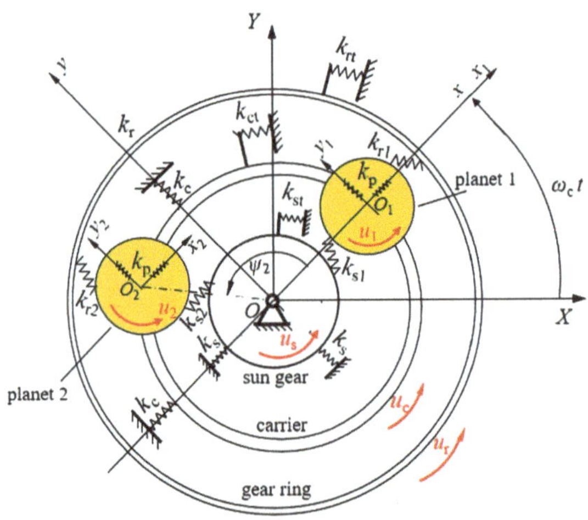

Figure 13.4 Translational–rotational model.

In the development, the following assumptions are made:

1) Given that spur gears are considered only, the planar motion of members are modeled while axial motion is not involved.
2) All members are represented by lumped parameters. Bearings and gear meshes are modeled as springs.
3) The planet gears are evenly distributed circumferentially and have identical masses and spring stiffness.
4) The time-varying meshing stiffness is simplified as a rectangular wave.

5) Damping is not included in the derivation.

With reference to Fig. 13.4, three coordinate frames are established. *OXY* is an absolute frame with its origin located at the center of the sun gear. *Oxy* is a rotating frame attached to the carrier with its origin coincident with the center of the carrier. $O_n x_n y_n$ is also a frame rotating with the carrier, the origin of which is at the center of the *n*-th planet, and its two coordinates are parallel to that of the *Oxy*.

Some parameters in Fig. 13.4 are explained as follows:

- x_i, y_i $(i = s, c, r, 1, \cdots, N)$: The translational displacements of the *i*-th member.
- θ_i $(i = s, c, r, 1, \cdots, N)$: The rotational displacement of the *i*-th member. It can be changed to a linear displacement along the tangential direction by:

$$u_i = r_i \theta_i \tag{13.26}$$

  r_i is the radius of either the base circle of a gear or the circle of planet centers. u_i is the linear displacement.
- ψ_n $(n = 1, 2 \cdots, N)$: The location angle of the *n*-th planet as shown in the figure. All are measured with regard to the first planet. Thus, we have

$$\psi_n = \frac{(n-1)2\pi}{N}. \tag{13.27}$$

- k_p: The stiffness of the planet bearing.
- k_c, k_r and k_s: The bearing stiffness of the carrier, the ring gear, and the sun gear, respectively, in the radial direction.
- k_{ct}, k_{rt}, and k_{st}: The bearing stiffness of the carrier, the ring gear, and the sun gear, respectively, in the tangential direction.

13.3.2 Acceleration transformation

Both *Oxy* and $O_n x_n y_n$ are attached to the rotating carrier so we need to transform the motion in these two frames to the absolute frame *OXY*. This can be done with reference to Fig. 13.5, in which *Oxy* is rotating and *OXY* is stationary.

$\boldsymbol{r}$ is a vector described in the rotating frame *Oxy*. The relationship between its coordinates in the two coordinate frames are given by:

$$\begin{cases} x = XC + YS \\ y = -XS + YC, \end{cases} \tag{13.28}$$

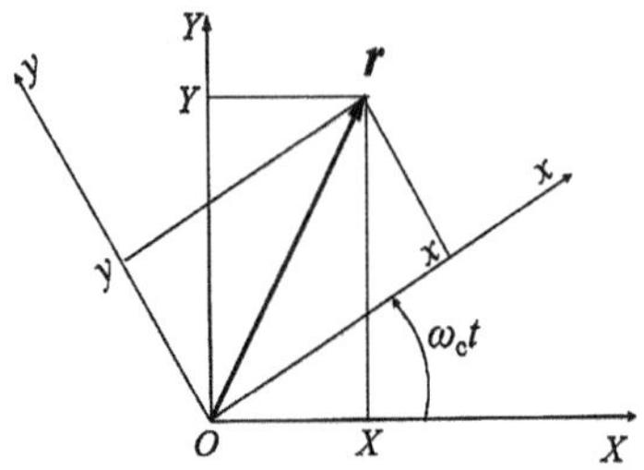

Figure 13.5 Coordinate transformation.

where $C = \cos\omega_c t$, $C = \sin\omega_c t$. ω_c is the angular velocity of the rotating frame, being the carrier angular velocity in the case of a planetary gear train.

Differentiating the above twice gives:

$$\begin{cases} \dot{x} = \dot{X}C + \dot{Y}S - \omega_c XS + \omega_c YC \\ \dot{y} = -\dot{X}S + \dot{Y}C - \omega_c XC - \omega_c YS \end{cases} \tag{13.29}$$

$$\begin{cases} \ddot{x} = \ddot{X}C + \ddot{Y}S - 2\omega_c \dot{X}S + 2\omega_c \dot{Y}C - \omega_c^2 XC - \omega_c^2 YS \\ \ddot{y} = -\ddot{X}S + \ddot{Y}C - 2\omega_c \dot{X}C + 2\omega_c \dot{Y}S + \omega_c^2 XS - \omega_c^2 YC. \end{cases} \tag{13.30}$$

After some manipulation and taking advantage of Eqs. (13.28) through (13.30), the following relationship can be obtained:

$$\begin{cases} a_x = \ddot{x} - 2\omega_c \dot{y} - \omega_c^2 x \\ a_y = \ddot{y} + 2\omega_c \dot{x} - \omega_c^2 y. \end{cases} \tag{13.31}$$

a_x and a_y are the absolute acceleration components along x and y, respectively. This is the general relation between the acceleration in a rotating coordinate frame and a stationary absolute coordinate frame. Applying this relationship to each of the rotating components, we have

$$\begin{cases} a_{ix} = \ddot{x}_i - 2\omega_c \dot{y}_i - \omega_c^2 x_i \\ a_{iy} = \ddot{y}_i + 2\omega_c \dot{x}_i - \omega_c^2 y_i, \end{cases} \tag{13.32}$$

where $(i = c, s, r, 1, \cdots, N)$.

13.3.3 Excitation

As in Chapter 12, the two types of internal excitation, namely, the time-varying meshing stiffness and the tooth profile error are considered in the modeling.

13.3.3.1 Time-varying meshing stiffness

The meshing stiffness of spur gears is simplified as a rectangular wave shown in Fig. 13.6. In these figures, k_{smax} and k_{smin} are the maximum and min-

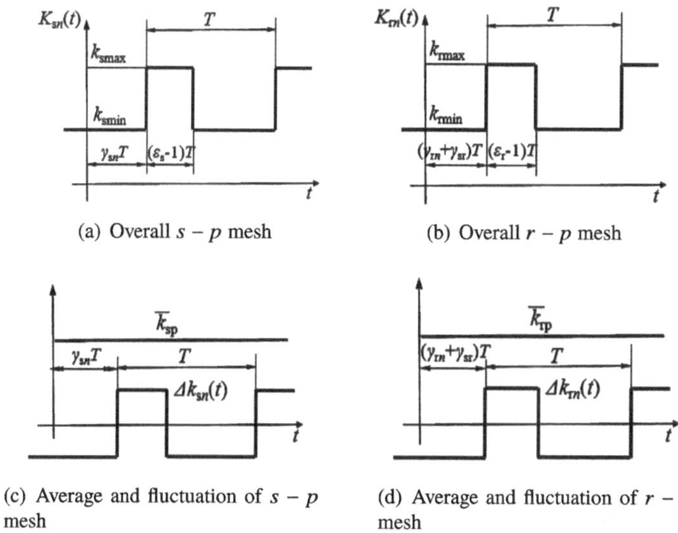

Figure 13.6 Time-varying meshing stiffness.

imum stiffness, respectively, of the $s-p$ mesh, while k_{rmax} and k_{rmin} are the maximum and minimum stiffness, respectively, of the $r-p$ mesh. T is the meshing period. ε_s and ε_r are the contact ratios of the $s-p$ and $r-p$ gear pairs, respectively. Once we have the meshing stiffness of a single gear mesh and the contact ratio, the time-varying rectangular wave can be constructed. The method to calculate the meshing stiffness of spur gear sets was already introduced in Chapter 12.

The overall time-varying meshing stiffness can be decomposed as an average component and a fluctuation component as shown in Figs. 13.6(c) and 13.6(d). Mathematically, this can be represented as:

$$\begin{cases} k_{sn}(t) = \overline{k}_{sp} + \Delta k_{sn}(t) \\ k_{rn}(t) = \overline{k}_{rp} + \Delta k_{rn}(t), \end{cases} \tag{13.33}$$

where $\overline{k}_{sp}$ and $\overline{k}_{rp}$ are the average stiffness of the $s-p$ and $r-p$ meshes, respectively. They are used in calculation of the natural frequency and are given by:

$$\begin{cases} \overline{k}_{rp} = k_{rmin}(2-\varepsilon_r) + k_{rmax}(\varepsilon_r - 1) \\ \overline{k}_{sp} = k_{smin}(2-\varepsilon_s) + k_{smax}(\varepsilon_s - 1). \end{cases} \tag{13.34}$$

$\Delta k_{sn}(t)$ and $\Delta k_{rn}(t)$ are the time-varying fluctuation components of the $s-p$ and $r-p$ meshes. Both vary in the *meshing frequency*, ω_m, which can be calculated through kinematic analysis as follows.

If observing from the rotating coordinate frame attached to the carrier, the angular velocity of the sun gear, ω_s^c, is given by:

$$\omega_s^c = \frac{Z_r}{Z_s + Z_r}\omega_s, \tag{13.35}$$

where ω_s is the absolute angular velocity of the sun gear. Z_r and Z_s are the tooth number of the ring gear and the sun gear, respectively.

The meshing frequency ω_m is then given by:

$$\omega_m = Z_s\omega_s^c = \frac{Z_s Z_r}{Z_s + Z_r}\omega_s. \tag{13.36}$$

The meshing period is, thus, given as

$$T = \frac{2\pi}{\omega_m}. \tag{13.37}$$

The time-varying meshing stiffness contributes excitation to the system with the meshing frequency. The fluctuation components, as shown in Figs. 13.6(c) and 13.6(d), can be expressed as Fourier series. γ_{rn} and γ_{sn} are the initial phases of the n-th $r-p$ and $s-p$ meshing stiffness. γ_{sr} is the phase difference of the $r-p$ and $s-p$. These will be used in the Fourier series expression given in Appendix E.

13.3.3.2 Tooth profile error

The tooth profile error is often simplified as a sinusoidal function having the meshing frequency. For the n-th $s-p$ and $r-p$ meshes, the tooth profile errors are then represented by:

$$\begin{cases} e_{sn} = E_{sn}\sin\left[\omega_m(t + \gamma_{sn}T\right] \\ e_{rn} = E_{rn}\sin\left\{\omega_m\left[t + (\gamma_{rn} + \gamma_{sr})T\right]\right\}, \end{cases} \tag{13.38}$$

where E_{sn} and E_{rn} are the amplitudes of tooth profile error of the n-th $s-p$ and $r-p$ meshes, respectively.

13.3.4 Relative displacements

The relative displacements can be analyzed as indicated in Fig. 13.7.

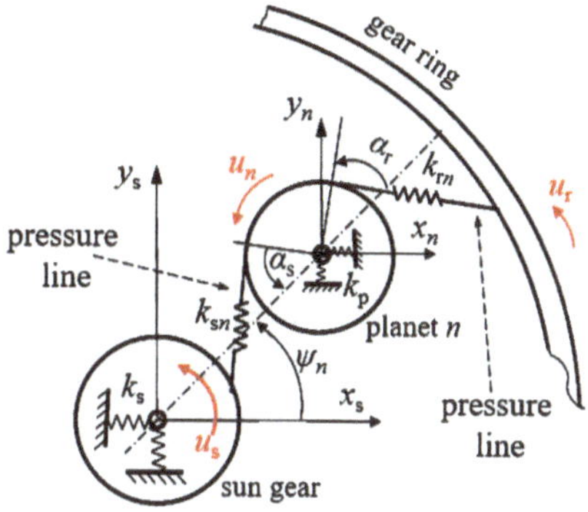

Figure 13.7 Relative displacement.

13.3.4.1 $s-p$ *mesh*

The relative displacement in the n-th $s-p$ mesh is calculated by projecting the displacements of relevant components to the line of action by:

$$\delta_{sn} = (x_n - x_s)\sin\psi_{sn} + (y_s - y_n)\cos\psi_{sn} + u_s + u_n + e_{sn}(t), \qquad (13.39)$$

where ψ_{sn} is shown in Fig. 13.7 and calculated by

$$\psi_{sn} = \psi_n - \alpha_s, \qquad (13.40)$$

where α_s is the pressure angle of this $s-p$ mesh.

13.3.4.2 $r-p$ *mesh*

Similarly, the relative displacement of the n-th $r-p$ mesh can be calculated by

$$\delta_{rn} = (x_n - x_r)\sin\psi_{rn} + (y_r - y_n)\cos\psi_{rn} + u_r - u_n + e_{rn}(t) \qquad (13.41)$$

and

$$\psi_{rn} = \psi_n - \alpha_r, \qquad (13.42)$$

where α_r is the pressure angle of this $r-p$ mesh.

13.3.4.3 Between carrier and planet

The two components along x and y directions of this relative displacement are

$$\begin{cases} \delta_{cnx} = x_c - x_n - u_c \sin\psi_n \\ \delta_{cny} = y_c - y_n + u_c \cos\psi_n. \end{cases} \tag{13.43}$$

Projecting the relative displacements along the tangential direction of the carrier at the center of the planet gives

$$\delta_{cnu} = (x_n - x_c)\sin\psi_n + (y_c - y_n)\cos\psi_n + u_c. \tag{13.44}$$

13.3.5 Governing equation of motion

Assuming the ring gear is fixed to the housing, the sun gear and the carrier work as the input and output, respectively. The equations of motion of each component can be derived according to Newton's second law of motion:

The carrier

$$\begin{cases} m_c\left(\ddot{x}_c - 2\omega_c\dot{y}_c - \omega_c^2 x_c\right) + \sum_{n=1}^{N} k_p\delta_{cnx} + k_c x_c = 0 \\ m_c\left(\ddot{y}_c + 2\omega_c\dot{x}_c - \omega_c^2 y_c\right) + \sum_{n=1}^{N} k_p\delta_{cny} + k_c y_c = 0 \\ \frac{J_c}{r_c^2}\ddot{u}_c + \sum_{n=1}^{N} k_p\delta_{cnu} + k_{ct}u_c = -\frac{T_c}{r_c}. \end{cases} \tag{13.45}$$

The ring gear

$$\begin{cases} m_r\left(\ddot{x}_r - 2\omega_c\dot{y}_r - \omega_c^2 x_r\right) + \sum_{n=1}^{N} k_{rn}\delta_{rn}\sin\psi_{rn} + k_r x_r = 0 \\ m_r\left(\ddot{y}_r + 2\omega_c\dot{x}_r - \omega_c^2 y_r\right) + \sum_{n=1}^{N} k_{rn}\delta_{rn}\cos\psi_{rn} + k_r y_r = 0 \\ \frac{J_r}{r_r^2}\ddot{u}_r + \sum_{n=1}^{N} k_{rn}\delta_{rn} + k_{rt}u_r = 0. \end{cases} \tag{13.46}$$

The sun gear

$$\begin{cases} m_s\left(\ddot{x}_s - 2\omega_c\dot{y}_s - \omega_c^2 x_s\right) - \sum_{n=1}^{N} k_{sn}\delta_{sn}\sin\psi_{sn} + k_s x_s = 0 \\ m_s\left(\ddot{y}_s + 2\omega_c\dot{x}_s - \omega_c^2 y_s\right) + \sum_{n=1}^{N} k_{sn}\delta_{sn}\cos\psi_{sn} + k_s y_s = 0 \\ \frac{J_s}{r_s^2}\ddot{u}_s - \sum_{n=1}^{N} k_{sn}\delta_{sn} + k_{st}u_s = \frac{T_s}{r_s}. \end{cases} \tag{13.47}$$

The n-th planet gear

$$\begin{cases} m_\mathrm{p}\left(\ddot{x}_n - 2\omega_\mathrm{c}\dot{y}_n - \omega_\mathrm{c}^2 x_n\right) + k_{\mathrm{s}n}\delta_{\mathrm{s}n}\sin\psi_{\mathrm{s}n} + k_{\mathrm{r}n}\delta_{\mathrm{r}n}\sin\psi_{\mathrm{r}n} - k_\mathrm{p}\delta_{\mathrm{c}n\mathrm{x}} = 0 \\ m_\mathrm{p}\left(\ddot{y}_n + 2\omega_\mathrm{c}\dot{x}_n - \omega_\mathrm{c}^2 y_n\right) - k_{\mathrm{s}n}\delta_{\mathrm{s}n}\cos\psi_{\mathrm{s}n} - k_{\mathrm{r}n}\delta_{\mathrm{r}n}\cos\psi_{\mathrm{r}n} - k_\mathrm{p}\delta_{\mathrm{c}n\mathrm{y}} = 0 \\ \dfrac{J_\mathrm{p}}{r_\mathrm{p}^2}\ddot{u}_n + k_{\mathrm{s}n}\delta_{\mathrm{s}n} - k_{\mathrm{r}n}\delta_{\mathrm{r}n} = 0. \end{cases} \tag{13.48}$$

where k_{sn} and k_{rn} are the time-varying meshing stiffness of the $s-p$ and $r-p$ meshes, respectively.

Eqs. (13.45) through (13.48) can be organized into more compact matrix form as:

$$\boldsymbol{M}\ddot{\boldsymbol{q}} + \omega_\mathrm{c}\boldsymbol{G}\dot{\boldsymbol{q}} + \left(\boldsymbol{K}_\mathrm{b} + \boldsymbol{K}_\mathrm{m} - \omega_\mathrm{c}^2\boldsymbol{K}_\Omega\right)\boldsymbol{q} = \boldsymbol{T} + \boldsymbol{F}. \tag{13.49}$$

The number of DOFs of this equation is $9+3N$. The vectors and matrices in this equation are given as follows:

$$\boldsymbol{q} = [x_\mathrm{c}\ y_\mathrm{c}\ u_\mathrm{c}\ x_\mathrm{r}\ y_\mathrm{r}\ u_\mathrm{r}\ x_\mathrm{s}\ y_\mathrm{s}\ u_\mathrm{s}\ x_1\ y_1\ u_1 \cdots x_N\ y_N\ u_N]^\mathrm{T}, \tag{13.50}$$

which is the generalized coordinate vector.

$$\boldsymbol{M} = \mathrm{diag}\left[\boldsymbol{M}_\mathrm{c}\quad \boldsymbol{M}_\mathrm{r}\quad \boldsymbol{M}_\mathrm{s}\quad \boldsymbol{M}_1\quad \cdots\quad \boldsymbol{M}_N\right] \tag{13.51}$$

is the mass matrix. Its submatrices are as follows:

$$\boldsymbol{M}_j = \mathrm{diag}\left[m_j\quad m_j\quad \frac{J_j}{r_j^2}\right] \quad (j = c, r, s, 1, \cdots, N). \tag{13.52}$$

$\mathbf{G}$ is the anti-symmetrical gyroscopic matrix given as follows:

$$\mathbf{G} = \mathrm{diag}\left[\mathbf{G}_\mathrm{c}\quad \mathbf{G}_\mathrm{r}\quad \mathbf{G}_\mathrm{s}\quad \mathbf{G}_1\quad \cdots\quad \mathbf{G}_N\right], \tag{13.53}$$

where the submatrices are given as:

$$\mathbf{G}_j = \begin{bmatrix} 0 & -2m_j & 0 \\ 2m_j & 0 & 0 \\ 0 & 0 & 0 \end{bmatrix} \quad (j = c, r, s, 1, \cdots, N). \tag{13.54}$$

$\boldsymbol{T}$ is the external excitation vector given by:

$$\boldsymbol{T} = \left[0\ 0\ -T_\mathrm{c}/r_\mathrm{c}\ 0\ 0\ 0\ 0\ 0\ T_\mathrm{s}/r_\mathrm{s}\ 0\ 0\ 0 \cdots 0\ 0\ 0\right]^\mathrm{T}. \tag{13.55}$$

$\boldsymbol{F}$ is the internal excitation vector given by:

$$\boldsymbol{F} = \begin{bmatrix} \mathbf{0} & \boldsymbol{F}_\mathrm{r} & \boldsymbol{F}_\mathrm{s} & \boldsymbol{F}_1 & \boldsymbol{F}_2 & \cdots & \boldsymbol{F}_N \end{bmatrix}^\mathrm{T}. \tag{13.56}$$

The submatrices are

$$\begin{cases} \boldsymbol{F}_\mathrm{s} = \sum\limits_{n=1}^{N} k_{\mathrm{s}n} e_{\mathrm{s}n}(t) \begin{bmatrix} \sin\psi_{\mathrm{s}n} & -\cos\psi_{\mathrm{s}n} & -1 \end{bmatrix} \\ \boldsymbol{F}_\mathrm{r} = \sum\limits_{n=1}^{N} k_{\mathrm{r}n} e_{\mathrm{r}n}(t) \begin{bmatrix} \sin\psi_{\mathrm{r}n} & -\cos\psi_{\mathrm{r}n} & -1 \end{bmatrix} \\ \boldsymbol{F}_n = k_{\mathrm{r}n} e_{\mathrm{r}n}(t) \begin{bmatrix} -\sin\psi_{\mathrm{r}n} & \cos\psi_{\mathrm{r}n} & 1 \end{bmatrix} + k_{\mathrm{s}n} e_{\mathrm{s}n}(t) \begin{bmatrix} -\sin\psi_{\mathrm{s}n} & \cos\psi_{\mathrm{s}n} & -1 \end{bmatrix}. \end{cases} \tag{13.57}$$

$\boldsymbol{K}_\mathrm{b}$ is the supporting stiffness matrix given as:

$$\boldsymbol{K}_\mathrm{b} = \mathrm{diag}\begin{bmatrix} \boldsymbol{K}_\mathrm{cb} & \boldsymbol{K}_\mathrm{rb} & \boldsymbol{K}_\mathrm{sb} & \mathbf{0} & \cdots & \mathbf{0} \end{bmatrix}. \tag{13.58}$$

The submatrices are:

$$\boldsymbol{K}_{j\mathrm{b}} = \mathrm{diag}\begin{bmatrix} k_{j\mathrm{x}} & k_{j\mathrm{y}} & k_{j\mathrm{u}} \end{bmatrix} \quad (j = c, r, s). \tag{13.59}$$

$\boldsymbol{K}_\Omega$ is the centripetal stiffness matrix in the following form:

$$\boldsymbol{K}_\Omega = [m_\mathrm{c}\ m_\mathrm{c}\ 0\ m_\mathrm{r}\ m_\mathrm{r}\ 0\ m_\mathrm{s}\ m_\mathrm{s}\ 0\ m_\mathrm{p}\ m_\mathrm{p}\ 0 \cdots m_\mathrm{p}\ m_\mathrm{p}\ 0]. \tag{13.60}$$

$\boldsymbol{K}_\mathrm{m}$ is the meshing stiffness matrix. It is symmetrical and given by:

$$\boldsymbol{K}_\mathrm{m} = \begin{bmatrix} \sum \boldsymbol{K}_\mathrm{c1}^n & 0 & 0 & \boldsymbol{K}_\mathrm{c2}^1 & \boldsymbol{K}_\mathrm{c2}^2 & \cdots & \boldsymbol{K}_\mathrm{c2}^N \\ & \sum \boldsymbol{K}_\mathrm{r1}^n & 0 & \boldsymbol{K}_\mathrm{r2}^1 & \boldsymbol{K}_\mathrm{r2}^2 & \cdots & \boldsymbol{K}_\mathrm{r2}^N \\ & & \sum \boldsymbol{K}_\mathrm{s1}^n & \boldsymbol{K}_\mathrm{s2}^1 & \boldsymbol{K}_\mathrm{s2}^2 & \cdots & \boldsymbol{K}_\mathrm{s2}^N \\ & & & \boldsymbol{K}_\mathrm{pp}^1 & 0 & \cdots & 0 \\ & & & & \boldsymbol{K}_\mathrm{pp}^2 & \cdots & 0 \\ & sym. & & & & \ddots & \vdots \\ & & & & & & \boldsymbol{K}_\mathrm{pp}^N \end{bmatrix}. \tag{13.61}$$

The submatrices are

$$\left\{\begin{aligned}
\boldsymbol{K}_{c1}^{n} &= k_{pn}\begin{bmatrix} 1 & 0 & -\sin\psi_n \\ & 1 & \cos\psi_n \\ sym. & & 1 \end{bmatrix} \\
\boldsymbol{K}_{c2}^{n} &= k_{pn}\begin{bmatrix} -\cos\psi_n & \sin\psi_n & 0 \\ -\sin\psi_n & -\cos\psi_n & 0 \\ 0 & -1 & 0 \end{bmatrix} \\
\boldsymbol{K}_{c3}^{n} &= k_{pn}\begin{bmatrix} 1 & 0 & 0 \\ 0 & 1 & 0 \\ 0 & 0 & 0 \end{bmatrix} \\
\boldsymbol{K}_{s1}^{n} &= k_{sn}\begin{bmatrix} \sin^2\psi_{sn} & -\cos\psi_{sn}\sin\psi_{sn} & -\sin\psi_{sn} \\ & \cos^2\psi_{sn} & \cos\psi_{sn} \\ sym. & & 1 \end{bmatrix} \\
\boldsymbol{K}_{s2}^{n} &= k_{sn}\begin{bmatrix} \sin\psi_{sn}\sin\alpha_s & \sin\psi_{sn}\cos\alpha_s & -\sin\psi_{sn} \\ -\cos\psi_{sn}\sin\alpha_s & -\cos\psi_{sn}\cos\alpha_s & \cos\psi_{sn} \\ -\sin\alpha_s & -\cos\alpha_s & 1 \end{bmatrix} \\
\boldsymbol{K}_{r1}^{n} &= k_{rn}\begin{bmatrix} \sin^2\psi_{rn} & -\sin\psi_{rn}\cos\psi_{rn} & -\sin\psi_{rn} \\ & \cos^2\psi_{rn} & \cos\psi_{rn} \\ sym. & & 1 \end{bmatrix} \\
\boldsymbol{K}_{r2}^{n} &= k_{rn}\begin{bmatrix} -\sin\psi_{rn}\sin\alpha_r & \sin\psi_{rn}\cos\alpha_r & \sin\psi_{rn} \\ \cos\psi_{rn}\sin\alpha_r & -\cos\psi_{rn}\cos\alpha_r & -\cos\psi_{rn} \\ \sin\alpha_r & -\cos\alpha_r & -1 \end{bmatrix} \\
\boldsymbol{K}_{pp}^{n} &= \boldsymbol{K}_{c3}^{n} + \boldsymbol{K}_{r3}^{n} + \boldsymbol{K}_{s3}^{n} \\
\boldsymbol{K}_{s3}^{n} &= k_{sn}\begin{bmatrix} \sin^2\alpha_s & \cos\alpha_s\sin\alpha_s & -\sin\alpha_s \\ & \cos^2\alpha_s & -\cos\alpha_s \\ sym. & & 1 \end{bmatrix} \\
\boldsymbol{K}_{r3}^{n} &= k_{rn}\begin{bmatrix} \sin^2\alpha_r & -\cos\alpha_r\sin\alpha_r & -\sin\alpha_r \\ & \cos^2\alpha_r & \cos\alpha_r \\ sym. & & 1 \end{bmatrix}.
\end{aligned}\right. \tag{13.62}$$

ψ_{sn} and ψ_{rn} are calculated by:

$$\begin{cases} \psi_{sn} = \psi_n - \alpha_s \\ \psi_{rn} = \psi_n + \alpha_r. \end{cases} \tag{13.63}$$

Solving Eq. (13.49) numerically gives the forced vibration response of the planetary gear train to both internal and external excitations. Due to the complexity, a closed-form solution is almost impossible. Various numerical methods for solving ODEs can be found in well-written textbooks (Chapra and Canale, 2011) or specialized papers (Kahraman, 1994b). Due to page limit, this book does not go further into this topic.

The free-vibration characteristics are discussed in the following section.

13.3.6 Free-vibration analysis

If the angular velocity of the carrier is not very high, the terms $\omega_c \boldsymbol{G}\dot{\boldsymbol{q}}$ and $\omega_c^2 \boldsymbol{K}_\Omega \boldsymbol{q}$ can be neglected in Eq. (13.49) without significantly influencing the accuracy. Then, the free-vibration equation is obtained by further dropping off the excitation on the right-hand side:

$$\boldsymbol{M}\ddot{\boldsymbol{q}} + \left(\boldsymbol{K}_b + \overline{\boldsymbol{K}}_m\right)\boldsymbol{q} = \boldsymbol{0}. \tag{13.64}$$

Given that the meshing stiffness, $\boldsymbol{K}_m$, is time-varying, it is replaced by its average $\overline{\boldsymbol{K}}_m$ in the free-vibration equation.

The eigenvalue problem is given by

$$\omega_i^2 \boldsymbol{M}\boldsymbol{\varphi}_i = \left(\boldsymbol{K}_b + \overline{\boldsymbol{K}}_m\right)\boldsymbol{\varphi}_i \quad (i = 1, 2, \cdots, 9 + 3N), \tag{13.65}$$

where ω_i and $\boldsymbol{\varphi}_i$ are the i-th natural frequency and the i-th vibration mode, respectively. $\boldsymbol{\varphi}_i$ is represented in the following form:

$$\boldsymbol{\varphi}_i = \begin{bmatrix} \boldsymbol{\varphi}_{ic} & \boldsymbol{\varphi}_{ir} & \boldsymbol{\varphi}_{is} & \boldsymbol{\varphi}_{i1} & \cdots & \boldsymbol{\varphi}_{iN} \end{bmatrix}^T, \tag{13.66}$$

where

$$\boldsymbol{\varphi}_{ij} = \begin{bmatrix} \varphi_{ijx} & \varphi_{ijy} & \varphi_{iju} \end{bmatrix} \quad (j = c, r, s, 1, \cdots, N). \tag{13.67}$$

A case study is conducted based on this analysis. The parameters in the case are given in Table 13.4.

Table 13.4 Planetary gear parameters.

Parameters	Carrier	Ring gear	Sun gear	Planet gear
m_j (kg)	5.43	2.35	0.4	0.66
J_j/r_j^2 (kg)	6.29	3.00	0.39	0.61
r_i (mm)	176.8	275.0	77.4	100.3
k_j (N/m)			10^8	
k_{jt} (N/m)	0	10^9	0	0
$\overline{k}_{sn}$, $\overline{k}_{rn}$ (N/m)			5×10^8	
α_r, α_s (°)			24.6	

Table 13.5 Natural frequencies (Hz).

N	3	4	5
$m=1$	0	0	0
	1475.7	1536.6	1567.4
	1930.3	1970.6	2006.1
	2658.3	2625.7	2614.8
	7462.8	7773.6	8065.4
	11,775.3	13,071.1	14,253.1
$m=2$	743.2	727.3	710.0
	1102.4	1091.2	1072.0
	1896.0	1892.8	1888.1
	2276.4	2342.5	2425.3
	6986.3	7189.9	7382.4
	9647.9	10,437.6	11,172.3
$m=N-3$		1808.2	1808.2
		5963.8	5963.8
		6981.7	6981.7
N_d	18	21	24

N_d is the total number of DOFs.

When the planet number $N \geq 3$, there are repeated natural frequencies. The calculation results of natural frequencies for the case with three–five planet gears are presented in Table 13.5, in which m is the multiplicity and N_d is the total number of DOFs.

The natural frequencies, also the vibration modes, are well-structured and can be grouped into three categories. For the case of $N=4$, the number of DOFs $N_d=21$. The features of the free vibration is summarized as follows:

1) **m = 1**
There are six unique frequencies in this group, among which one is the zero frequency corresponding to rigid motion and the other five are nonzero.
The vibration modes corresponding to the five nonzero natural frequencies have a common feature:

$$\varphi_{\mathrm{cx}} = \varphi_{\mathrm{cy}} = \varphi_{\mathrm{rx}} = \varphi_{\mathrm{ry}} = \varphi_{\mathrm{sx}} = \varphi_{\mathrm{sy}} = 0. \tag{13.68}$$

In other words, in these vibration modes, the three main members, including the carrier, the ring gear, and the sun gear, have no translational motion, but nonzero rotation. Thus, they are referred to as *rotation modes of main members.*

2) **m = 2**
There are six natural frequencies with a multiplicity of $m = 2$. Due to the multiplicity, there are 12 vibration modes in this group. The common feature of vibration modes in this group is

$$\varphi_{\mathrm{cu}} = \varphi_{\mathrm{ru}} = \varphi_{\mathrm{su}} = 0. \tag{13.69}$$

This means that, in all these modes, the main members have no rotational, but only translational motion. Thus, they are referred to as *translational modes of main members.*

3) **m = N-1**
There are three frequencies in this group. For $N = 4$, the multiplicity is 1; thus, there are three vibration modes in this group. Their common feature is

$$\varphi_{\mathrm{ci}} = \varphi_{\mathrm{ri}} = \varphi_{\mathrm{si}} = 0 \quad (i = x, y, u). \tag{13.70}$$

This means that in these modes all the main members do not move and only the planets move. Thus, they are referred to as *planet modes*. It is noted that, for $N = 3$, this group of vibration modes disappear.

The vibration modes corresponding to the natural frequencies 1536 Hz, 2342.5 Hz, and 1808.2 Hz are given as representatives of the three types of vibration modes in Table 13.6. They are also schematically illustrated in Fig. 13.8.

13.4 Planet phasing and parameter selection

The symmetrical arrangement of multiple planet gears in a planetary gear train makes it possible to have the meshing forces of different planet gears

Table 13.6 Vibration modes.

Vibration modes	$f_0 = 1536.6$ **Hz**	$f_0 = 2342.5$ **Hz**	$f_0 = 1808.2$ **Hz**
φ_{cx}	0.0000	0.0983	0.0000
φ_{cy}	0.0000	0.1857	0.0000
φ_{cu}	0.2147	0.0000	0.0000
φ_{rx}	0.0000	−0.0311	0.0000
φ_{ry}	0.0000	−0.0659	0.0000
φ_{ru}	0.1358	0.0000	0.0000
φ_{sx}	0.0000	−0.1475	0.0000
φ_{sy}	0.0000	−0.4975	0.0000
φ_{su}	−0.3732	0.0000	0.0000
φ_{1x}	0.2278	−0.3550	0.4556
φ_{1y}	−0.1000	−0.3015	0.0000
φ_{1u}	0.3704	0.0986	0.2059
φ_{2x}	0.2278	−0.3266	−0.4556
φ_{2y}	−0.1000	−0.0226	0.000
φ_{2u}	0.3704	−0.0825	−0.2059
φ_{3x}	0.2278	0.3550	0.4556
φ_{3y}	−0.1000	−0.3015	0.0000
φ_{3u}	0.3704	−0.0986	0.2059
φ_{4x}	0.2278	0.3266	−0.4556
φ_{4y}	−0.1000	0.0226	0.0000
φ_{4u}	0.3704	0.0825	−0.2059

canceled out, at least partially, by each other on the sun and ring gears through proper selection of parameters. Thus, the resultant force on them, or the dynamic excitation, could be reduced. This fact potentially reduces the vibration through proper selection of parameters in the design stage. The idea is referred to as *planet phasing*. The earliest report of this concept was made by Schlegel and Mard (1967), who successfully reduced the noise level of a helicopter by 11 dB. Later, Parker (2000) put the concept on the a rigorous mathematical basis by using the Fourier series. More recently, Wang (2005) validated the effectiveness of planet phasing in reduction of vibration through experiments. Based on this work, Zhang et al. (2005) proposed new guidelines for choosing parameters of planetary gear trains.

In the section following, the concept of planet phasing is discussed based on the work by Parker (2000) and Wang (2005).

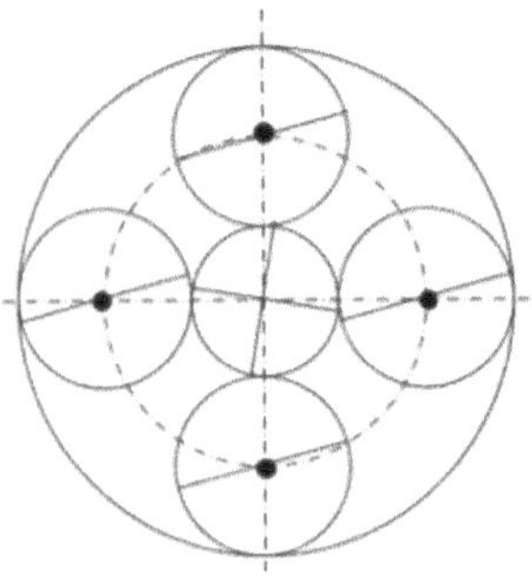

(a) Rotational modes of main members

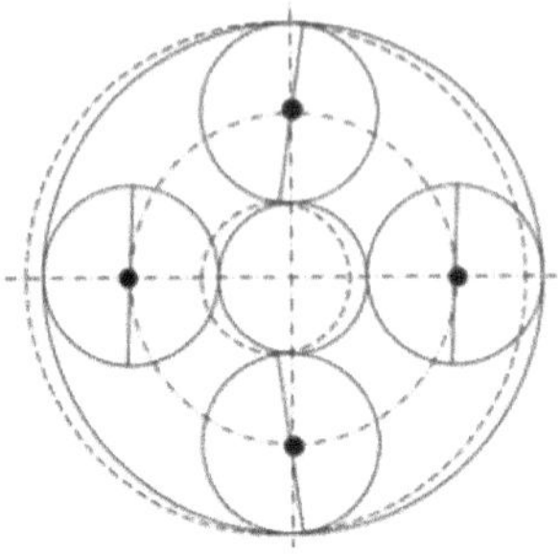

(b) Translational modes of main members

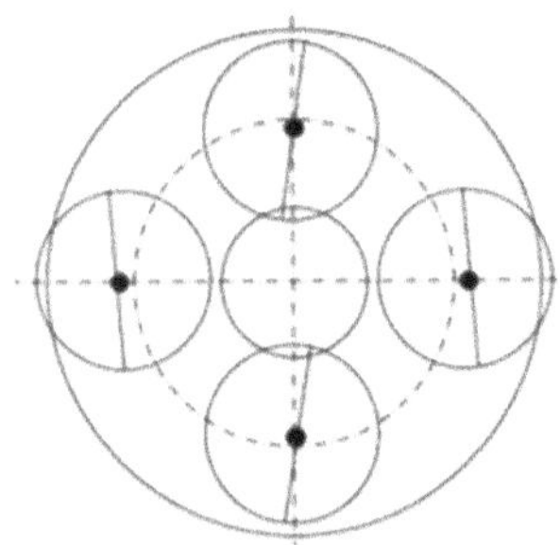

(c) Planet modes

Figure 13.8 Characteristic vibration modes.

13.4.1 Planet phasing

The rational of planet phasing is explained with reference to Fig. 13.9.

To derive the meshing forces in closed-form for different gear meshes, we establish coordinate frames $Oe_1^{(n)}e_2^{(n)}$, $(n = 1, 2, \cdots, N)$, attached to each individual planet gear with the origin at O and $e_1^{(n)}$ passing the center of the planet gear as shown in Fig. 13.9. The position angle of the n-th planet is

$$\psi_n = \frac{2\pi(n-1)}{N}. \tag{13.71}$$

In the following analysis, we take the $s - p$ meshes as an example, and the $r - p$ meshes can be analyzed in a similar way.

For the n-th $s - p$ mesh, the meshing force for spur gears can be represented as components as below:

$$\boldsymbol{F}_n = F_{n1}\boldsymbol{e}_1^{(n)} + F_{n2}\boldsymbol{e}_2^{(n)}, \tag{13.72}$$

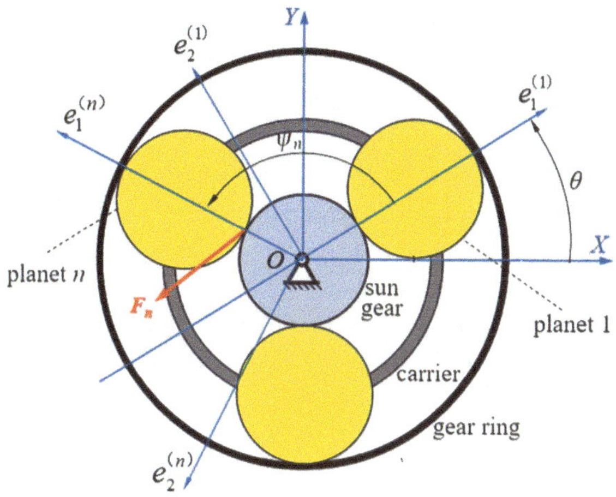

Figure 13.9 Force analysis of planetary gear trains.

where $\boldsymbol{e}_1^{(n)}$ and $\boldsymbol{e}_2^{(n)}$ are the unit vector along the $e_1^{(n)}$ and $e_2^{(n)}$ axes, respectively. F_{n1} and F_{n2} are the components of $\boldsymbol{F}_n$ along the two axes.

It is reasonable to assume $\boldsymbol{F}_n$ is a periodic function of time with the meshing frequency ω_{m}. So are F_{n1} and F_{n2}. Thus, we can expand them into Fourier series as:

$$\begin{cases} F_{n1} = \sum\limits_{l=0}^{\infty} \left[a_n^l \sin(l\omega_{\text{m}} t + l z_{\text{s}} \psi_n) + b_n^l \cos(l\omega_{\text{m}} t + l z_{\text{s}} \psi_n) \right] \\ F_{n2} = \sum\limits_{l=0}^{\infty} \left[c_n^l \sin(l\omega_{\text{m}} t + l z_{\text{s}} \psi_n) + d_n^l \cos(l\omega_{\text{m}} t + l z_{\text{s}} \psi_n) \right]. \end{cases} \tag{13.73}$$

Decomposing $\boldsymbol{F}_n$ into components along X and Y, we have:

$$\begin{bmatrix} F_{nx} \\ F_{ny} \end{bmatrix} = \begin{bmatrix} \cos\psi_n & \sin\psi_n \\ -\sin\psi_n & \cos\psi_n \end{bmatrix} \begin{bmatrix} F_{n1} \\ F_{n1} \end{bmatrix}. \tag{13.74}$$

The resultant force acting on the sun gear from all the N planet gears are the summation of all the forces of the $s-p$ meshes:

$$\boldsymbol{F}_{\text{sun}} = F_{\text{x}} \boldsymbol{i} + F_{\text{y}} \boldsymbol{j} = \sum_{n=1}^{N} \left(F_{n\text{x}} \boldsymbol{i} + F_{n\text{y}} \boldsymbol{j} \right), \tag{13.75}$$

where $\boldsymbol{i}$ and $\boldsymbol{j}$ are the unit vectors along X and Y, respectively.

The resultant torque acting on the sun gear is given by:

$$T_{\text{sun}} = r_s \sum_{n=1}^{N} F_{n2}. \tag{13.76}$$

Now let's take a look at F_x, which can be represented as follows by combining Eqs. (13.74) and (13.75):

$$F_x = \sum_{n=1}^{N} F_{nx} = \sum_{n=1}^{N} (F_{n1} \cos\psi_n + F_{n2} \sin\psi_n) = \sum_{l=1}^{\infty} F_x^l, \tag{13.77}$$

where F_x^l is the l-th harmonic of F_x. It is given by:

$$\begin{aligned} F_x^l = {} & \overbrace{\sum_{n=1}^{N} a_n^l \cos\psi_n \sin\left(l\omega_m t + lz_s\psi_n\right)}^{I} + \overbrace{\sum_{n=1}^{N} b_n^l \cos\psi_n \cos\left(l\omega_m t + lz_s\psi_n\right)}^{II} \\ & + \overbrace{\sum_{n=1}^{N} c_n^l \sin\psi_n \sin\left(l\omega_m t + lz_s\psi_n\right)}^{III} + \overbrace{\sum_{n=1}^{N} d_n^l \sin\psi_n \cos\left(l\omega_m t + lz_s\psi_n\right)}^{IV} \Bigg]. \end{aligned} \tag{13.78}$$

Clearly F_x^l consists of the sum of four terms as marked. If we can make the terms zero through proper selection of the parameter, F_x^l can be made zero.

Taking the term I as an example, it can be cast into the following form through trigonometric identities:

$$\begin{aligned} I &= \sum_{n=1}^{N} a_n^l \cos\psi_n \sin\left(l\omega_m t + lz_s\psi_n\right) \\ &= \sum_{n=1}^{N} a_n^l \left[\cos\psi_n \cos(lz_s\psi_n) \sin(l\omega_m t) + \cos\psi_n \sin(lz_s\psi_n) \cos(l\omega_m t)\right] \\ &= \frac{1}{2} \sum_{n=1}^{N} a_n^l \Bigg\{ \Big[\cos(\psi_n lz_s + \psi_n) + \cos(\psi_n lz_s - \psi_n)\Big] \sin(l\omega_m t) \\ &\quad + \Big[\sin(\psi_n lz_s + \psi_n) + \sin(\psi_n lz_s - \psi_n)\Big] \cos(l\omega_m t) \Bigg\}. \end{aligned} \tag{13.79}$$

It should be noted that trigonometric identities have been used in arriving at the previous notations.

If the planet gears are identical and evenly distributed along the circumference, it is reasonable to assume that the a_n^l, b_n^l, c_n^l and d_n^l for all meshing forces are identical. For brevity, letting $a_l = a_n^l$ and substituting Eq. (13.71) into the previous formulation, we have:

$$I = \frac{1}{2}a^l \sum_{n=1}^{N} \left\{ \left[\cos\frac{2\pi(n-1)(lz_s+1)}{N} + \cos\frac{2\pi(n-1)(lz_s-1)}{N} \right] \sin(l\omega_m t) + \left[\sin\frac{2\pi(n-1)(lz_s+1)}{N} + \sin\frac{2\pi(n-1)(lz_s-1)}{N} \right] \cos(l\omega_m t) \right\}. \tag{13.80}$$

We define

$$k = \mathrm{mod}\left(\frac{lz_s}{N}\right), \tag{13.81}$$

which is referred to as the *phasing factor* in this book.

Eq. (13.80) can be cast into the following form:

$$I = \frac{1}{2}a^l \sum_{n=1}^{N} \left\{ \left[\cos\frac{2\pi(n-1)(k+1)}{N} + \cos\frac{2\pi(n-1)(k-1)}{N} \right] \sin(l\omega_m t) + \left[\sin\frac{2\pi(n-1)(k+1)}{N} + \sin\frac{2\pi(n-1)(k-1)}{N} \right] \cos(l\omega_m t) \right\}. \tag{13.82}$$

It can be proved that the following relationships exist for an integer m:

$$\sum_{n=1}^{N} \cos\frac{2\pi(n-1)m}{N} = \begin{cases} 0 & \text{if } m/N \neq \text{integer} \\ N & \text{if } m/N = \text{integer} \end{cases} \tag{13.83}$$

$$\sum_{n=1}^{N} \sin\frac{2\pi(n-1)m}{N} = 0. \tag{13.84}$$

When $k \neq 1$ and $N-1$, I in Eq. (13.82) reduces to zero.

Similarly, we can conclude that F_y^l is zero. The terms II, III, and IV in Eq. (13.78) become zero as well when $k \neq 1$ and $N-1$. In summary, the l-th harmonic of the resultant force acting on the sun gear disappears when $k \neq 1$ and $N-1$.

Now, let's look at the resultant torque acting on the sun gear, which is

$$T_{\text{sun}} = r_{\text{s}} \sum_{n=1}^{N} \sum_{l=0}^{\infty} \left[c_n^l \sin(l\omega_{\text{m}} t + l z_{\text{s}} \psi_n) + d_n^l \cos(l\omega_{\text{m}} t + l z_{\text{s}} \psi_n) \right] = \sum_{l=0}^{\infty} T^l, \tag{13.85}$$

where T^l is the l-th harmonic of the torque given by:

$$T^l = r_{\text{s}} \sum_{n=1}^{N} \left[c_n^l \sin(l\omega_{\text{m}} t + l z_{\text{s}} \psi_n) + d_n^l \cos(l\omega_{\text{m}} t + l z_{\text{s}} \psi_n) \right]. \tag{13.86}$$

For the case of N identical planet gears uniformly distributed circumferentially, we have identical c_n^l and d_n^l. Letting $c^l = c_n^l$, $d^l = d_n^l$ and substituting Eq. (13.81) into the previous equation yield:

$$\begin{aligned} \frac{T^l}{r_{\text{s}}} &= \left[c^l \sum_{n=1}^{N} \cos \frac{2\pi(n-1)k}{N} - d^l \sum_{n=1}^{N} \sin \frac{2\pi(n-1)k}{N} \right] \sin(l\omega_{\text{m}} t) \\ &+ \left[c^l \sum_{n=1}^{N} \sin \frac{2\pi(n-1)k}{N} - d^l \sum_{n=1}^{N} \cos \frac{2\pi(n-1)k}{N} \right] \cos(l\omega_{\text{m}} t). \end{aligned} \tag{13.87}$$

Obviously, when $k \neq 0$, T^l becomes zero.

To this point, the relationship between k and F^l and between k and T^l is clear. This is summarized in Table 13.7.

Table 13.7 Relationship between k and F_i, as well as T^l.

k	F^l and T^l	Features of vibration mode
0	$F^l = 0,\ T^l \neq 0$	translational motion eliminated, rotational motion excited
$1, N-1$	$F^l \neq 0,\ T^l = 0$	translational motion excited, rotational motion eliminated
$2, 3, \cdots, N-2,$	$F^l = 0,\ T^l = 0$	both rotational and translational motion eliminated

In the previous content, we have analyzed the sun gear. Likewise, we can analyze the force on the ring gear.

In a planetary gear train, the selection of tooth number of gears is constrained by several conditions, among which the *mounting condition* is given by

$$\frac{z_{\text{s}} + z_{\text{r}}}{N} = m' \quad (m' \text{is a positive integer}). \tag{13.88}$$

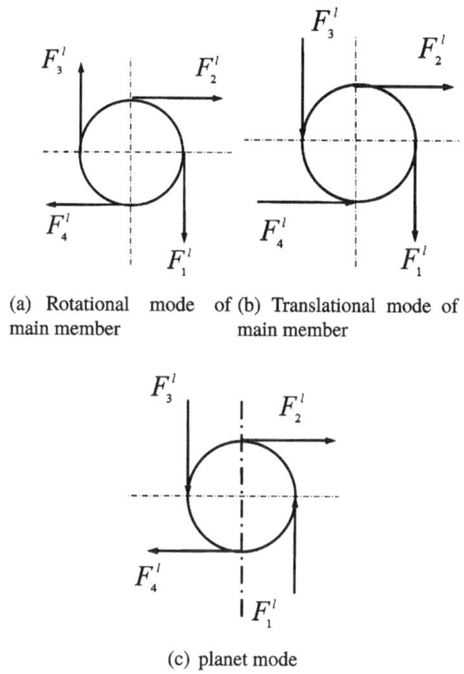

Figure 13.10 Three vibration modes.

Obviously, the phasing factor k for the sun gear and the ring gear is the same. In fact, the sun gear, the ring gear, and the carrier behave exactly the same.

13.4.2 Physical illustration

In the following the conclusion in the previous subsection for the case $N = 4$ is illustrated in Fig. 13.10.

In Fig. 13.10(a) the resultant force acting on the sun gear is zero, but the torque is not. In this case, translation is eliminated, but rotation is excited. Thus, this is the rotational mode of the main members discussed in the previous subsection.

In Fig. 13.10(b) the resultant force is not zero, but the torque is. This indicates that the rotation is eliminated while the translation is excited. Thus, this represents the translational mode.

However, in Fig. 13.10(c), both the resultant force and resultant torque are zero, meaning there is no motion of the main members. In another words, only the planet gears move. Obviously, this is the planet mode.

13.4.3 Experimental validation

To validate the effectiveness of planet phasing, Wang (2005) conducted experimental studies of two planetary gear trains (reducers). The two reducers have identical dimensions, transmission ratios, and natural frequencies. Both have three planet gears, and gear tooth numbers as shown in Table 13.8.

Table 13.8 Parameters of planetary gear trains.

Parameter	Reducer I	Reducer II
Tooth number of ring gear	66	67
Tooth number of sun gear	21	23
Tooth number of planet gear	22	22
Gear module (mm)	3	3
Nominal pressure angle (°)	20	20
Actual $s-p$ pressure angle (°)	26.11	20
Actual $r-p$ pressure angle (°)	23.25	20
$s-p$ contact ratio	1.23	1.59
$r-p$ contact ratio	1.57	1.79
Transmission ratio	4.14	3.91

The two reducers are intentionally made with the following difference:

- **Reducer I:** The tooth number of sun gear is an integer multiple of the number of planet gears ($Z_s/N = 21/3 = 7$).
- **Reducer II:** The tooth number of sun gear is not an integer multiple of the number of planet gears ($Z_s/N = 23/3 \neq \text{integer}$).

Analysis of planet phasing indicates that reducer I has only the rotational mode of the three main members excited, whereas reducer II has both rotational and translational modes excited. Consequently, reducer II would be expected to have more severe vibration and generate higher level of noise than reducer I if the excitation is the same.

The noise spectra of the two reducers are measured and calculated, which are given in Fig. 13.11. Obviously reducer I has a lower level of noise than reducer II, validating the analysis of the planet phasing.

13.4.4 Parameter selection

Tooth number selection of gears in planetary gear trains is much more complicated than ordinary gear sets. Some general suggestions for the selection of tooth number in planetary gear trains include:

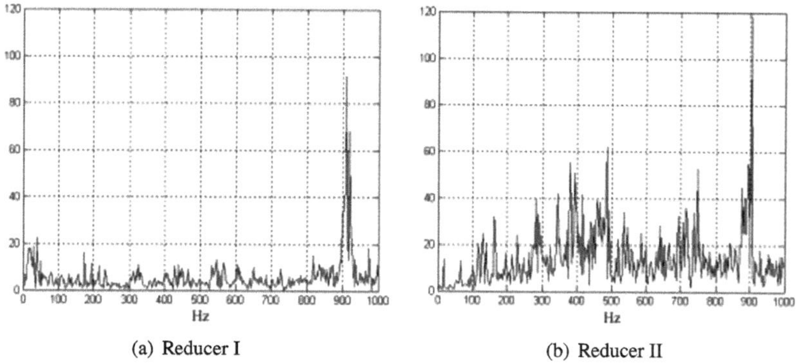

Figure 13.11 Noise spectra of reducers.

- In general, the numbers of different gears have to be matched to meet the *coaxility condition*, *mounting condition* and *adjacency condition* (Grote and Antonsson, 2009).
- In addition, it is also desirable to make the tooth numbers of the sun gear, the planet gear, and the ring gear incommensurate so that more uniform wear is expected.

It is noted that these suggestions are made without accounting for dynamics.

For high-speed planetary gear trains, however, dynamic analysis plays an important role in design. Planet phasing provides potentials to suppress certain type of vibration through proper selection of the number of gear teeth and the number of planet gears. After systematic analysis, Wang (2005) summarized some general recommendations in choosing the gear tooth number and planet number N from the viewpoint of dynamics. These, as outlined next, can be regarded as an amendment to the previous suggestions:

1) The selection should be made with consideration of the specific application. For example, when noise is a main concern, the translational vibration of the main members should be suppressed because the translational vibration will be transmitted to the housing through bearings, causing a high level of noise. However, if the motion accuracy is the main objective, such as in a robotic manipulator, the rotational vibration of the main members should be controlled because the rotational vibration deteriorates the output accuracy.
2) The harmonics contribute to the vibration unevenly. Focus should be placed on the harmonics that affect the most. Assuming the planetary gear train operates at a constant speed and the natural frequency of the

<table>
<tr><td>N=2</td><td>$Z_s=2m$</td><td>$Z_s=2m+1$</td></tr>
<tr><td>$l=2n$</td><td rowspan="2">◎</td><td>◎</td></tr>
<tr><td>$l=2n+1$</td><td>+</td></tr>
</table>

<table>
<tr><td>N=3</td><td>$Z_s=3m$</td><td>$Z_s=3m+1$</td><td>$Z_s=3m+2$</td></tr>
<tr><td>$l=3n$</td><td colspan="3">◎</td></tr>
<tr><td>$l=3n+1$</td><td rowspan="2">◎</td><td colspan="2" rowspan="2">+</td></tr>
<tr><td>$l=3n+2$</td></tr>
</table>

<table>
<tr><td>N=4</td><td>$Z_s=4m$</td><td>$Z_s=4m+1$</td><td>$Z_s=4m+2$</td><td>$Z_s=4m+3$</td></tr>
<tr><td>$l=4n$</td><td colspan="4">◎</td></tr>
<tr><td>$l=4n+1$</td><td rowspan="3">◎</td><td>★</td><td>+</td><td>★</td></tr>
<tr><td>$l=4n+2$</td><td>+</td><td>◎</td><td>+</td></tr>
<tr><td>$l=4n+3$</td><td>★</td><td>+</td><td>★</td></tr>
</table>

<table>
<tr><td>N=5</td><td>$Z_s=5m$</td><td>$Z_s=5m+1$</td><td>$Z_s=5m+2$</td><td>$Z_s=5m+3$</td><td>$Z_s=5m+4$</td></tr>
<tr><td>$l=5n$</td><td colspan="5">◎</td></tr>
<tr><td>$l=5n+1$</td><td rowspan="4">◎</td><td>+</td><td colspan="2">★</td><td>+</td></tr>
<tr><td>$l=5n+2$</td><td rowspan="2">★</td><td colspan="2" rowspan="2">+</td><td rowspan="2">★</td></tr>
<tr><td>$l=5n+3$</td></tr>
<tr><td>$l=5n+4$</td><td>+</td><td colspan="2">★</td><td>+</td></tr>
</table>

<table>
<tr><td>N=6</td><td>$Z_s=6m$</td><td>$Z_s=6m+1$</td><td>$Z_s=6m+2$</td><td>$Z_s=6m+3$</td><td>$Z_s=6m+4$</td><td>$Z_s=6m+5$</td></tr>
<tr><td>$l=6n$</td><td colspan="6">◎</td></tr>
<tr><td>$l=6n+1$</td><td rowspan="5">◎</td><td>+</td><td colspan="3">★</td><td>+</td></tr>
<tr><td>$l=6n+2$</td><td rowspan="3">★</td><td>★</td><td>◎</td><td>★</td><td rowspan="3">★</td></tr>
<tr><td>$l=6n+3$</td><td>◎</td><td>★</td><td>◎</td></tr>
<tr><td>$l=6n+4$</td><td>★</td><td>◎</td><td>★</td></tr>
<tr><td>$l=6n+5$</td><td>+</td><td colspan="3">★</td><td>+</td></tr>
</table>

Note: "◎" —rotational vibration mode of central members,
"+"—translation vibration mode of central members,
"★"— vibration mode of planet gears.

Figure 13.12 Parameter selection chart.

system can be estimated, the harmonic order l that has the largest effect on dynamics can be determined based on Eq. (13.81).

3) For high-speed planetary gear trains, sealing and lubrication are generally well-designed and maintained. In general, wear is not the main concern. Thus, dynamics should be the primary consideration in the selection of the gear tooth number and the planet number. However, if the dynamics and the uniform wear considerations do not conflict, uniform wear can be considered by making the tooth numbers incommensurable.

Wang (2005) also systematically analyzed the detailed vibration modes with N ranging from two through six. The results are summarized into a design chart in Fig. 13.12. From this chart, the following observations can be made:

1) When $Z_s = Nm$ (m is an integer number), the translational modes of all harmonics are suppressed.
2) Based on observation 1), to control the translational vibration, we can choose $Z_s = Nm$ for all cases.
3) However, for the case $N = 6$, it is even more desirable to have $Z_s = 6m + 2, 6m + 3$, or $6m + 4$ because, in these cases, some harmonics of the rotational vibration modes are suppressed in addition to the translational mode.

CHAPTER 14

Elasto-dynamics of mechanical systems

14.1 Introduction

In Chapters 3 and 4, the dynamics of rigid systems was already discussed. Based on these two chapters, the velocity of machines during the starting, stable operation, and braking stages can be analyzed. The results are then fed to the next stage for the design of the relevant components. However, the analysis of the two chapters is based on the assumption of rigid components without accounting for the components' elasticity and system vibration. Thus, the results have limited accuracy for high-speed machine systems. In Chapters 10 through 13, although the component elasticity is considered, focus is placed on the mechanism itself without consideration of the interaction between the driver and the mechanism. Thus, they are referred to as *mechanism dynamics*.

A machine always has at least one driver. The combination of the mechanism and the driver is referred to as a *mechanical system* in this book. For more realistic dynamic analysis, both the component elasticity and the driver–machine interaction are incorporated into a dynamic model in this chapter. To differentiate from the mechanism dynamics, this type of dynamic analysis is here termed as the *elasto-dynamics of mechanical systems*.

The chapter is organized as follows:

- In Sections 14.2 and 14.3, a crane machine and a rolling mill are analyzed, respectively. These two machines both have constant speed ratios.
- Section 14.4 deals with cam mechanisms having variable speed ratios.

14.2 Bridge crane systems

In Chapter 3, an electric hoisting machine based on the rigid-body assumption (see p. 65) was investigated. We illustrated through an example that large dynamic loads are generated within the system when the object is suddenly lifted up or stopped. This generated load generated needs to be properly accounted for in the strength calculation. For this purpose, we will develop a dynamic model considering both the component elasticity and the driving motor of an overhead bridge crane in the following section.

Machinery Dynamics
https://doi.org/10.1016/B978-0-12-815785-5.00025-X

14.2.1 Dynamic model

A bridge crane hoisting system is shown in Fig. 14.1, where 1 is the driv-

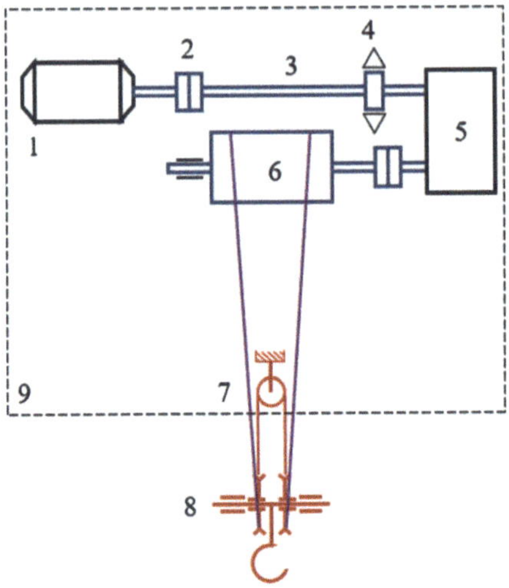

Figure 14.1 Bridge crane.

ing motor; 2 is a coupling; 3 is the main shaft; 4 is the braking drum; 5 is a reducer; 6 is the lifting drum; 7 is a fixed pulley; and 8 is the moving pulley and the picking-up hook. Components 1–7 are installed into a trolley, shown as a dashed box in the figure, which moves along the track on the bridge. In the modeling, the elasticity of the following components is considered:

1) **The bridge:** This is modeled as a simply supported beam.
2) **The main shaft 3:** This is modeled as a shaft subjected to a torque.
3) **The ropes:** This includes the ropes between the lift drum 6 and the moving pulley 8, between the fixed pulley 7 and the moving pulley 8, as well as between the moving pulley and the picking-up hook within 8.

Then, the crane system can be abstracted into a dynamic model shown in Fig. 14.2. The parameters in this model are explained as follows:

M_d: driving torque of the motor.

M_f: resisting torque of the system.

J_1: moment of inertia of the motor rotor and coupling.

J_2: equivalent moment of inertia converted to the main shaft from the braking drum, the reducer, and the lifting drum,

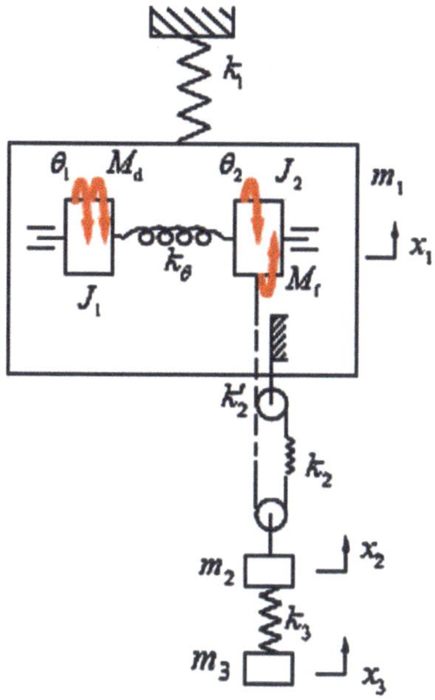

Figure 14.2 Dynamic model.

m_1: equivalent mass, including $m_1^{(1)}$ and $m_1^{(2)}$. $m_1^{(1)}$, that is converted from the bridge at the lifting point, and $m_1^{(2)}$ that is converted from the trolley.

m_2: mass of the moving pulley and the picking-up hook.

m_3: mass of the object to be lifted.

k_1: stiffness of the bridge at the lifting point.

k_2: stiffness of the lifting rope.

k_3: stiffness of the rope between the hook and the moving trolley.

k_θ: torsional stiffness of the main shaft.

θ_1: torsional displacement of J_1.

θ_2: torsional displacement of J_2.

x_1, x_2, x_3: linear displacements of mass m_1, m_2 and m_3, respectively.

In addition, the following parameters, which are not shown in Fig. 14.2, are also used in the model:

c_j: damping coefficient corresponding to k_j ($j = 1, 2, 3, \theta$).

i: transmission ratio of the reducer.

r: radius of the lifting drum.

n: mechanical advantage of the lifting rope system.

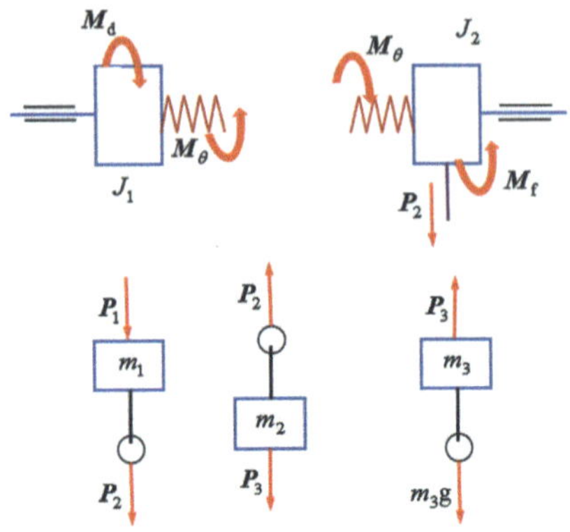

Figure 14.3 FBD of parts.

14.2.2 Equation of motion

The FBD of the m_j and J_i are shown in Fig. 14.3. P_i $(i=1,2,3)$ and M_θ in these diagrams include both the contribution from the spring and damping. The equation of motion of each lumped mass and moment of inertia is derived from Newton's second law of motion:

$$\begin{cases} J_1\ddot{\theta}_1 = M_\mathrm{d} - M_\theta \\ J_2\ddot{\theta}_2 = M_\theta - \dfrac{P_2 r}{ni} - M_\mathrm{f} \\ m_1\ddot{x}_1 = -P_1 - P_2 \\ m_2\ddot{x}_2 = P_2 - P_3 \\ m_3\ddot{x}_3 = P_3 - m_3 g, \end{cases} \tag{14.1}$$

where

$$\begin{cases} P_1 = k_1 x_1 + c_1\dot{x}_1 \\ P_2 = k_2\left[x_1 + \dfrac{\theta_2 r}{ni} - x_2\right] + c_2\left[\dot{x}_1 + \dfrac{\dot{\theta}_2 r}{ni} - \dot{x}_2\right] \\ P_3 = k_3(x_2 - x_3) + c_3(\dot{x}_2 - \dot{x}_3) \\ M_\theta = k_\theta(\theta_1 - \theta_2) + c_\theta(\dot{\theta}_1 - \dot{\theta}_2). \end{cases} \tag{14.2}$$

The motor torque M_d varies with the speed of the rotor. As discussed in Chapter 3, it can be approximated by a linear function of the rotation speed around the rated working point:

$$M_\mathrm{d} = M_0 - b\dot{\theta}_1, \tag{14.3}$$

where b and M_0 are given by:

$$\begin{cases} b = \dfrac{M_k}{\omega_0 - \omega_k} \\ M_0 = b\omega_0. \end{cases} \tag{14.4}$$

M_k is the maximum torque of the motor, and ω_k is the angular velocity at the maximum torque M_k. ω_0 is the synchronous angular velocity of the motor.

The resistance torque of the system, M_f, is calculated in the following, considering the mechanical efficiency η:

$$M_f = (1 - \eta)M_d. \tag{14.5}$$

Substituting Eqs. (14.2), (14.3) and (14.5) into Eq. (14.1) gives

$$\boldsymbol{M}\ddot{\boldsymbol{X}} + (\boldsymbol{C} + \boldsymbol{D}) + \boldsymbol{K}\boldsymbol{X} = \mathbf{Q}, \tag{14.6}$$

where the generalized coordinate vector $\boldsymbol{X}$ is given as

$$\boldsymbol{X} = \begin{bmatrix} \theta_1 & \theta_2 & x_1 & x_2 & x_3 \end{bmatrix}^{\mathrm{T}}. \tag{14.7}$$

$\boldsymbol{M}$ is the mass matrix in the following form:

$$\boldsymbol{M} = \mathrm{diag}\begin{bmatrix} J_1 & J_2 & m_1 & m_2 & m_3 \end{bmatrix}. \tag{14.8}$$

$\mathbf{Q}$ is the generalized force vector

$$\mathbf{Q} = \begin{bmatrix} M_0 & -(1-\eta)M_0 & 0 & 0 & -m_3 g \end{bmatrix}^{\mathrm{T}}. \tag{14.9}$$

$\boldsymbol{K}$ is the stiffness matrix

$$\boldsymbol{K} = \begin{bmatrix} k_\theta & -k_\theta & 0 & 0 & 0 \\ & k_\theta + k_2\left(\dfrac{r}{ni}\right)^2 & \dfrac{k_2 r}{ni} & -\dfrac{k_2 r}{ni} & 0 \\ & & k_1 + k_2 & -k_2 & 0 \\ & sym. & & k_2 + k_3 & -k_3 \\ & & & & k_3 \end{bmatrix}. \tag{14.10}$$

$\boldsymbol{C}$ is the damping matrix

$$\boldsymbol{C} = \begin{bmatrix} c_\theta & -c_\theta & 0 & 0 & 0 \\ & c_\theta + c_2\left(\dfrac{r}{ni}\right)^2 & \dfrac{c_2 r}{ni} & -\dfrac{c_2 r}{ni} & 0 \\ & & c_1 + c_2 & -c_2 & 0 \\ & sym. & & c_2 + c_3 & -c_3 \\ & & & & c_3 \end{bmatrix}. \quad (14.11)$$

$\boldsymbol{D}$ is an auxiliary matrix caused by the motor characteristics curve:

$$\boldsymbol{D} = \begin{bmatrix} b & 0 & 0 & 0 & 0 \\ -(1-\eta)b & 0 & 0 & 0 & 0 \\ 0 & 0 & 0 & 0 & 0 \\ 0 & 0 & 0 & 0 & 0 \\ 0 & 0 & 0 & 0 & 0 \end{bmatrix}. \quad (14.12)$$

14.2.3 Solution of the equation

Eq. (14.6) is a linear ODE set with constant coefficients. Due to the existence of matrix $\boldsymbol{D}$, it is hard to solve analytically. Thus, numerical methods are often used for its solution.

The maximum dynamic load generally occurs upon two cases, namely, abrupt starting and emergency braking.

14.2.3.1 Abrupt starting

It can be seen from Example 3.6 that an AC motor reaches the stable speed before the rope is completely tightened. Neglecting the friction, this stable motor speed is approximately equal to the synchronous speed. Thus, the initial condition for the numerical solution is given by:

$$\begin{cases} \theta_1(0) = \theta_2(0) = x_1(0) = x_2(0) = x_3(0) = 0 \\ \dot{\theta}_1(0) = \omega_0. \end{cases} \quad (14.13)$$

14.2.3.2 Emergency braking

In the case of emergency braking when lowering the object, the braking drum is stopped suddenly. The object, however, keeps moving, and the

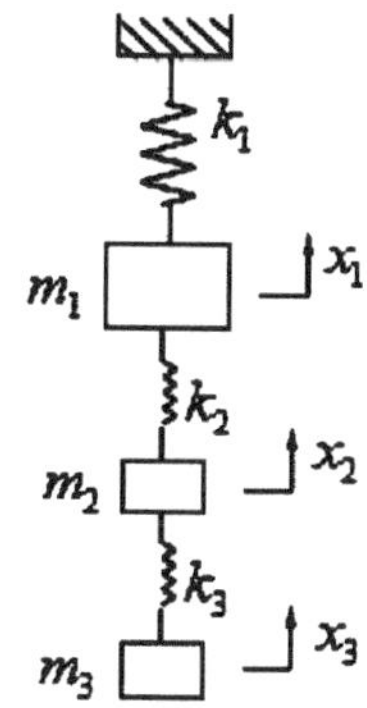

Figure 14.4 Three-DOF mass–spring model.

lifting rope works as a spring. This is a free-vibration problem, which is represented as a three-DOFs mass–spring system as shown in Fig. 14.4. The initial conditions for the three masses are:

$$\begin{cases} x_1(0) = x_2(0) = x_3(0) = 0 \\ \dot{x}_3(0) = v, \end{cases} \tag{14.14}$$

where v is the velocity of the object right before the braking is applied.

With the initial conditions of either Eq. (14.13) or Eq. (14.14), Eq. (14.6) can be solved numerically. Then the dynamic loads within the lifting rope and within the system can be calculated. Given the page limit, the detailed process for the numerical solution is not given here; it can be found in well-written texts on numerical methods.

It is worth pointing out that in most design handbooks detailed dynamic analysis has not been adopted yet. Design based on statics with an estimated dynamic load factor is still prevalent. The detailed dynamic analysis, however, provides a verification to the estimation of the dynamic load in design.

14.3 Rolling mill system

A typical rolling mill system is schematically shown in Fig. 14.5 (Kashay et al., 1972). The main components in this system include two motors, 1 and 2; three couplings 3, 5 and 8; two gears 4 and 9; and two rolls 6 and 7, as well as transmitting shaft segments.

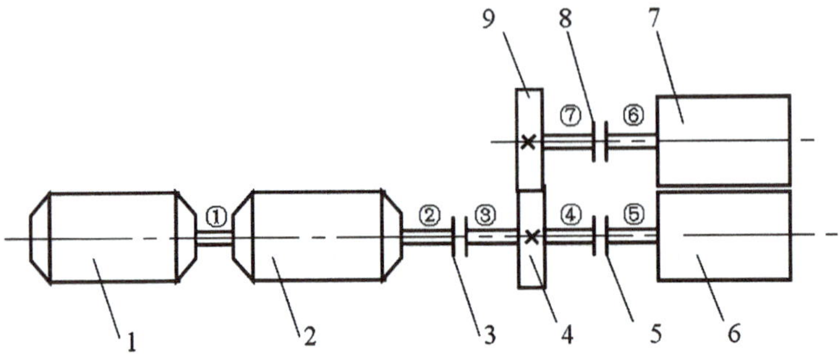

Figure 14.5 Rolling mill.

When the raw material enters the mill, the load on the two rolls increases suddenly, causing the system to vibrate torsionally. The dynamic torque may be as large as several times of the static value; thus, analysis on the transient dynamics is necessary.

14.3.1 Dynamic model

A lumped-parameter model is developed for the system. The shaft segments are converted to equivalent moments of inertia at both ends and massless torsional springs connecting components at both ends. Then the system can be represented by the torsional dynamic model in Fig. 14.6.

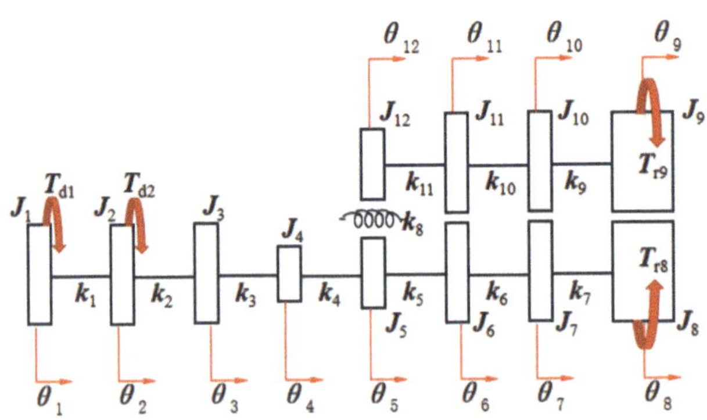

Figure 14.6 Dynamic model of a rolling mill.

The parameters in Fig. 14.6 are defined as follows:

J_1: sum of the moment of inertia of the motor 1 and the equivalent moment of inertia converted from shaft segment ①.

J_2: sum of the moment of inertia of motor 2 and the equivalent moment of inertia converted from shaft segments ① and ②.

J_3: sum of the moment of inertia of the left flange of coupling 3 and the equivalent moment of inertia converted from shaft segment ②.

J_4: sum of the moment of inertia of the right flange of coupling 3 and the equivalent moment of the inertia converted from shaft segment ③.

J_5: sum of the moment of inertia of gear 4 and the equivalent moment of inertia converted from shaft segments ③ and ④.

J_6: sum of the moment of inertia of the left flange of coupling 5 and the equivalent moment of inertia converted from shaft segment ④.

J_7: sum of the moment of inertia of the right flange of coupling 5 and the equivalent moment of inertia converted from shaft segment ⑤.

J_8: sum of the moment of inertia of roll 6 and the equivalent moment of inertia converted from shaft segment ⑤.

J_9: sum of the moment of inertia of roll 7 and the equivalent moment of inertia converted from shaft segment ⑥.

J_{10}: sum of the moment of inertia of the right flange of coupling 8 and the equivalent moment of inertia converted from shaft segment ⑥.

J_{11}: sum of the moment of inertia of the left flange of coupling 8 and the equivalent moment of inertia converted from shaft segment ⑦.

J_{12}: sum of the moment of inertia of gear 9 and the equivalent moment of inertia converted from shaft segment ⑦.

θ_i: torsional displacement of J_i $(i = 1, 2, \cdots, 12)$.

T_{di}: driving torques of motor i $(i = 1, 2)$.

T_{r8}, T_{r9}: resisting torques on J_8 and J_9, respectively.

$k_1, k_2, k_4, k_5, k_7, k_9, k_{11}$: torsional stiffness of shaft segments ①, ②, ③, ④, ⑤, ⑥, ⑦, respectively.

k_3, k_6, k_{10}: torsional stiffness of couplings 3, 5, and 8, respectively.

k_8: meshing stiffness of the gear mesh.

14.3.2 Equation of motion

In the rolling operation, the material's internal damping is significant; thus, it is included in the modeling. However, for the sake of clarity, damping is not shown in Fig. 14.6. According to Newton's second law of motion, we have the following equations for the lumped moments of inertia:

$$
\begin{cases}
J_1\ddot{\theta}_1 + c_1\left(\dot{\theta}_1 - \dot{\theta}_2\right) + k_1\left(\theta_1 - \theta_2\right) = T_{\mathrm{d}1} \\
J_2\ddot{\theta}_2 + c_1\left(\dot{\theta}_2 - \dot{\theta}_1\right) + c_2\left(\dot{\theta}_2 - \dot{\theta}_3\right) + k_1\left(\theta_1 - \theta_2\right) \\
+k_2\left(\theta_2 - \theta_3\right) = T_{\mathrm{d}2} \\
J_3\ddot{\theta}_3 + c_2\left(\dot{\theta}_3 - \dot{\theta}_2\right) + c_3\left(\dot{\theta}_3 - \dot{\theta}_4\right) + k_2\left(\theta_3 - \theta_2\right) \\
+k_3\left(\theta_4 - \theta_4\right) = 0 \\
J_4\ddot{\theta}_4 + c_3\left(\dot{\theta}_4 - \dot{\theta}_3\right) + c_4\left(\dot{\theta}_4 - \dot{\theta}_5\right) + k_3\left(\theta_4 - \theta_3\right) \\
+k_4\left(\theta_4 - \theta_5\right) = 0 \\
J_5\ddot{\theta}_5 + c_4\left(\dot{\theta}_5 - \dot{\theta}_4\right) + c_5\left(\dot{\theta}_5 - \dot{\theta}_6\right) + c_8 r_{04}\left(r_{04}\dot{\theta}_5 - r_{09}\dot{\theta}_{12}\right) \\
+k_4\left(\theta_5 - \theta_4\right) + k_5\left(\theta_5 - \theta_6\right) + k_8 r_{04}\left(r_{04}\theta_5 - r_{09}\theta_{12}\right) = 0 \\
J_6\ddot{\theta}_6 + c_5\left(\dot{\theta}_6 - \dot{\theta}_5\right) + c_6\left(\dot{\theta}_6 - \dot{\theta}_7\right) + k_5\left(\theta_6 - \theta_5\right) \\
+k_6\left(\theta_6 - \theta_7\right) = 0 \\
J_7\ddot{\theta}_7 + c_6\left(\dot{\theta}_7 - \dot{\theta}_6\right) + c_7\left(\dot{\theta}_7 - \dot{\theta}_8\right) + k_6\left(\theta_7 - \theta_6\right) \\
+k_7\left(\theta_7 - \theta_8\right) = 0 \\
J_8\ddot{\theta}_8 + c_7\left(\dot{\theta}_8 - \dot{\theta}_7\right) + k_7\left(\theta_8 - \theta_7\right) = -T_{\mathrm{r}8} \\
J_9\ddot{\theta}_9 + c_9\left(\dot{\theta}_9 - \dot{\theta}_{10}\right) + k_9\left(\theta_9 - \theta_{10}\right) = -T_{\mathrm{r}9} \\
J_{10}\ddot{\theta}_{10} + c_9\left(\dot{\theta}_{10} - \dot{\theta}_9\right) + c_{10}\left(\dot{\theta}_{10} - \dot{\theta}_{11}\right) + k_9\left(\theta_{10} - \theta_9\right) \\
+k_{10}\left(\theta_{10} - \theta_{11}\right) = 0 \\
J_{11}\ddot{\theta}_{11} + c_{10}\left(\dot{\theta}_{11} - \dot{\theta}_{10}\right) + c_{11}\left(\dot{\theta}_{11} - \dot{\theta}_{12}\right) + k_{10}\left(\theta_{11} - \theta_{10}\right) \\
+k_{11}\left(\theta_{11} - \theta_{12}\right) = 0 \\
J_{12}\ddot{\theta}_{12} + c_{11}\left(\dot{\theta}_{12} - \dot{\theta}_{11}\right) + c_8 r_{09}\left(r_{09}\dot{\theta}_{12} - r_{04}\dot{\theta}_5\right) \\
+k_{11}\left(\theta_{12} - \theta_{11}\right) + k_8 r_{09}\left(r_{09}\theta_{12} - r_{04}\theta_5\right) = 0,
\end{cases}
\tag{14.15}
$$

where r_{04} and r_{09} are the radii of base circles of gear 4 and 9, respectively.

The equation can be organized into matrix form as:

$$
\boldsymbol{M}\ddot{\boldsymbol{\theta}} + \boldsymbol{C}\dot{\boldsymbol{\theta}} + \boldsymbol{K}\boldsymbol{\theta} = \boldsymbol{F}, \tag{14.16}
$$

where $\boldsymbol{M}$, $\boldsymbol{C}$, and $\boldsymbol{K}$ are the mass, damping, and stiffness matrices, respectively. They are not given here for the sake of space. The generalized coordinate and the excitation vectors are given as follows:

$$\begin{cases} \boldsymbol{\theta} = \begin{bmatrix} \theta_1 & \theta_2 & \cdots & \theta_{12} \end{bmatrix}^{\mathrm{T}} \\ \boldsymbol{F} = \begin{bmatrix} T_{\mathrm{d1}} & T_{\mathrm{d2}} & 0 & \cdots & 0 & -T_{\mathrm{r8}} & -T_{\mathrm{r9}} & 0 & 0 & 0 \end{bmatrix}^{\mathrm{T}}. \end{cases} \tag{14.17}$$

Kashay et al. (1972) numerically solved the equation with parameters of an actual rolling mill. The driving torque of the motor was obtained through tests and is given in Fig. 14.7. The resistance toque from the rolls is taken as a step function, as shown in Fig. 14.8.

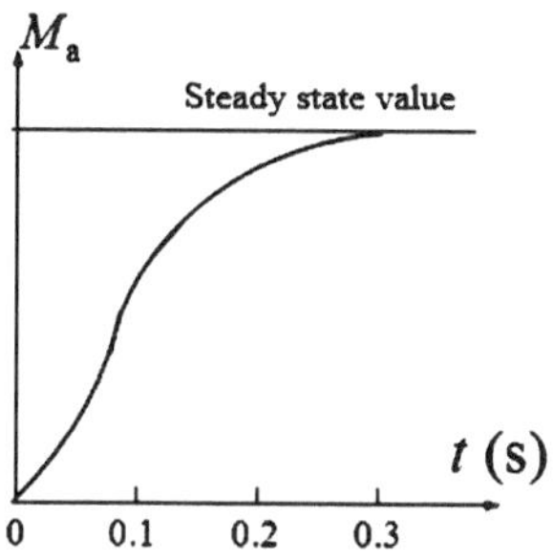

Figure 14.7 Motor torque.

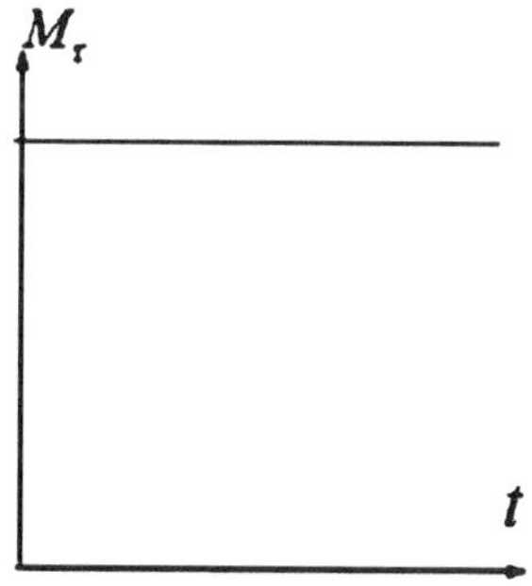

Figure 14.8 Resistance torque.

The numerical simulation results are compared with the measured data. The comparison of the torque between the two motors is presented in Fig. 14.9, in which the two-step resistance torques are taken as 2.03×10^6 and 2.19×10^6 lb-in. Clearly, the simulation results agree well with the measured data.

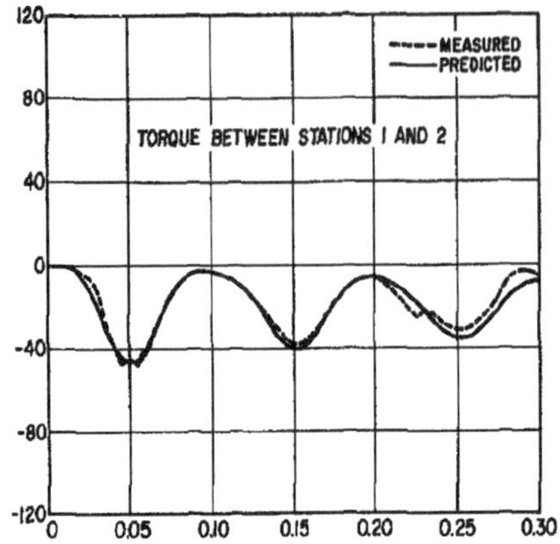

Figure 14.9 Comparison of torques. *(Courtesy of Kashay et al. (1972))*

14.4 Polydyne servo-cam design

Stoddart (1953) first proposed the concept of a polydyne cam, which takes into dynamics into consideration in cam design. However, one issue of the polydyne cam system is the high sensitivity to parameter variations. Small changes in dynamic parameters may significantly degrade the cam performance. Yan et al. (2001) developed a method to improve the dynamic performance of the polydyne cam through servo-control of the input speed. This method, integrating the polydyne cam design and servo-control system, is introduced in the following section.

14.4.1 Principle

The cam profile in this method is designed through the ordinary polydyne method. A microprocessor-controlled servomotor works as the actuator to drive the cam mechanism: It alters the cam input speed as required to achieve a desired polynomial follower motion. In essence, the method changes the *rigid, static polydyne cam* into a *flexible, dynamic polydyne cam* system.

Two concepts need to be clearly defined before introducing the design procedures. The *design speed* is the constant cam speed with which the cam profile is synthesized. The *off-design speed* is a speed which is different from the design speed.

The procedures of this method are given in the flow-chart in Fig. 14.10 and summarized as follows:

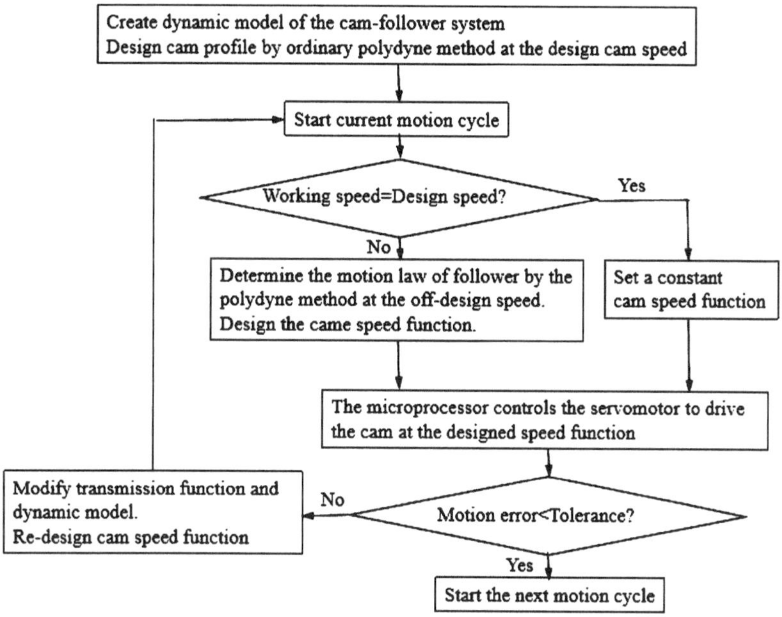

Figure 14.10 Flow chart.

Step 1: Create the dynamic model of the system and design the cam profile by the ordinary polydyne method at the design speed.
Step 2: Start one motion cycle.
Step 3: If the cam operates at the design speed, then set a constant cam-speed function; otherwise, determine the motion of the follower first by the polydyne method at the off-design speed, and then design the corresponding nonconstant cam-speed function.
Step 4: The microprocessor controls the servomotor to drive the cam at the designed speed function.
Step 5: If the output motion error exceeds a preset tolerance, go to step 6; otherwise, go to step 7.
Step 6: Modify the transmission function and the dynamic model of the cam-follower system and redesign the cam speed function; then go to step 2.
Step 7: Start the next motion cycle.

According to control theory, it is easy to implement a particular input speed function by a microprocessor-controlled servomotor. Lower vibration response at high speed is the primary design objective of the servo-control system that has periodic and variable inertia loading. A measuring system (exclude the measuring components in the servo loop) is used to detect the actual motion parameters of the cam and follower, and the data are fed back to the microprocessor for modifying the cam profile function and dynamic model of the elastic cam-follower system.

The process of design is described in detail in the following.

14.4.2 Design process

The D–R–D (dwell–rise–dwell) motion is selected as an example. The parameters are explained as follows:

θ_r, θ_d: The angular displacements of cam in rise and dwell periods, respectively.

θ_{hr}, θ_{hd}: The total angular displacement of cam in rise and dwell periods, respectively.

h, h_a: The total displacements of the follower and output, respectively.

s, s_a: The displacements of the follower and output, respectively.

t, t_r, t_d: Time; total rise time; and total dwell time, respectively.

As usual, we first normalize the parameters as follows:

$$\begin{cases} T = \dfrac{t}{t_r} \\ \Theta = \dfrac{\theta}{\theta_r} \\ S = \dfrac{s}{h} \\ S_a = \dfrac{s_a}{h_a}. \end{cases} \tag{14.18}$$

14.4.2.1 Cam profile designed at design speed in rise

The dynamics of the system can be represented by the following single-DOF vibration equation:

$$\ddot{S}_a + 2\xi\,(2\pi\lambda)\,\dot{S}_a + (2\pi\lambda)^2\,S_a = (2\pi\lambda)^2\,S, \tag{14.19}$$

where ξ is the damping ratio and λ is the period ratio defined as $\lambda = t_r/t_0$. t_0 is the natural period of the system.

The desired displacement output can be represented by the polynomial

$$S_a = S_a(T) = \sum_{r=0}^{N} a_r T^r, \tag{14.20}$$

where N is an integer representing the order of the polynomial.

Substituting S_a, $\dot{S}_a$, and $\ddot{S}_a$ into Eq. (14.19) and taking advantage of Eq. (14.20), we obtain

$$S = S(\lambda, T) = \sum_{r=0}^{N} a_r \left[T^r + 2r\xi\, (2\pi\lambda)^{-1}\, T^{r-1} + r(r-1)\, (2\pi\lambda)^{-2}\, T^{r-2} \right]. \tag{14.21}$$

If the cam rotates at a constant speed, we have $\Theta = T$ and Eq. (14.21) will become

$$S = S(\lambda, \Theta) = \sum_{r=0}^{N} a_r \left[\Theta^r + 2r\xi\, (2\pi\lambda)^{-1}\, \Theta^{r-1} + r(r-1)\, (2\pi\lambda)^{-2}\, \Theta^{r-2} \right]. \tag{14.22}$$

Eq. (14.21) represents the follower displacement as a function of the angular displacement of the cam if the cam rotates at constant speed. It is the basis for making the cam profile.

Eq. (14.22) is true regardless of the input speed of the cam. However, in the case of variable input speed, the output obtained by the cam profile determined on the basis of Eq. (14.22) will depart from that of Eq. (14.20).

Letting ω_0 and λ_0 represent the cam design speed and the period ratio, the follower displacement at the design sped ω_0 (or λ_0) can be obtained by Eq. (14.22):

$$S_0 = S(\lambda_0, \Theta). \tag{14.23}$$

14.4.2.2 Off-design speed in rise

Denoting ω_p as the cam speed corresponding to λ_p, we assume $\omega_p \neq \omega_0$, $\lambda_p \neq \lambda_0$. The follower displacement at λ_p is represented as a function of T according to Eq. (14.21):

$$S_p = S(\lambda_p, T). \tag{14.24}$$

Without modifying the polydyne cam profile designed at λ_0, a cam-speed function $\Theta(T)$ can be designed to make the follower have the same motion as determined at λ_0. This is represented by

$$S(\lambda_0, \Theta) = S(\lambda_p, T). \tag{14.25}$$

Eq. (14.25) forms a nonlinear algebra equation about $\Theta(T)$. It can be solved by the following methods:

1) **Analytic method**
It can be solved analytically as a problem of inverse kinematics. This method is generally tedious and not easy to be implemented.
2) **Leaning control method**
The learning control approach needs solution of the forward kinematics and dynamics repeatedly to find the solution of the inverse problem. It is powerful, but is could not deal with design constraints.
3) **Optimization method**
In this method, the problem is formulated into an optimization problem. Optimal control theory is a powerful optimization tool for solving this type of problem.

14.4.2.3 Off-design speed in dwell

For simplicity, an indexing cam is considered. The indexing motion consists of a rise and a dwell period. To maintain continuity of speed, the dimensionless displacement of the cam in the dwell period needs to satisfy the following boundary conditions:

$$\begin{cases} \theta_d(0) = 0; & \theta_d(t_d) = \theta_{hd} \\ \dot{\theta}_d(0) = \dot{\theta}_r(t_r); & \dot{\theta}_d(t_d) = \dot{\theta}_r(0) \\ \ddot{\theta}_d(0) = \ddot{\theta}_r(t_r); & \ddot{\theta}_d(t_d) = \ddot{\theta}_r(0) \\ \dddot{\theta}_d(0) = \dddot{\theta}_r(t_r); & \dddot{\theta}_d(t_d) = \dddot{\theta}_r(0). \end{cases} \tag{14.26}$$

In dimensionless form they become:

$$\begin{cases} \Theta_d(0) = 0; & \Theta_d(1) = \Theta_r(0) \\ \dot{\Theta}_d(0) = \dot{\Theta}_r(1); & \dot{\Theta}_d(1) = \dot{\Theta}_r(0) \\ \ddot{\Theta}_d(0) = (\theta_{hd}/\theta_{hr})\,\ddot{\Theta}_r(1); & \ddot{\Theta}_d(1) = (\theta_{hd}/\theta_{hr})\,\ddot{\Theta}_r(0) \\ \dddot{\Theta}_d(0) = (\theta_{hd}/\theta_{hr})^2\,\dddot{\Theta}_r(1); & \dddot{\Theta}_d(1) = (\theta_{hd}/\theta_{hr})^2\,\dddot{\Theta}_r(0). \end{cases} \tag{14.27}$$

The cam displacement function minimizing input speed variation can be written as

$$\Theta_d = C_1 T + C_2 T^2 + C_3 T^3 + C_4 T^4 + C_5 T^5 + C_6 T^6 + C_7 T^7. \tag{14.28}$$

The coefficients are as follows:

$$
\begin{cases}
C_1 = & \Theta_r(1) \\
C_2 = & \dfrac{1}{2}\dfrac{\theta_{hd}}{\theta_{hr}}\ddot{\Theta}_r(1) \\
C_3 = & \dfrac{1}{6}\left(\dfrac{\theta_{hd}}{\theta_{hr}}\right)^2 \dddot{\Theta}_r(1) \\
C_4 = & -\dfrac{1}{6}\left(\dfrac{\theta_{hd}}{\theta_{hr}}\right)^2 \dddot{\Theta}_r(0) - \dfrac{1}{6}\dfrac{\theta_{hd}}{\theta_{hr}}\ddot{\Theta}_r(0) - 15\dot{\Theta}_r(0) - 20\dot{\Theta}_r(1) \\
 & -5\dfrac{\theta_{hd}}{\theta_{hr}}\ddot{\Theta}_r(1) - \dfrac{2}{3}\left(\dfrac{\theta_{hd}}{\theta_{hr}}\right)^2 \dddot{\Theta}_r(1) + 35 \\
C_5 = & -\dfrac{1}{2}\left(\dfrac{\theta_{hd}}{\theta_{hr}}\right)^2 \dddot{\Theta}_r(0) + \dfrac{5}{3}\dfrac{\theta_{hd}}{\theta_{hr}}\ddot{\Theta}_r(0) + 39\dot{\Theta}_r(0) + 45\dot{\Theta}_r(1) \\
 & +10\dfrac{\theta_{hd}}{\theta_{hr}}\ddot{\Theta}_r(1) + \left(\dfrac{\theta_{hd}}{\theta_{hr}}\right)^2 \dddot{\Theta}_r(1) - 84 \\
C_6 = & -\dfrac{1}{2}\left(\dfrac{\theta_{hd}}{\theta_{hr}}\right)^2 \dddot{\Theta}_r(0) + \dfrac{13}{6}\dfrac{\theta_{hd}}{\theta_{hr}}\ddot{\Theta}_r(0) - 34\dot{\Theta}_r(0) - 36\dot{\Theta}_r(1) \\
 & -\dfrac{15}{2}\dfrac{\theta_{hd}}{\theta_{hr}}\ddot{\Theta}_r(1) - \dfrac{2}{3}\left(\dfrac{\theta_{hd}}{\theta_{hr}}\right)^2 \dddot{\Theta}_r(1) + 70 \\
C_7 = & \dfrac{1}{6}\left(\dfrac{\theta_{hd}}{\theta_{hr}}\right)^2 \dddot{\Theta}_r(0) - 2\dfrac{\theta_{hd}}{\theta_{hr}}\ddot{\Theta}_r(0) + 10\dot{\Theta}_r(0) + 10\dot{\Theta}_r(1) \\
 & +2\dfrac{\theta_{hd}}{\theta_{hr}}\ddot{\Theta}_r(1) + \dfrac{1}{6}\left(\dfrac{\theta_{hd}}{\theta_{hr}}\right)^2 \dddot{\Theta}_r(1) - 20.
\end{cases}
\tag{14.29}
$$

14.4.3 Design example

In this example the follower is supposed to have an indexing motion. We assume $t_r = t_d$. The procedures for the design are illustrated as follows:

Step 1: Select the standard 5–6–7–8–9 polynomial as the desired motion of the output. The dimensionless acceleration of the output is shown in Fig. 14.11.

Step 2: Letting the period ratios $\lambda_1 = 2.0$, $\lambda_2 = 2.5$, and $\lambda_3 = 3.0$, design the follower motion functions by the polydyne approach. The resulting dimensionless accelerations of the follower are shown in Fig. 14.12.

Step 3: Letting one of the λ_i $(i = 1, 2, 3)$ as the λ_0 (the period ratio at the design speed), solve Eq. (14.25) to determine the off-design speeds at

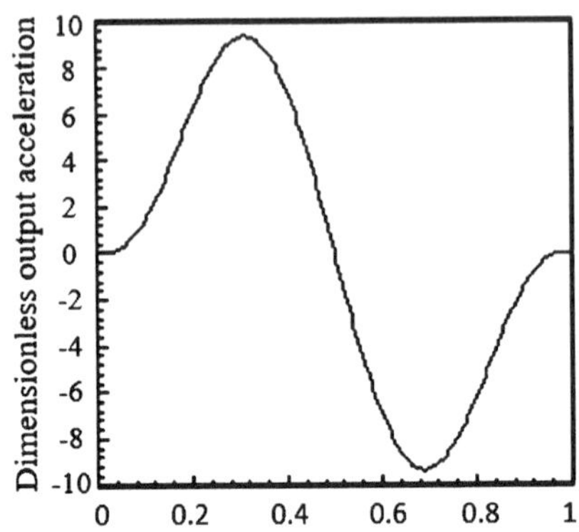

Figure 14.11 Desired output acceleration 5–6–7–8–9 polynomial.

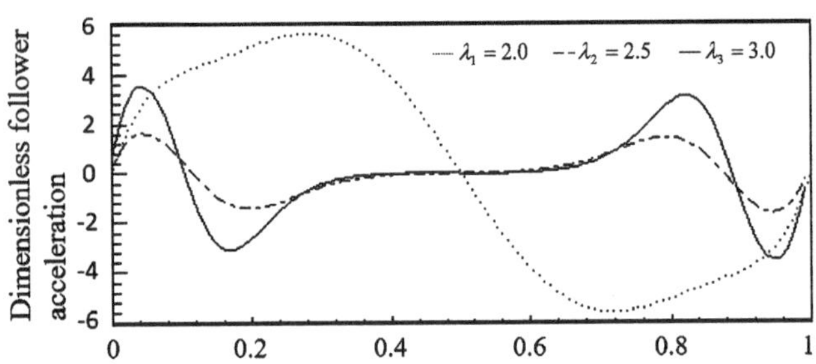

Figure 14.12 Dimensionless accelerations of the follower.

the other two λ_i. The learning control approach is used, and the results are shown in Fig. 14.13.

It can be seen from Figs. 14.11 through 14.13 that the results of this example verify the feasibility of the method of polydyne servo-cam design. As to the selection of design cam speed (or design period ratio), Fig. 14.13 indicates that it has direct influence on the variation of the resulting input speed function. Furthermore, the conditions of cam fabricating and forcing can also be improved by selecting an appropriate design cam speed.

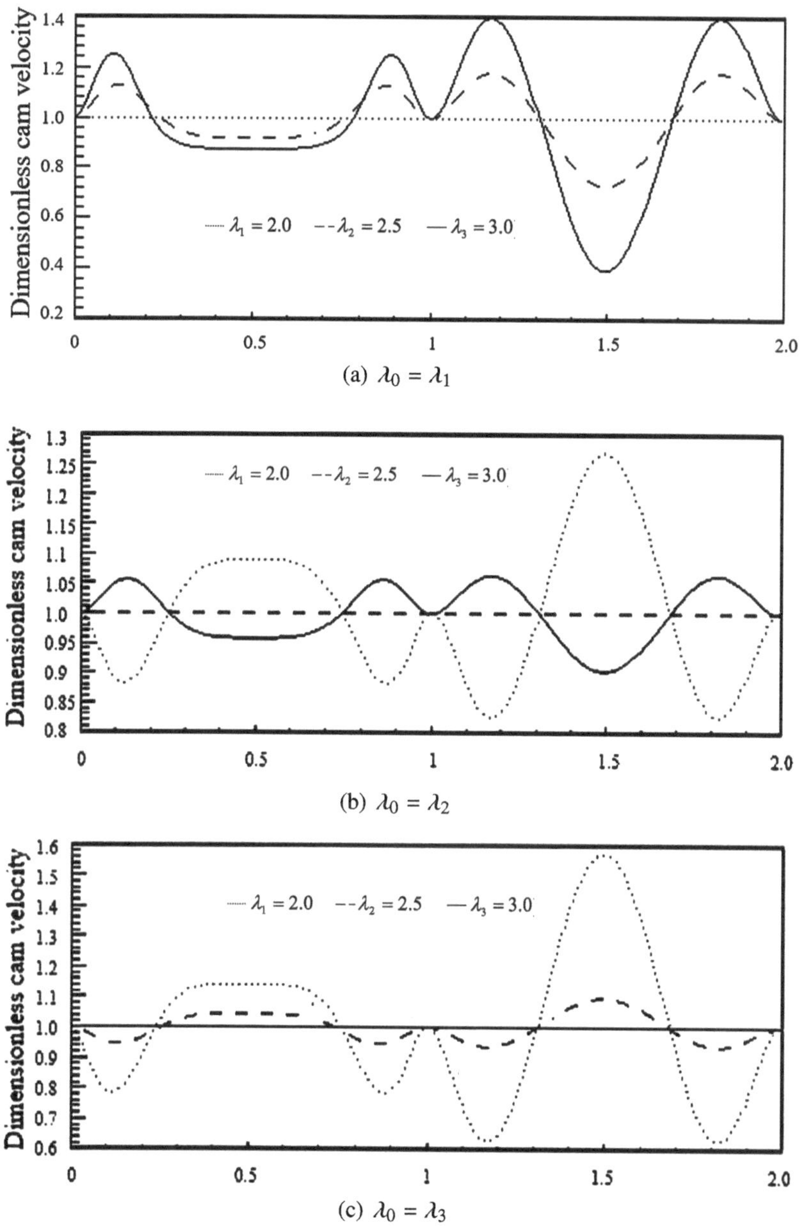

Figure 14.13 Dimensionless cam speed.

CHAPTER 15

Dynamics of machinery with joint clearance

15.1 Introduction

Up to this point of this book, clearance has not been considered in the modeling and analysis of dynamics. However, clearance exists in almost all kinematic pairs (joints) or between mating parts primarily for two reasons. First, clearance may be purposely allocated for fabrication and lubrication purposes. Second, wear generates clearance between originally tightly mated parts and enlarges existing clearance. Clearance has significant effect on dynamics. For example, it causes impact and quickens fatigue. It also deteriorates the precision of machines. In some cases, completely chaotic behavior may be caused by clearance. Researchers started to study the effect of joint clearance on the dynamics of machines around the late 1960s (Chace, 1967). In this paper, Lagrange's equation was utilized in combination with second-order constraints. Since then, many methods and models have been developed to tackle the clearance dynamic problem. Publications have appeared formed on the topic mainly in the form of journal and conference papers scattered in various journals and proceedings. Some well-written review papers and monographs, such as those by (Tian et al., 2018; Flores et al., 2008), are also available. Readers who want to throughly understand dynamics with clearance are advised to consult these relevant materials.

Clearance brings a strong nonlinearity into the dynamic equation and causes very complicated nonlinear response phenomena. The basic theory of nonlinear vibration was already introduced in Chapter 8. In this chapter we will focus on the modeling of clearance in dynamics of specific mechanisms and basic analysis of typical nonlinear responses. In addition, the discussion is limited to only planar mechanisms.

The chapter is organized as follows:

- In Section 15.2, the general models of joint clearance are first introduced with examples of revolute joints. The three possible modes and relevant treatment in dynamics are presented.
- Section 15.3 focuses on linkages. Various treatments of joint clearance are presented with the specific mechanisms and modeling methods.

Machinery Dynamics
https://doi.org/10.1016/B978-0-12-815785-5.00026-1

- An indexing cam and a gear pair are presented in Sections 15.4 and 15.5. Some typical response features are analyzed.
- In the final Section 15.6, some general features of nonlinear machinery dynamics are outlined.

15.2 Three modes of clearance

For a planar revolute joint having clearance in Fig. 15.1(a), the clearance can be defined in radius as

$$c = r_{\mathrm{j}} - r_{\mathrm{b}}, \tag{15.1}$$

where c is the clearance of the joint. r_{j} and r_{b} are the radii of the journal and the bearing, respectively.

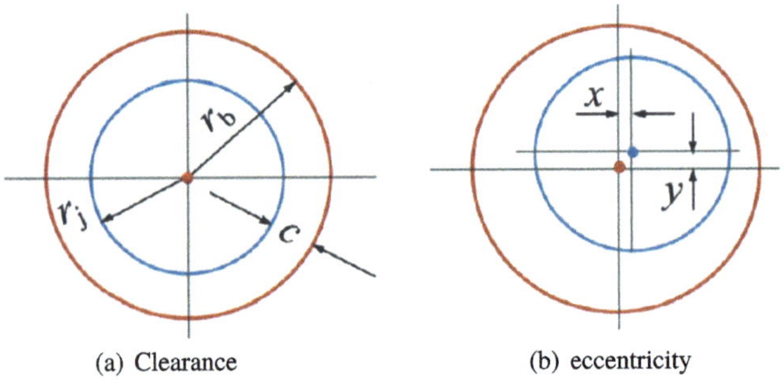

Figure 15.1 Modes of joint clearance.

Three possible relative statuses, called *modes*, may exist between the journal and the bearing. For convenient description of the modes, we define a variable δ as

$$\delta = \sqrt{x^2 + y^2} - c, \tag{15.2}$$

where x and y are the Cartesian components of the distance between the two centers as shown in Fig. 15.1(b). The three modes can then be described as:

Free flight: In this mode $\delta < 0$, and the journal is not in contact with the bearing. Thus, no constraint is applied by the joint.

Impact: In this mode $\delta = 0$, and the journal impacts the bearing with some relative velocity before being bounced back and forth. The contact between the journal and the bearing is momentary. It is commonly modeled with a restitution coefficient as

$$v_i^+ = -s v_i^-, \tag{15.3}$$

where s is called the restitution coefficient. v_i^+ and v_i^- are the (relative) impact velocity right after and before the impact moment. If the impact is completely elastic, $s = 1$; otherwise, $0 < s < 1$.

Continuous contact: In this mode, the bearing and the journal are in continuous contact. If the joint is considered rigid, $\delta = 0$, and continuous sliding happens between the journal and the bearing. If the joint is elastic, $\delta > 0$, and the joint is often modeled as a pair of spring and damper as

$$F_N = k\delta^n + c\dot{\delta}, \tag{15.4}$$

where F_N is the normal contact force. k and c are the spring stiffness and the damping coefficient, respectively. $n = 3/2$ if Hertzian contact is considered. For simplicity, a linear spring, $n = 1$, is often used.

The most comprehensive model considers all three modes. Some others consider two modes or even only one mode. These models will be applied to various mechanisms in the following.

15.3 Linkage mechanism

Many models of various levels of complexity have been proposed in dynamics of linkages containing clearance. In some works (Miedema and Mansour, 1976), all three modes were considered. One difficulty in the three-mode model is the determination of the mode. Some other researchers (Dubowsky and Freudenstein, 1971; Tang et al., 2013) considered two modes. The simplest case is to consider only the rigid-contact mode; in this case, the clearance becomes a massless link (Feng et al., 1991). In the following section, a two-mode model and a massless link model are introduced focusing on the modeling procedure.

15.3.1 Two-mode model

Tang et al. (2013) investigated the four-bar mechanism shown in Fig. 15.2. In this mechanism, there is clearance at one revolute joint as shown. Two modes, free flight, and continuous contact, are considered. The contact is modeled as a pair of linear spring and a damper as shown in Fig. 15.3. The

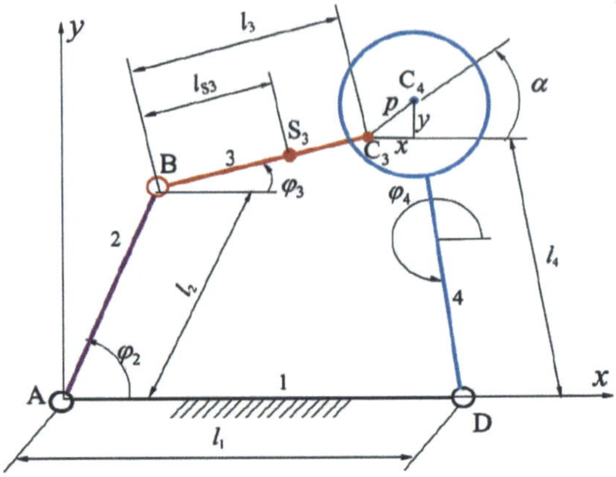

Figure 15.2 Four-bar linkage.

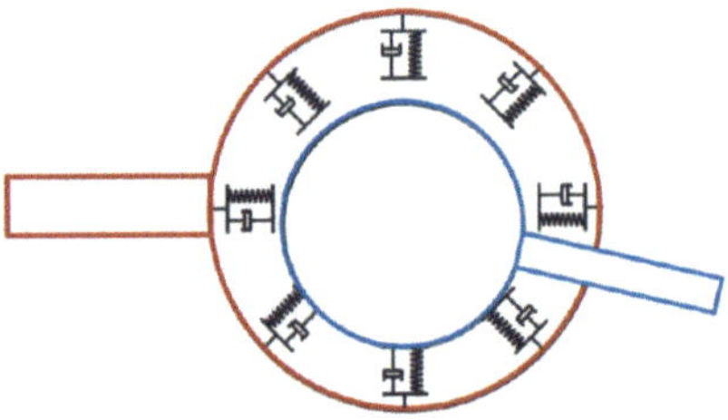

Figure 15.3 Clearance model.

contact force between the journal and the bearing is therefore calculated by:

$$F = \begin{cases} k\delta + c\dot{\delta} & \text{if } \delta \geq 0 \\ 0 & \text{if } \delta < 0, \end{cases} \tag{15.5}$$

where δ is defined as

$$\delta = \sqrt{x^2 + y^2} - c_0. \tag{15.6}$$

c_0 is the clearance. k and c are the stiffness and damping coefficient, respectively. It should be noted that, in arriving at relationship just stated, the friction force between the journal and the bearing is not considered. In general the friction is insignificant compared to the normal force.

15.3.1.1 Equation of motion

If no clearance exists in the joint, the following relationship can be obtained from the vector loop:

$$\begin{cases} l_2 \cos\varphi_2 + l_3 \cos\varphi_{30} + l_4 \cos\varphi_{40} - l_1 = 0 \\ l_2 \sin\varphi_2 + l_3 \sin\varphi_{30} + l_4 \sin\varphi_{40} = 0, \end{cases} \tag{15.7}$$

where φ_2 is the position angle of the crank. φ_{30} and φ_{40} are the position angles of links 3 and 4, respectively, without clearance. $l_i(i = 1 \cdots 4)$ are the lengths of the links. If the clearance at the revolute joint between links 3 and 4 is considered, the vector loop equation becomes:

$$\begin{cases} l_2 \cos\varphi_2 + l_3 \cos\varphi_3 + x + l_4 \cos\varphi_4 - l_1 = 0 \\ l_2 \sin\varphi_2 + l_3 \sin\varphi_3 + y + l_4 \sin\varphi_4 = 0, \end{cases} \tag{15.8}$$

where φ_3 and φ_4 are the actual position angles of links 3 and 4, respectively, with clearance.

φ_3 and φ_4 can be represented as a small disturbance to the position angles without clearance:

$$\begin{cases} \varphi_3 \approx \varphi_{30} + \Delta\varphi_3 \\ \varphi_4 \approx \varphi_{40} + \Delta\varphi_4. \end{cases} \tag{15.9}$$

Taking advantage of the following trigonometric identities

$$\begin{cases} \sin(\alpha + \beta) = \sin\alpha \cos\beta + \cos\alpha \sin\beta \\ \cos(\alpha + \beta) = \cos\alpha \cos\beta - \sin\alpha \sin\beta, \end{cases} \tag{15.10}$$

$\Delta\varphi_3$ and $\Delta\varphi_4$ are small disturbances; thus, the following approximation is true:

$$\begin{cases} \sin(\Delta\varphi_3) \approx 0; \quad \cos(\Delta\varphi_3) \approx \Delta\varphi_3 \\ \sin(\Delta\varphi_4) \approx 0; \quad \cos(\Delta\varphi_4) \approx \Delta\varphi_4. \end{cases} \tag{15.11}$$

Eq. (15.8) can be expanded to

$$\begin{cases} l_2 \cos\varphi_2 + l_3 \cos\varphi_{30} - l_3 \sin\varphi_{30}\Delta\varphi_3 + x + l_4 \cos\varphi_{40} - l_4 \sin\varphi_{40}\Delta\varphi_3 - l_1 = 0 \\ l_2 \sin\varphi_2 + l_3 \sin\varphi_{30} + l_3 \cos\varphi_{30}\Delta\varphi_3 + y + l_4 \sin\varphi_{40} + l_4 \cos\varphi_{40}\Delta\varphi_4 = 0. \end{cases} \tag{15.12}$$

With Eqs. (15.7) and (15.12), the following relationship is established:

$$\begin{cases} \Delta\varphi_3 = \dfrac{x\cos\varphi_{40} + y\sin\varphi_{40}}{l_3\sin(\varphi_{30} - \varphi_{40})} \\ \Delta\varphi_4 = \dfrac{x\cos\varphi_{30} + y\sin\varphi_{30}}{l_4\sin(\varphi_{40} - \varphi_{30})}. \end{cases} \tag{15.13}$$

By defining the following coefficients:

$$\begin{cases} a_1 = \dfrac{\cos\varphi_{40}}{\sin(\varphi_{30} - \varphi_{40})} \\ a_2 = \dfrac{\sin\varphi_{40}}{\sin(\varphi_{30} - \varphi_{40})} \\ b_1 = \dfrac{\cos\varphi_{30}}{\sin(\varphi_{40} - \varphi_{30})} \\ a_2 = \dfrac{\sin\varphi_{30}}{\sin(\varphi_{40} - \varphi_{30})}, \end{cases} \tag{15.14}$$

Eq. (15.13) can be shortened to:

$$\begin{cases} \Delta\varphi_3 = \dfrac{xa_1 + ya_2}{l_3} \\ \Delta\varphi_4 = \dfrac{xb_1 + yb_2}{l_4}. \end{cases} \tag{15.15}$$

Assuming the crank, link 2, rotates at a constant angular velocity, the kinetic energy without inclusion of the crank is

$$E_K = \frac{1}{2}I_{S3}\dot{\varphi}_3^2 + \frac{1}{2}m_3\left[l_2\dot{\varphi}_2^2 + l_{S3}^2\dot{\varphi}_3^2 + 2l_2l_{S3}\dot{\varphi}_2\dot{\varphi}_3\cos(\varphi_3 - \varphi_2)\right] + \frac{1}{2}I_4\dot{\varphi}_4^2, \tag{15.16}$$

where I_2 and I_4 are the moments of inertia of link 2 and 4 about the pivots to the ground. I_{S3} is the moment of inertia of link 3 about its center of gravity S_3.

The potential energy, if taking the position of the fixed link 1 as the zero potential, is

$$E_P = \frac{1}{2}k_e\delta^2 + m_3g(l_2\sin\varphi_2 + l_{S3}\sin\varphi_3), \tag{15.17}$$

where k_e is the effective stiffness of the joint having the following form:

$$k_e = \begin{cases} k & \text{if } \delta \geq 0 \\ 0 & \text{if } \delta < 0. \end{cases} \tag{15.18}$$

Taking the generalized coordinate vector as

$$q = \begin{bmatrix} x & y \end{bmatrix}^{\mathrm{T}} \tag{15.19}$$

and substituting Eqs. (15.16) and (15.17) into the Lagrange's equation (see Eq. (3.8) on p. 61), the governing equation of motion can be obtained as follows:

$$\left\{\begin{aligned}
&\ddot{x} + 2\dot{x}\left(\dot{a}_1 \sin\varphi_{30} + \dot{b}_1 \sin\varphi_{40}\right)\\
&+2\dot{y}\left(\dot{a}_2 \sin\varphi_{30} + \dot{b}_2 \sin\varphi_{40}\right)\\
&+x\left(\ddot{a}_1 \sin\varphi_{30} + \ddot{b}_1 \sin\varphi_{40}\right) + y\left(\ddot{a}_2 \sin\varphi_{30} + \ddot{b}_2 \sin\varphi_{40}\right)\\
&+kM_1\left(M_3 \sin^2\varphi_{30} + M_4 \sin^2\varphi_{40}\right)x\\
&-kM_1\left(M_3 \sin\varphi_{30}\cos\varphi_{30} + M_4 \sin\varphi_{40}\cos\varphi_{40}\right)y\\
&+M_2 \sin\varphi_{30}\left[\ddot{\varphi}_2 \cos(\varphi_3 - \varphi_2) + \varphi_2^2 \sin(\varphi_{30} - \varphi_2)\right]\\
&+l_3\ddot{\varphi}_{30}\sin\varphi_{30} + l_4\ddot{\varphi}_{40}\sin\varphi_{40} = 0\\
&\ddot{y} - 2\dot{x}\left(\dot{a}_1 \cos\varphi_{30} + \dot{b}_1 \cos\varphi_{40}\right)\\
&-2\dot{y}\left(\dot{a}_2 \cos\varphi_{30} + \dot{b}_2 \cos\varphi_{40}\right)\\
&-x\left(\ddot{a}_1 \cos\varphi_{30} + \ddot{b}_1 \cos\varphi_{40}\right)\\
&-y\left(\ddot{a}_2 \cos\varphi_{30} + \ddot{b}_2 \cos\varphi_{40}\right)\\
&-kM_1\left(M_3 \sin\varphi_{30}\cos\varphi_{30} + M_4 \sin\varphi_{40}\cos\varphi_{40}\right)x\\
&+kM_1\left(M_3 \cos^2\varphi_{30} + M_4 \cos^2\varphi_{40}\right)y\\
&-M_2 \cos\varphi_{30}\left[\ddot{\varphi}_2 \cos(\varphi_3 - \varphi_2) + \varphi_2^2 \sin(\varphi_{30} - \varphi_2)\right]\\
&-l_3\ddot{\varphi}_{30}\cos\varphi_{30} - l_4\ddot{\varphi}_{40}\cos\varphi_{40} = 0,
\end{aligned}\right. \tag{15.20}$$

where M_1 through M_4 are coefficients defined as follows:

$$\begin{aligned}
M_1 &= \left(1 - \frac{c_0}{\sqrt{x^2 + y^2}}\right)\\
M_2 &= \frac{m_3 l_2 l_{S3} l_3}{m_3 l_{S3}^2 + I_{S3}}\\
M_3 &= \frac{l_3^2}{m_3 l_{S3}^2 + I_{S3}}\\
M_4 &= \frac{l_4^2}{I_4}.
\end{aligned} \tag{15.21}$$

Obviously, Eq. (15.20) are coupled nonlinear ODEs with time-varying coefficients.

15.3.1.2 Solution and result analysis

Eq. (15.20) can only be solved numerically. Generally the equation of nonlinear dynamics belongs to the type of stiff equation, which requires excessively small integration time steps in some region (Chapra and Canale, 2011). In addition, some commonly used numerical methods, such as the Runge–Kutta method, for solving the equation are unstable, thus, could not provide satisfactory results. For this reason, special methods and techniques are usually required to obtain adequate solutions. Gear's method is an implicit scheme, which has very good stability, so it is probably the most widely used numerical method to solve stiff equations. In view of this fact, the modified Gear's method is used to solve the equation. The parameters of the linkage are given in Table 15.1.

Table 15.1 Parameters of linkage.

Link No.	1	2	3	4
l_i (m)	0.2	0.03	0.181	0.1175
I_i (kg·m^2)	–	–	0.001244	0.001340
I_{Si} (kg·m^2)	–	–	0.0905	0.05875
m_i (kg)	–	0.4489	0.1743	0.2916

The angular velocity of the crank is taken as $\omega_2 = 320$ rpm. At the clearance $c_0 = 0.1$ mm, the simulation results of displacement of the journal center, the trajectory of the journal center and the contact force vector at the joint are presented in Fig. 15.4.

Clearance is a strong nonlinear factor; it brings a few important nonlinear characteristics into the response, such as chaos and bifurcation, etc. For example, Fig. 15.5 can be drawn by gradually changing the clearance value and solving the equation repeatedly. It shows the changes of the solution, especially some fundamental changes at some points. Clearly, periodic responses, such as 2-subharmonic response and 4-subharmonic response, are observed in this figure.

15.3.2 One-mode model

Feng et al. (1991) studied the dynamics of a four-bar linkage having a clearance in each of the four revolute joints. In this study, the links are treated as elastic with the finite element method introduced in Chapter 11.

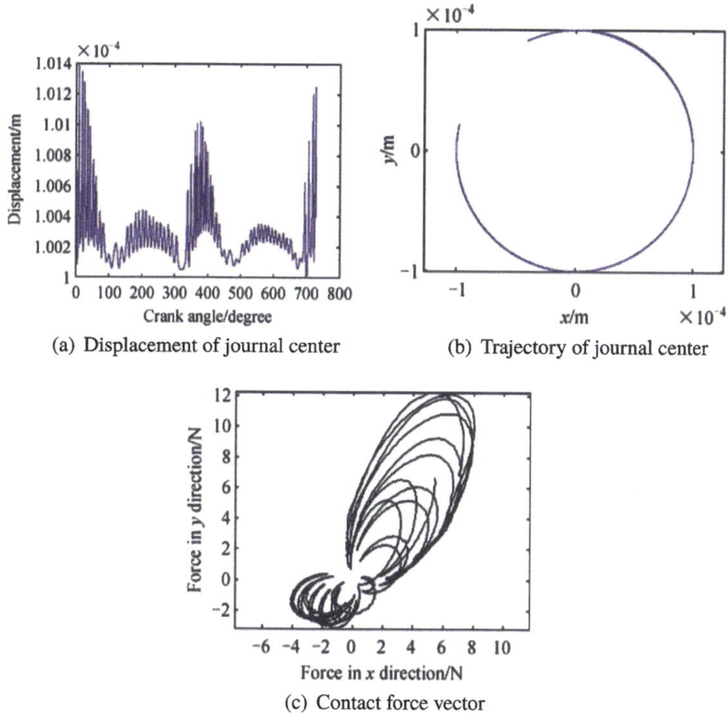

Figure 15.4 Response at $c_0 = 0.01$ mm.

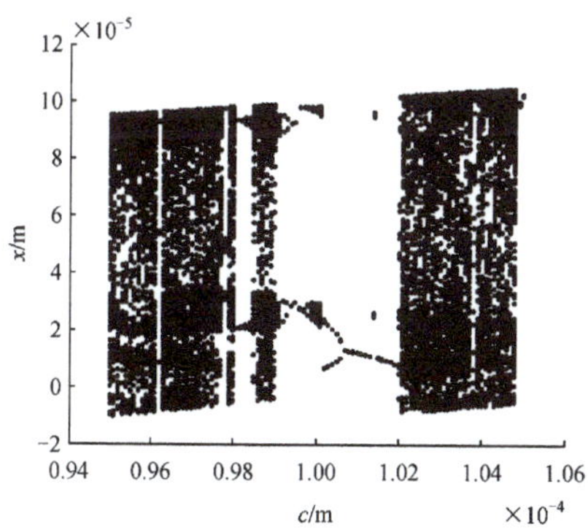

Figure 15.5 Bifurcation diagram.

The journal and bearing are considered to be in continuous contact, and each clearance is treated as a massless link, also called a *pseudo element*. As shown in Fig. 15.6(a), the generalized coordinates are numbered directly

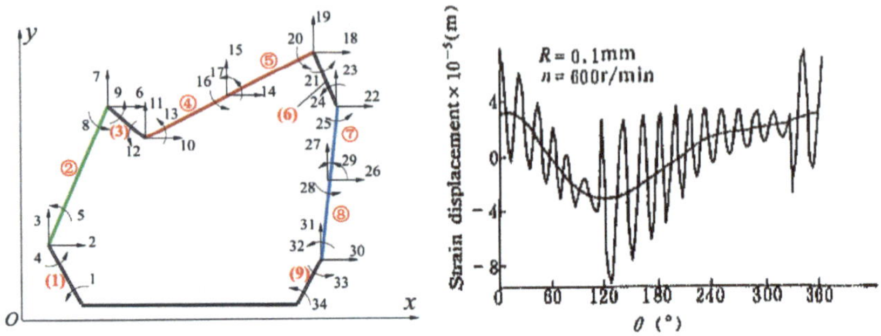

(a) FE model (b) Strain at midpoint of coupler

Figure 15.6 FE model with joint clearance.

with a number, as $1, 2, \cdots, N_u$; the normal element is numbered with a circle ⓘ, while the pseudo element number is placed in parenthesis, (j), for distinction. (1), (3), (6), and (9) are the pseudo elements of clearance.

Following the procedure in Chapter 11, the governing equation of motion is formed as follows (see Eq. (11.77) on p. 314):

$$\boldsymbol{M}\ddot{\boldsymbol{U}} + \boldsymbol{C}\dot{\boldsymbol{U}} + \boldsymbol{K}\boldsymbol{U} = \boldsymbol{F} - \boldsymbol{M}\ddot{\boldsymbol{U}}_r, \tag{15.22}$$

where $\boldsymbol{M}$, $\boldsymbol{C}$, and $\boldsymbol{K}$ are the mass, damping, and stiffness matrices, respectively. $\boldsymbol{U}$, $\dot{\boldsymbol{U}}$, and $\ddot{\boldsymbol{U}}$ are the generalized coordinate, generalized velocity, and generalized acceleration vectors, respectively. $\boldsymbol{F}$ is the generalized force vector. $\ddot{\boldsymbol{U}}_r$ is the rigid acceleration vector.

Solving Eq. (15.22) gives the response based on which the dynamic stress and strain at any point can be calculated. One example of the strain at the midpoint of the coupler is given in Fig. 15.6(b) under a particular set of parameters.

15.4 Cam mechanisms

Clearance also affects dynamics of cam mechanisms. Indexing cam mechanisms are generally of form-closed and are more sensitive to clearance in terms of dynamics and motion accuracy. The dynamics of cam mechanisms was already discussed in Chapter 10, however, clearance was not considered. In the following section, the dynamics of an indexing cam mechanism incorporating the clearance effect is introduced based on Chang et al. (2001).

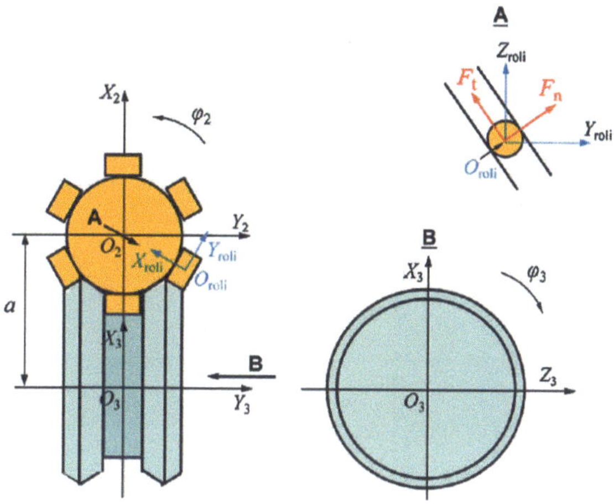

Figure 15.7 Indexing cam mechanism with clearance.

15.4.1 Dynamic model

The indexing cam mechanism under investigation is shown in Fig. 15.7. Clearance exists between the cam groove and the roller turret as shown.

Two modes, free flight and linear elastic contact, are considered in the clearance model. The contact force is then represented as

$$P = kf(\theta); \tag{15.23}$$

k is the contact stiffness, and $f(\theta)$ is defined as

$$f(\theta) = \begin{cases} \theta - \epsilon & \text{if } \theta > \epsilon \\ 0 & \text{if } -\epsilon \le \theta \le \epsilon \\ \theta + \epsilon & \text{if } \theta < -\epsilon, \end{cases} \tag{15.24}$$

where ϵ is the clearance and θ is the rotation angle. Choosing the generalized coordinate vector as

$$\boldsymbol{q} = \begin{bmatrix} \varphi_1 & \varphi_2 & \varphi_3 & \varphi_4 & x_1 & y_1 & z_1 & x_2 & y_2 & z_2 & x_3 & y_3 & z_3 \end{bmatrix}^{\mathrm{T}}, \tag{15.25}$$

where φ is the rotation displacement, x, y and z are the linear displacements. The subscripts 1, 2, 3, and 4 represent the rotary table, the turret, the cam, and the motor, respectively. To account for the interaction between the driving motor and the cam mechanism, the motor characteristics is approximated by the following linear function of the angular velocity:

$$M_0 = a_0 - a_1\omega. \tag{15.26}$$

The governing equation of motion is given as

$$\begin{cases} I_1\ddot{\varphi}_1 + k_1(\varphi_1 - \varphi_2) = 0 \\ I_2\ddot{\varphi}_2 + k_1(\varphi_2 - \varphi_1) - \sum\Big[k_{c2}f(\varphi_{21} - \varphi_2)r_2 \\ -\mu k_{c2}f(\varphi_{21} - \varphi_2)r_2\dfrac{\tau r_2}{a - r_2\cos\varphi_{2i}} - c\dot{\varphi}_2 r_2\Big]r_2 = 0 \\ I_3\ddot{\varphi}_3 + k_2(\varphi_3 - \varphi_4) + \sum\Big[k_{c2}f(\varphi_{21} - \varphi_2)r_2\dfrac{\tau r_2}{a - r_2\cos\varphi_{2i}} \\ -\mu k_{c2}f(\varphi_{21} - \varphi_2)r_2 - c\dot{\varphi}_3(a - r_2\cos\varphi_2)\Big](a - r_2\cos\varphi_{2i}) = 0 \\ I_4\ddot{\varphi}_4 + k_2(\varphi_4 - \varphi_3) = 0 \\ m_1\ddot{x}_1 + k_{w11}x_1 + k_{w12}x_2 = 0 \\ m_1\ddot{y}_1 + k_{w21}y_1 + k_{w22}y_2 = 0 \\ m_1\ddot{z}_1 + k_{l11}z_1 + k_{l12}z_2 = 0 \\ m_2\ddot{x}_2 + k_{w11}x_1 + k_{w12}x_2 - \sum\Big[k_{c2}f(\varphi_{21} - \varphi_2)r_2\cos\Big(\dfrac{\pi}{2} + \varphi_{2i}\Big)\Big] = 0 \\ m_2\ddot{y}_2 + k_{w21}y_1 + k_{w22}y_2 - \sum\Big[k_{c2}f(\varphi_{21} - \varphi_2)r_2\cos\varphi_{2i}\Big] = 0 \\ m_2\ddot{z}_2 + k_{l21}z_1 + k_{l22}z_2 + \sum\Big[k_{c2}f(\varphi_{21} - \varphi_2)r_2\dfrac{\tau r_2}{a - r_2\cos\varphi_{2i}}\Big] = 0 \\ m_3\ddot{x}_3 + k_{w3}x_3 + \sum\Big[k_{c2}f(\varphi_{21} - \varphi_2)r_2\cos\Big(\dfrac{\pi}{2} + \varphi_{2i}\Big)\Big] = 0 \\ m_3\ddot{y}_3 + k_{l3}y_3 + \sum\Big[k_{c2}f(\varphi_{21} - \varphi_2)r_2\cos\varphi_{2i}\Big] = 0 \\ m_3\ddot{z}_3 + k_{w3}z_3 - \sum\Big[k_{c2}f(\varphi_{21} - \varphi_2)r_2\dfrac{\tau r_2}{a - r_2\cos\varphi_{2i}}\Big] = 0. \end{cases} \tag{15.27}$$

15.4.2 Solution

As stated in the previous section, Eq. (15.27) is stiff, thus, solved numerically with Gear's method. The parameters in the simulation can be found in Chang et al. (2001). For the sake of space saving, they are not given here. The results of simulation with and without a clearance are given in Figs. 15.8 and 15.9, respectively.

Obviously, the clearance affects the response significantly. In particular, the responses of the turret (Figs. 15.8(b) and 15.9(b)) and the camshaft

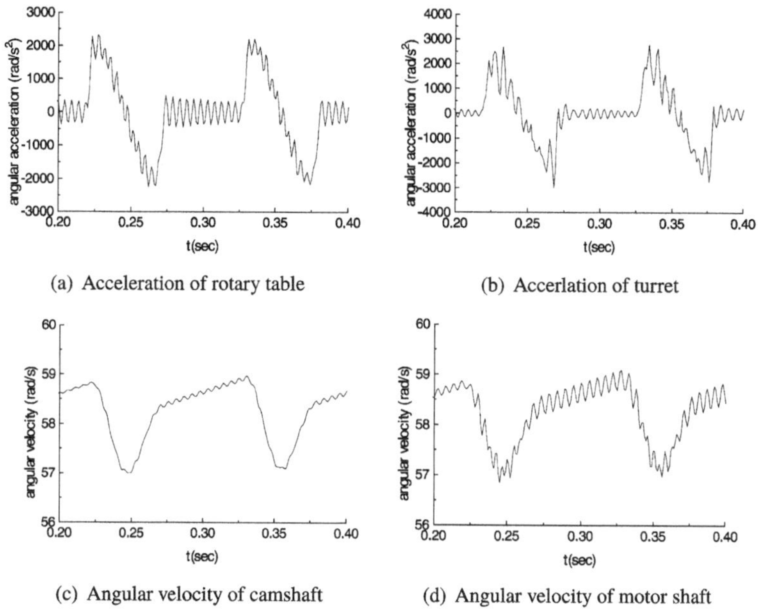

Figure 15.8 Response without clearance.

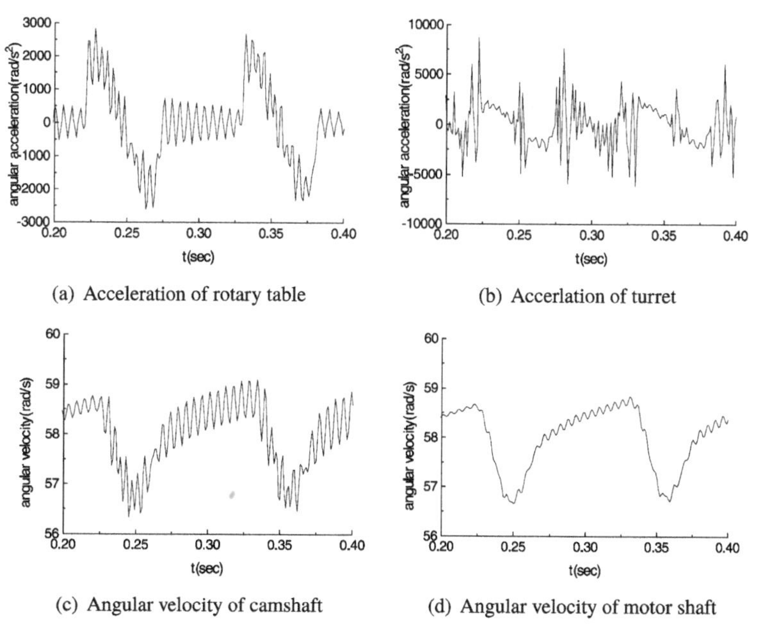

Figure 15.9 Response with clearance.

(Figs. 15.8(c) and 15.9(c)) are influenced by the clearance more obviously than rotary table and the motor shaft.

15.5 Gears

Clearance exists in all gear meshes mainly for lubrication purposes. The methods for modeling clearance in gear dynamics are roughly classified into two groups, namely, the impact model and the contact model. In each case, two motion modes are considered. In the impact model the two modes considered are the free flight and impact. The impact is modeled as an instantaneous momentum exchange with a restitution coefficient. This type of model is primarily used in the investigation of gear rattle (Pfeiffer and Kunert, 1990). The contact model considers the free flight and the continuous contact. The relationship between the penetration depth and the contact force during the contact is generally simplified as linear in nature as shown in Fig. 15.10(b). This contact model is commonly used in the general gear dynamics (Wang et al., 2003). This book introduces only the contact model and focuses on the simplest case in which only the rotation of a single gear pair is considered based on the work by Theodossiades and Natsiavas (2000).

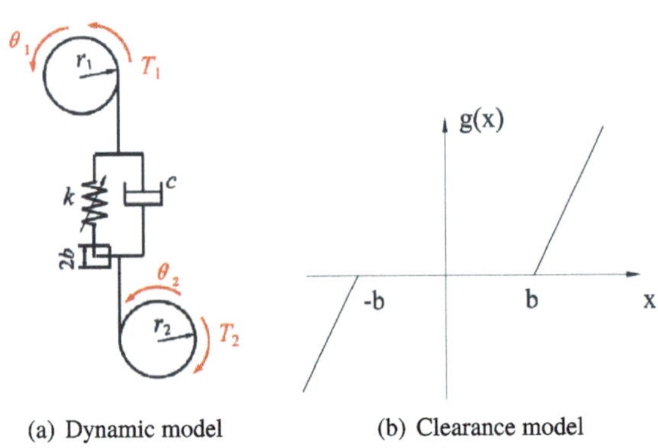

Figure 15.10 Gear-pair model.

15.5.1 Dynamic model

The model of a pair of spur gears is illustrated in Fig. 15.10. The two gears are represented by two moments of inertia J_1 and J_2, and the meshing gear tooth pair is modeled as a spring k and a damper c. The clearance between the meshing gear teeth is b.

Define a linear relative displacement $x(t)$ as

$$x(t) = r_1\theta_1(t) - r_2\theta_2(t) - e(t), \tag{15.28}$$

where r_1 and r_2 are the radii of the gears, and θ_1 and θ_2 are the elastic rotational displacements of the gears. $e(t)$ is the profile error defined in Chapter 12.

Following the procedures in Section 12.3 (see p. 335), the equation of motion of single DOF for the gear pair has the following form:

$$m\ddot{x} + c\dot{x}(t) + k(t)g(x) = f - m\ddot{e}(t), \tag{15.29}$$

where

$$\begin{cases} m = \dfrac{J_1 J_2}{J_1 r_2^2 + J_2 r_1^2} \\ f = \dfrac{r_1 J_2 T_1 - r_2 J_1 T_2}{J_1 r_2^2 + J_2 r_1^2}. \end{cases} \tag{15.30}$$

T_1 and T_2 are the torques applied on the two gears. The clearance is illustrated in Fig. 15.10(b) and can be represented mathematically as

$$g(x) = \begin{cases} x - b & \text{if } x \geq b, \\ 0 & \text{if } -b < x < b, \\ x + b & \text{if } x \leq -b. \end{cases} \tag{15.31}$$

$k(t)$, as stated in Chapter 12, changes periodically with time t. It can be expanded into Fourier series as

$$k(t) = k_0 + \sum_{i=1}^{\infty} [a_i \cos(i\omega_m t - \varphi_i)]. \tag{15.32}$$

Common practice in nonlinear vibration study is to normalize the variables with characteristic quantities. For this purpose define the following parameters

$$\begin{cases} \omega_0 = \sqrt{\dfrac{k_0}{m}} \\ \xi = \dfrac{c}{2\sqrt{mk_0}} \\ \Omega_\mathrm{m} = \dfrac{\omega_\mathrm{m}}{\omega_0} \\ w(\bar{t}) = \dfrac{k(t)}{k_0} \\ \bar{t} = \omega_0 t \\ u(\bar{t}) = \dfrac{x(t)}{b} \\ \bar{e}(\bar{t}) = \dfrac{e(\bar{t})}{n}. \end{cases} \tag{15.33}$$

Eq. (15.29) can then be written in a dimensionless form as

$$\ddot{u} + 2\xi\dot{u} + w(\bar{t})g(u) = f_0 - \ddot{\bar{e}}(\bar{t}), \tag{15.34}$$

where $g(u)$ has the form

$$g(u) = \begin{cases} u - 1 & \text{if } u \geq 1, \\ 0 & \text{if } -1 < u < 1, \\ u + 1 & \text{if } u \leq -1, \end{cases} \tag{15.35}$$

$w(\bar{t})$ and $\ddot{\bar{e}}(\bar{t})$ both change with the meshing frequency periodically. If only the first harmonic is retained for both and defined as small parameter ϵ, Eq. (15.34) can be simplified to

$$\ddot{u} + 2\xi\dot{u} + [1 + 2\epsilon\cos(\Omega_\mathrm{m} t)]g(u) = f_0 + \epsilon f_1 \cos(\Omega_\mathrm{m} t + \phi). \tag{15.36}$$

It should be noted that $\bar{t}$ has been replaced by t for simplicity in the above. f_0 and f_1 represent the constant and time varying excitation, and ϕ represents a phase difference angle between the stiffness and the profile error. This equation is the basis for the subsequent analysis.

15.5.2 Solution and result analysis

Theodossiades and Natsiavas (2000) solved Eq. (15.36) both analytically and numerically. It was found that there was a very complicated response in the nonlinear gear pair, including primary resonance, jump phenomenon, and chaos. Systematical parametric study was also conducted. In the following, we present only two of the response diagrams.

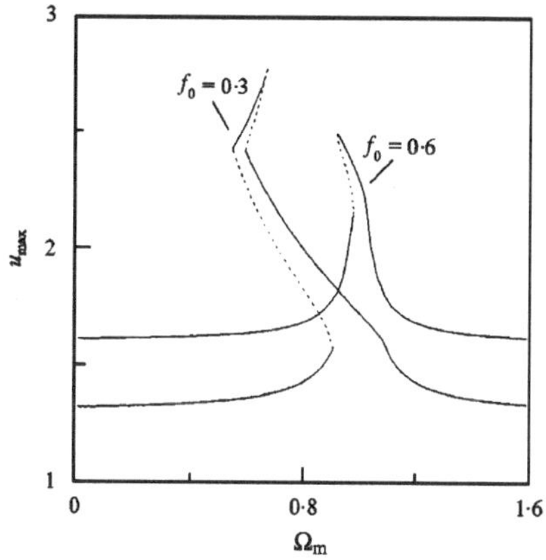

Figure 15.11 Response for different f_0.

With the parameters $\epsilon = 0.03$, $\xi = 0.8$, $\phi = 0$, and $f_1 = 2.5$, the effect of the constant excitation f_0 on the response was calculated with $f_0 = 0.3$ and 0.6. The result is shown in Fig. 15.11. The following observations can be made from this figure:

- The amplitude–frequency response at the primary resonance bends towards the left with a small vibration, indicating that the clearance is characterized clearly as a *softening spring*. However, with the increase of the vibration amplitude, the curve bends toward the right, behaving as a *hardening spring*. This is caused by the contact on the opposite side of the gear teeth.
- Because of the bending of the curve, jump phenomenon will exist if sweeping the excitation frequency occurs at a certain range.
- Increasing the constant force f_0 will alleviate the nonlinearity caused by the clearance. It is expected that the system will become linear after f_0 exceeds a certain threshold value.

To verify the analytical method, a numerical solution was also conducted. A comparison of the results is presented in Fig. 15.12, in which the dotted line results from the numerical simulation and the solid line is obtained through the analytic method. It can be seen that both the numerical simulation and the analytic method capture the primary resonance. The two results agree well on the primary resonance. However, the two

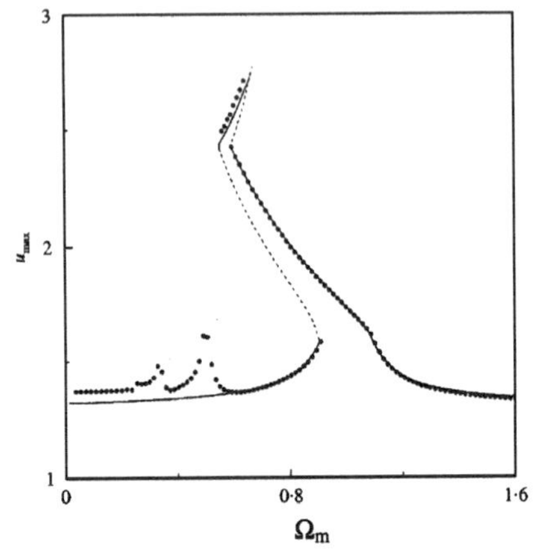

Figure 15.12 Results of analysis and simulation.

superharmonic resonances, which are obvious in the numerical results, do not show up in the analytic result. The reason for this is that the analytic method assumes a specific form of response that does not meet the requirement of both primary and superharmonic resonance .

15.6 Features of dynamics with joint clearance

Clearance exists in almost all joints and introduces strong nonlinearity into the dynamics. In this chapter, we present several examples of various mechanisms having joint clearance. However, this fails to give a complete picture of the nonlinear dynamics with clearance because of two reasons. First, the nonlinear dynamics has a very complicated response, to name a few, primary resonance, superharmonic resonance, subharmonic resonance, jump, bifurcation, chaos, etc. Second, the limited space makes it impossible to comprehensively include all of the very diversified mechanisms and models. Here we summarize some common features in nonlinear dynamics of mechanisms with joint clearance:

- Model selection: The three-mode model is literally the most comprehensive one, but it requires the determination of the modes in the process. The continuous contact model (massless link) is the simplest; it is good for KED analysis. In the selection of a specific model, a trade-off should be considered between the tractability and model complexity.
- Analytic and numerical: In nonlinear dynamic analysis, analytic and numerical methods are often used in combination. Analytic methods

provide more insight, but sometimes miss revealing an overall picture. In addition, they can be used for single-DOF problems in most cases; thus, simplification of the model is required. On the other hand, numerical methods yield the results, but they couldn't explain the mechanism behind the results. The advantage of numerical methods is that they can handle systems having many DOFs with modern computers.

- Qualitative and quantitative: Qualitative and quantitative methods are employed often in combination in nonlinear dynamics.

APPENDIX A

Crossover shock

In form-closed cams, crossover shock happens between the roller and the groove when the acceleration changes sign. The severity of the shock is related to the velocity of impact. In this following, the calculation of the crossover shock based on Koster (1974a) is introduced with Fig. A.1.

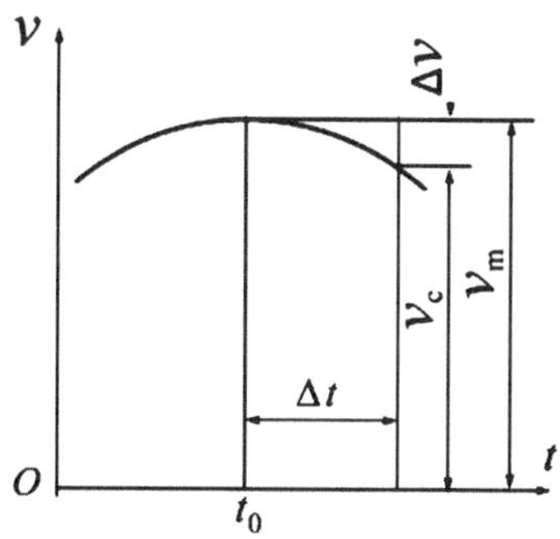

Figure A.1 Crossover shock.

At time $t = t_0$, the acceleration changes from + to −. The velocity reaches peak v_m at this moment. We assume it takes Δt for the roller to cross over the groove. The roller flies across the groove at t_0 with a velocity $v_{\rm m}$. After Δt, the roller impacts the groove, and its velocity becomes $v_{\rm c}$. The impact velocity is then

$$v_{\rm imp} = v_{\rm m} - v_{\rm c}. \tag{A.1}$$

$v_{\rm c}$ can be approximately obtained from a Taylor expansion:

$$v_{\rm c} = v(t_0 + \Delta t) \approx v(t_0) + a(t_0)\Delta t + \frac{1}{2} j(t_0)\Delta t^2 + \cdots, \tag{A.2}$$

where $a(t_0)$ and $j(t_0)$ are the acceleration and jerk of the follower at t_0. Substituting Eq. (A.2) into (A.1) and taking advantage of $v(t_0) = v_{\rm m}$, $a(t_0) = 0$ give

$$v_{\rm imp} = \frac{1}{2} j(t_0)\Delta t^2. \tag{A.3}$$

The clearance between the roller and the groove width is designated as δ, which can be calculated with the following integral:

$$\delta = \int_{t_0}^{t_0+\Delta t} (v_m - v)\,dt. \tag{A.4}$$

δ can also be expanded into a Taylor series as

$$\begin{aligned}\delta =& v_m \Delta t - [s(t_0 + \Delta t) - s(t_0)] \\ \approx& v_m \Delta t - \left[s(t_0) + v(t_0)\Delta t) + \frac{1}{2}a(t_0)\Delta t^2 + \frac{1}{6}j(t_0)\Delta t^3 + \cdots - s(t_0) \right].\end{aligned} \tag{A.5}$$

This reduces to

$$\delta = -\frac{1}{6}j(t_0)\Delta t^3. \tag{A.6}$$

Substituting Eq. (A.6) into (A.3) and considering that $j(t_0)$ is negative at the crossover point, we have

$$v_{imp} = \sqrt[3]{4.5|j(t_0)|\delta^2}. \tag{A.7}$$

Clearly, the impact velocity is proportional to the cubic root of the jerk. In practice, dimensionless quantities are more often used. We introduce the following transformations:

$$V_{imp} = \frac{v_{imp}}{h/t_h}, \tag{A.8}$$

where h and t_h are the rise displacement and the rising time, and

$$J_{imp} = \frac{j_{imp}}{h/t_h^3}. \tag{A.9}$$

Substituting Eqs. (A.8) and (A.9) into Eq. (A.7) gives

$$V_{imp} = \sqrt[3]{4.5}\sqrt[3]{J_{imp}}\sqrt[3]{\Delta^2}, \tag{A.10}$$

where $\Delta = \delta/h$. It can be concluded from Eqs. (A.7) and (A.10) that the velocity of impact is proportional to: 1) the clearance between the roller and the groove, and 2) the cubic root of jerk at the crossover point. This is the reason we used $\sqrt[3]{J_{imp}}$ in Chapter 10 as one of the characterized quantities.

APPENDIX B

Common motions of harmonic combinations

The most commonly-used follower motions constructed from harmonic combinations are given in this appendix. In view of the page limit, only S is given except in the first motion. If needed, V, A, and J can be obtained by differentiating the displacement function with respect to time for other cases.

1. **Sinusoidal acceleration** (see Fig. 10.1 on p. 252)

$$\begin{cases} S = T - \dfrac{1}{2\pi}\sin(2\pi T) \\ V = 1 - \cos(2\pi T) \\ A = 2\pi \sin(2\pi T) \\ J = 4\pi^2 \cos(2\pi T). \end{cases} \tag{B.1}$$

2. **Modified trapezoidal acceleration** (see Fig. 10.5(a) on p. 257)

$$S = \begin{cases} \dfrac{2T_a A_m}{\pi} T - \dfrac{4T_a^2 A_m}{\pi^2} \sin \dfrac{\pi T}{2T_a} & \text{if } 0 \le T \le T_a \\ \dfrac{A_m}{2}(T - T_a)^2 + V_a(T - T_a) + S_a & \text{if } T_a < T \le 0.5 - T_a \\ \dfrac{4T_a^2 A_m}{\pi^2}\left[1 - \cos \dfrac{\pi(T - 0.5 + T_a)}{2T_a}\right] & \\ \qquad + V_b(T - 0.5 + T_a) + S_b & \text{if } 0.5 - T_a < T \le 0.5 + T_a \\ -\dfrac{A_m}{2}(T - 0.5 - T_a)^2 & \\ \qquad + V_b(T - 0.5 - T_a) + S_c & \text{if } 0.5 + T_a < T \le 1 - T_a \\ \dfrac{4T_c^2 A_m}{\pi^2}\left[\cos \dfrac{\pi(T - 1 + T_a)}{2T_a} - 1\right] & \\ \qquad + V_a(T - 1 + T_a) + S_d & \text{if } 1 - T_a < T \le 1, \end{cases} \tag{B.2}$$

where

$$\begin{cases} A_{\mathrm{m}} = \dfrac{1}{1/4 - T_a + 2T_a/\pi} \\ V_a = \dfrac{2T_a A_{\mathrm{m}}}{\pi} \\ V_b = A_{\mathrm{m}}(0.5 - 2T_a) + V_a \\ S_a = \dfrac{2T_a^2 A_{\mathrm{m}}}{\pi} - \dfrac{4T_a^2 A_{\mathrm{m}}}{\pi^2} \\ S_b = \dfrac{A_{\mathrm{m}}}{2}(0.5 - 2T_a)^2 + V_a(0.5 + 2T_a) + S_a \\ S_c = 1 - S_b \\ S_d = 1 - S_a. \end{cases} \tag{B.3}$$

T_a can be any value between 0 and 1/4. When $T_a = 1/8$, it is the commonly used modified trapezoidal acceleration motion. When $T_a = 1/4$, it gives the cycloidal motion.

3. **Modified sinusoidal acceleration** (see Fig. 10.5(b) on p. 257)

$$S = \begin{cases} \dfrac{2T_a A_{\mathrm{m}}}{\pi} T - \dfrac{4T_a^2 A_{\mathrm{m}}}{\pi^2} \sin \dfrac{\pi T}{2T_a} & \text{if } 0 \le T \le T_a \\ \dfrac{(1 - 2T_a)^2 A_{\mathrm{m}}}{\pi^2} \left[1 - \cos \dfrac{\pi(T - T_a)}{1 - 2T_a} \right] + \\ \qquad V_a(T - T_a) + S_a & \text{if } T_a < T \le 1 - T_a \\ \dfrac{4T_a^2 A_{\mathrm{m}}}{\pi^2} \left[\cos \dfrac{\pi(T - 1 + T_a)}{2T_a} - 1 \right] + \\ \qquad V_a(T - 1 + T_a) + S_b & \text{if } 1 - T_a < T \le 1, \end{cases} \tag{B.4}$$

where

$$\begin{cases} A_{\mathrm{m}} = \dfrac{1}{2T_a/\pi + (2 - 8T_a)/\pi^2} \\ V_a = \dfrac{2T_a A_{\mathrm{m}}}{\pi} \\ S_a = \dfrac{2T_a^2 A_{\mathrm{m}}}{\pi} - \dfrac{4T_a^2 A_{\mathrm{m}}}{\pi^2} \\ S_b = 1 - S_a. \end{cases} \tag{B.5}$$

T_a can be arbitrarily taken between $0 - 1/2$. When $T_a = 1/8$, it is the most-widely used modified sinusoidal acceleration.

4. **Modified constant velocity** (see Fig. 10.5(c) on p. 257)

$$S=\begin{cases} \dfrac{2T_aA_m}{\pi}T-\dfrac{4T_a^2A_m}{\pi^2}\sin\dfrac{\pi T}{2T_a} & \text{if } 0\le T\le T_a \\ \dfrac{4(T_b-T_a)^2A_m}{\pi^2}\left[1-\cos\dfrac{\pi(T-T_a)}{2(T_b-T_a)}\right]+ & \\ \qquad V_a(T-T_a)+S_a & \text{if } T_a<T<T_b \\ V_m(T-T_b)+S_b & \text{if } T_b<T\le 1-T_b \\ \dfrac{4(T_b-T_a)^2A_m}{\pi^2}\sin\dfrac{\pi(T-1+T_b)}{2(T_b-T_a)}- & \\ \dfrac{2(T_b-T_a)A_m}{\pi}(T-1+T_b)+ & \\ \qquad V_m(T-1+T_b)+S_c & \text{if } 1-T_b<T\le 1-T_a \\ \dfrac{4T_a^2A_m}{\pi^2}\left[\cos\dfrac{\pi(T-1+T_a)}{2T_a}-1\right]+ & \\ \qquad V_a(T-1+T_a)+S_d & \text{if } 1-T_a<T\le 1, \end{cases} \tag{B.6}$$

where

$$\begin{cases} A_m=\dfrac{1}{\frac{2}{\pi}\left[(2-\frac{8}{\pi})T_aT_b+(\frac{4}{\pi}-2)T_b^2+T_b\right]} \\ V_a=\dfrac{2T_aA_m}{\pi} \\ V_m=\dfrac{2T_bA_m}{\pi} \\ S_a=\dfrac{2T_a^2A_m}{\pi}-\dfrac{4T_a^2A_m}{\pi^2} \\ S_b=\dfrac{4(T_b-T_a)^2A_m}{\pi^2}+V_a(T_b-T_a)+S_a \\ S_c=1-S_b \\ S_d=1-S_a. \end{cases} \tag{B.7}$$

The length of constant velocity period is controlled by T_a and T_b. When $T_a=1/16$, $T_b=1/4$, the constant velocity time is half the rising time. Choosing different values of T_a and T_b, the constant velocity period can be made shorter or longer.

APPENDIX C

Calculation of deformation of gear teeth

The following calculation is based on the mechanics of material.

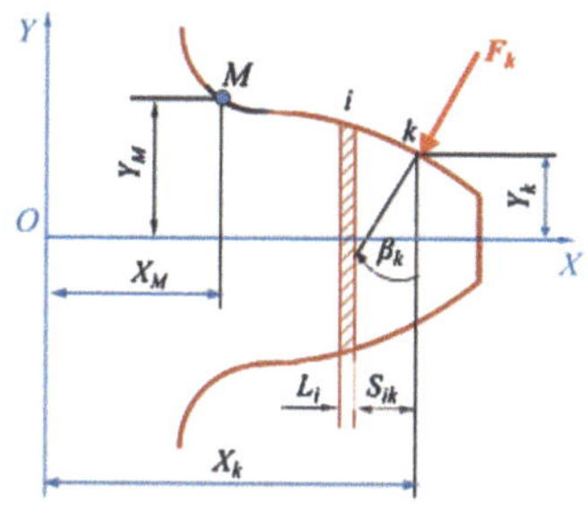

Figure C.1 Tooth deformation.

For a gear tooth as shown in Fig. C.1, we want to calculate the deformation of point k, to which a force F_k is applied, along the line of action.

1. **Bending and shear deformation**

   The tooth can be treated as a nonuniform cantilever beam. To calculate the deformation using the mechanics of materials, we divide the gear tooth into a series of small segments, each of which is approximated as a constant cross-section beam. For the i-th segment, the total deformation along the line of action is calculated as follows (Lin et al., 1988):

$$\delta_{\mathrm{B}ik} = \frac{F_k}{E_{\mathrm{e}}}\left\{\cos^2\beta_k\left[\frac{L_i^3+3L_i^2S_{ik}^2+3L_iS_{ik}^2}{3I_i}\right]\right.$$
$$-\cos\beta_k\sin\beta_k\left[\frac{L_i^2Y_k+2Y_kL_iS_{ik}}{2I_i}\right]$$
$$\left.+\cos^2\beta_k\left[\frac{12(1+\nu)L_i}{5A_i}\right]+\sin^2\beta_k\frac{L_i}{A_i}\right\}, \tag{C.1}$$

where E_{e} is the effective modulus of elasticity, which has the following form:

$$E_e = \begin{cases} \dfrac{E}{1-\nu^2} & \text{if } B/H_p \geq 5 \\ E & \text{if } B/H_p < 5, \end{cases} \tag{C.2}$$

where B is the facial width of the gear and H_p is the tooth thickness. E is the modulus of elasticity. ν is the Poisson ratio. The total bending and shear deformation is the summation of all the segments:

$$\delta_{Bk} = \sum_{i=1}^{n} \delta_{Bik}. \tag{C.3}$$

2. **Deformation caused by elasticity of the tooth root**
In the previous, the gear tooth is assumed to be a cantilever beam. However, the material is elastic, and the tooth root is not a fixed end. This elasticity causes a deformation at point k as well. It is calculated as follows:

$$\delta_{Fk} = \begin{cases} \dfrac{F_k \cos^2 \beta_k (1-\nu^2)}{BE} \left[5.306 \left(\dfrac{L_f}{H_f} \right)^2 + 2 \dfrac{L_f (1-\nu-2\nu^2)}{H_f (1-\nu^2)} \right. \\ \left. \quad + 1.534 \left(1 + \dfrac{0.4167 \tan^2 \beta_k}{1+\nu} \right) \right] & \text{if } B/H_p \geq 5 \\ \dfrac{F_k \cos^2 \beta_k}{BE} \left[5.306 \left(\dfrac{L_f}{H_f} \right)^2 + 2 \dfrac{L_f (1-\nu)}{H_f} \right. \\ \left. \quad + 1.534 \left(1 + \dfrac{0.4167 \tan^2 \beta_k}{1+\nu} \right) \right] & \text{if } B/H_p < 5, \end{cases}$$

where

$$\begin{cases} L_f = X_k - X_M - Y_k \tan \beta_k \\ H_f = 2Y_M. \end{cases} \tag{C.4}$$

X_M and Y_M are the coordinates of point M as shown in Fig. C.1.

3. **Deformation caused by surface contact**
The contact deformation at point k is evaluated using the following formula:

$$\delta_{Ck} = 1.275 E^{-0.9} B^{-0.8} F_k^{-0.1}. \tag{C.5}$$

The total deformation is the sum of the three types of deformation:

$$\delta_k = \delta_{Bk} + \delta_{Fk} + \delta_{Ck}. \tag{C.6}$$

APPENDIX D

Matrices in gear dynamics

The matrices in Eq. (12.46) are given as follows: $\boldsymbol{K}$ is the supporting stiffness matrix. It is symmetrical having the nonzero entries as follows:

$$\begin{cases} K_{11} = K_{22} = k_{S1} \\ K_{19} = K_{2,10} = -\zeta_1 k_{S1} \\ K_{17} = K_{28} = -\zeta_2 k_{S1} \\ K_{33} = k_{T1} \\ K_{44} = K_{55} = k_{S2} \\ K_{4,11} = K_{5,12} = -\zeta_4 k_{S2} \\ K_{4,13} = K_{5,14} = -\zeta_3 k_{S2} \\ K_{77} = \zeta_2^2 k_{S1} + k_{by1} \\ K_{79} = K_{8,10} = \zeta_2 \zeta_1 k_{S1} \\ K_{88} = \zeta_2^2 k_{S1} + k_{bx1} \\ K_{99} = \zeta_1^2 k_{S1} + k_{by2} \\ K_{10,10} = \zeta_1^2 k_{S1} + k_{bx2} \\ K_{11,11} = \zeta_4^2 k_{S2} + k_{by3} \\ K_{11,13} = K_{12,14} = \zeta_4 \zeta_3 k_{S2} \\ K_{12,12} = \zeta_4^2 k_{S2} + k_{bx3} \\ K_{13,13} = \zeta_3^2 k_{S2} + k_{by4} \\ K_{14,14} = \zeta_3^2 k_{S2} + k_{bx4} \\ K_{15,15} = k_{T2}. \end{cases} \tag{D.1}$$

$\boldsymbol{K}'$ is the meshing stiffness. It is asymmetrical, and its nonzero elements are:

$$\begin{cases} K'_{11} = k_v; & K'_{14} = -k_v; & K'_{13} = k_v r_{b1}; & K'_{16} = -k_v r_{b2} \\ K'_{31} = k_v r_{b1}; & K'_{33} = k_v r_{b1}^2; & K'_{34} = -k_v r_{b1}; & K'_{36} = -k_v r_{b1} r_{b2} \\ K'_{41} = -k_v; & K'_{43} = -k_v r_{b1}; & K'_{44} = k_v; & K'_{46} = k_v r_{b2} \\ K'_{61} = -k_v r_{b2}; & K'_{63} = -k_v r_{b1} r_{b2}; & K'_{64} = k_v r_{b2}; & K'_{66} = k_v r_{b2}^2. \end{cases} \tag{D.2}$$

$\boldsymbol{C}$ and $\boldsymbol{C}'$ are the supporting damping matrix and meshing damping matrix, respectively. They have the same structure with $\boldsymbol{K}$ and $\boldsymbol{K}'$, respectively, except that the stiffness elements are replaced with corresponding damping elements. Some damping elements are given as follows:

The meshing damping c_{vi} is

$$c_{vi} = 2\zeta_v \sqrt{\frac{k_{vi} r_{b1}^2 r_{b2}^2 J_1 J_2}{r_{b1}^2 J_1 + r_{b2}^2 J_2}}. \tag{D.3}$$

The torsional damping of the shafts c_{Ti} is

$$\begin{cases} c_{T1} = 2\zeta_T \sqrt{\dfrac{k_{T1} J_M J_1}{J_M + J_1}} \\ c_{T2} = 2\zeta_T \sqrt{\dfrac{k_{T2} J_L J_2}{J_L + J_2}}. \end{cases} \tag{D.4}$$

The bending damping c_{Si} is

$$\begin{cases} c_{S1} = 2\zeta_S \sqrt{\dfrac{k_{S1} J_M J_1}{J_M + J_1}} \\ c_{S2} = 2\zeta_S \sqrt{\dfrac{k_{S2} J_L J_2}{J_L + J_2}}, \end{cases} \tag{D.5}$$

where ζ_v, ζ_T and ζ_S are damping ratios. Experiments show that their values are in the ranges 0.03–0.17, 0.005–0.075, and 0.03–0.1, respectively.

The mass matrix $\boldsymbol{M}$ is symmetrical. Its nonzero entries are.

$$\begin{cases} M_{11} = M_{22} = m_1; & M_{44} = M_{55} = m_2; & M_{33} = J_1 \\ M_{66} = J_2; & M_{77} = M_{88} = m_{b1}; & M_{99} = M_{10,10} = m_{b2} \\ M_{11,11} = M_{12,12} = m_{b3}; & M_{13,13} = M_{14,14} = m_{b4}; & M_{15,6} = M_{15,15} = J_L. \end{cases} \tag{D.6}$$

The meshing error excitation $\boldsymbol{E}$ has the following nonzero entries:

$$\begin{cases} E_1 = \sum_i \left(k_{vi} e_i + c_{vi} \dot{e}_i\right) \\ E_3 = r_{b1} E_1 \\ E_4 = -E_1 \\ E_6 = -r_{b2} E_1. \end{cases} \tag{D.7}$$

The external force vector $\boldsymbol{T}$ has only one nonzero element:

$$T_{15} = -T_L. \tag{D.8}$$

APPENDIX E

Meshing stiffness calculation of a planetary gear train

The calculation can be made with reference to Fig. E.1. As stated in Chapter 13, γ_{rn} and γ_{sn} are the initial phases of the n-th $r-p$ and $s-p$ meshing stiffnesses. γ_{sr} is the phase difference of the $r-p$ and $s-p$.

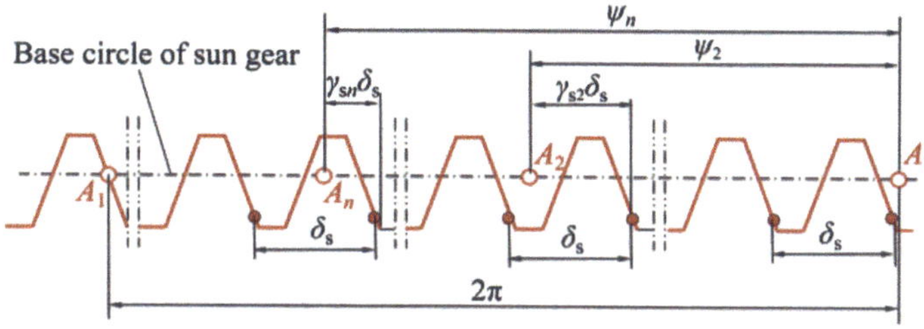

Figure E.1 Meshing phase at various planets.

We assume the multiple $s-p$ (or $r-p$) meshes have a meshing stiffness wave of the same shape but a phase difference. If we straighten the pitch circle of the sun gear into a straight line, it would appear as shown in Fig. E.1.

In this figure A_i is the tangential point of the pitch circle of the i-th planet gear and the pitch circle of the sun gear. ψ_n is the orientation angle of the n-th planet as shown in Fig. 13.4 and given by Eq. (13.27). The small solid circles on the teeth represent the points entering meshing. δ_s, the distance between adjacent solid circles is $2\pi/Z_s$.

From this figure, the phase difference is given by

$$\gamma_{sn} = \quad \mathrm{mod}\left(\frac{\psi_n}{\delta_s}\right) = \quad \mathrm{mod}\left(\frac{Z_s\psi_n}{2\pi}\right), \tag{E.1}$$

where mod represents the modulo operation (remainder after division). Because of the periodicity of stiffness, this formula can also be recast as:

$$\gamma_{sn} = \frac{Z_s\psi_n}{2\pi}. \tag{E.2}$$

Similarly, we can have

$$\gamma_{sn} = \frac{Z_r \psi_n}{2\pi}. \tag{E.3}$$

The phase difference between the $r-p$ and $s-p$ meshes is given by:

$$\gamma_{sr} = \begin{cases} \approx 0 & \text{if } Z_p \text{ is an odd number} \\ \approx 1/2 & \text{if } Z_p \text{ is an even number.} \end{cases} \tag{E.4}$$

The fluctuation components in Figs. 13.6(c) and 13.6(d) can be represented as a Fourier series:

$$\begin{cases} \Delta k_{rn}(t) = \sum_{l=1}^{\infty} \left(a_{rn}^l \cos l\omega_m t + b_{rn}^l \sin l\omega_m t \right) \\ \Delta k_{sn}(t) = \sum_{l=1}^{\infty} \left(a_{sn}^l \cos l\omega_m t + b_{sn}^l \sin l\omega_m t \right). \end{cases} \tag{E.5}$$

The coefficients are:

$$\begin{cases} b_{sn}^l = k^* \{ 2 \sin \pi l (2\gamma_{sn} + \epsilon_s - 1) \sin \pi l (\epsilon_s - 1) \\ \qquad + (1 - \epsilon_s) \left[\cos 2\pi l (2 - \gamma_{sn} - \epsilon_s) - 1 \right] \} \\ a_{sn}^l = k^* \{ 2 \cos \pi l (2\gamma_{sn} + \epsilon_s - 1) \sin \pi l (\epsilon_s - 1) \\ \qquad + (1 - \epsilon_s) \left[\sin 2\pi l (2 - \gamma_{sn} - \epsilon_s) - 1 \right] \} \\ b_{rn}^l = k^* \{ 2 \sin \pi l (2\gamma_{rn} + 2\gamma_{sr} + \epsilon_r - 1) \sin \pi l (\epsilon_r - 1) \\ \qquad + (1 - \epsilon_r) \left[\cos 2\pi l (2 - \gamma_{rn} - \gamma_{sr} - \epsilon_r) - 1 \right] \} \\ a_{rn}^l = k^* \{ 2 \cos \pi l (2\gamma_{rn} + 2\gamma_{sr} + \epsilon_r - 1) \sin \pi l (\epsilon_r - 1) \\ \qquad + (1 - \epsilon_r) \sin 2\pi l (2 - \gamma_{rn} - \gamma_{sr} - \epsilon_r) \}, \end{cases}$$

where

$$k^* = \frac{k_{smax} - k_{smin}}{l\pi}. \tag{E.6}$$

References

Arakelian, V., Briot, S., 2015. Balancing of Linkages and Robot Manipulators: Advanced Methods with Illustrative Examples, vol. 27. Springer.

Arakelian, V., Dahan, M., Smith, M., 2000. A historical review of the evolution of the theory on balancing of mechanisms. In: International Symposium on History of Machines and Mechanisms Proceedings HMM 2000. Springer, pp. 291–300.

Berkof, R., Lowen, G., 1969. A new method for completely force balancing simple linkages. Journal of Engineering for Industry 91 (1), 21–26.

Berkof, R.S., 1973. Complete force and moment balancing of inline four-bar linkages. Mechanism and Machine Theory 8 (3), 397–410.

Buckingham, E., 1931. Dynamic Loads on Gear Teeth. American Society of Mechanical Engineers.

Buckingham, E., 1949. Analytical Mechanics of Gears. McGraw-Hill.

Chace, M.A., 1967. Analysis of the time-dependence of multi-freedom mechanical systems in relative coordinates. ASME Journal of Manufacturing Science and Engineering 89 (1), 119–125.

Chang, Z., Zhang, C., Yang, Y., Wang, Y., 2001. A study on dynamics of roller gear cam system considering clearances. Mechanism and Machine Theory 36 (1), 143–152.

Chapra, S.C., Canale, R.P., 2011. Numerical Methods for Engineers, vol. 2. Mcgraw-Hill, New York.

Chew, M., Chuang, C.H., 1995. Minimizing residual vibrations in high-speed cam-follower systems over a range of speeds. Journal of Mechanical Design 117 (1), 166–172.

Clough, R.W., 1960. The finite element method in plane stress analysis. In: Proceedings of 2nd ASCE Conference on Electronic Computation. Pittsburgh Pa., Sept. 8 and 9, 1960.

Dresig, H., Holzweißig, F., 2010. Dynamics of Machinery: Theory and Applications. Springer Science & Business Media.

Dubowsky, S., Freudenstein, F., 1971. Dynamic analysis of mechanical systems with clearances—part 1: formation of dynamic model. ASME Journal of Engineering for Industry 93 (1), 305–309.

Emőd, I., Jurek, A., 1967. Massenausgleich am kurbelgetriebe von sechszylinder-viertakt-v-motoren mit 6 kurbeln und 60 zylinderwinkeln. Periodica Polytechnica. Mechanical Engineering 11 (3–4), 205–211.

Erdman, A.G., Sandor, G., 1972. Kineto-elastodynamics–a review of the state of the art and trends. Mechanism and Machine Theory 7 (1), 19–33.

Erdman, A., Sandor, G., Oakberg, R., 1972. A general method for kineto-elastodynamic analysis and synthesis of mechanisms. ASME Journal of Manufacturing Science and Engineering 94 (4), 1193–1205.

Esat, I., Bahai, H., 1999. A theory of complete force and moment balancing of planer linkage mechanisms. Mechanism and Machine Theory 34 (6), 903–922.

Feng, Z., Sun, X., Zhang, C., 1991. Kineto-elastodynamic analysis of four-bar linkages incorporating the effects of clearances in kinematic pairs. Chinese Journal of Mechanical Engineering 4 (3), 1.

Fischer, O., 1902. Über die reduzierten systeme und die hauptpunkte der glieder eines gelenkmechanismus. Zeitschrift für Angewandte Mathematik und Physik 47, 429–466.

Flores, P., Ambrósio, J., Claro, J., Lankarani, M., 2008. Kinematics and Dynamics of Multibody Systems with Imperfect Joints: Models and Case Studies, vol. 34. Springer.

Grote, K.-H., Antonsson, E.K., 2009. Springer Handbook of Mechanical Engineering. Springer.

Harris, S.L., 1958. Dynamic loads on the teeth of spur gears. Proceedings of the Institution of Mechanical Engineers 172 (1), 87–112.
Holzer, H., 1921. Die Berechnung der Drehschwingungen. Springer.
Hrones, J.A., et al., 1948. An analysis of the dynamic forces in a cam-driven system. Transactions of the American Society of Mechanical Engineers 70, 473–482.
Huang, Z., Shi, Z., 1990. Theory of mini-flywheel with high efficiency. Journal of Northeast Institute of Heavy Machinery 14 (1), 1–7.
Hughes, T.J., 2012. The Finite Element Method: Linear Static and Dynamic Finite Element Analysis. Courier Corporation.
Imam, I., Sandor, G.N., Gramor, S., 1979. Deflection and stress analysis in high speed planar mechanisms with elastic links. ASME Journal of Engineering for Industry 101 (1), 154–162.
Jazar, R.N., 2010. Theory of Applied Robotics: Kinematics, Dynamics, and Control. Springer Science & Business Media.
Kahraman, A., 1994a. Natural modes of planetary gear trains. Journal of Sound and Vibration 173 (1), 125–130.
Kahraman, A., 1994b. Planetary gear train dynamics. ASME Journal of Mechanical Design 116, 713–720.
Kashay, A.M., Voelker, F.C., Smalley, A.J., 1972. Dynamic shock phenomena in rolling mills. ASME Journal of Manufacturing Science and Engineering 94 (2), 647–657.
Kochev, I., 2000. General theory of complete shaking moment balancing of planar linkages: a critical review. Mechanism and Machine Theory 35 (11), 1501–1514.
Koster, M., 1974a. The effects of backlash and shaft flexibility on the dynamic behaviour of a cam mechanism. In: Jones, J. Rees (Ed.), Cams and Cam Mechanisms. MEP, London and Birmingham, Alabama.
Koster, M.P., 1974b. Vibrations of Cam Mechanisms. Macmillan International Higher Education.
Leissa, A.W., Qatu, M.S., 2011. Vibrations of Continuous Systems. McGraw-Hill Education.
Lin, H.H., Huston, R.L., Coy, J.J., 1988. On dynamic loads in parallel shaft transmissions, part 1: modeling and analysis. ASME Journal of Mechanisms, Transmissions, and Automation in Design 110, 221–225.
Lin, J., Parker, R.G., 1999. Analytical characterization of the unique properties of planetary gear free vibration. ASME Journal of Vibration and Acoustics 121 (3), 316–321.
Loewy, R.G., Piarulli, V.J., 1970. Dynamics of Rotating Shafts. Shock and Vibration Information Center.
Lutes, L.D., Sarkani, S., 2004. Random Vibrations: Analysis of Structural and Mechanical Systems. Butterworth-Heinemann.
Mata, A.S., Torras, A.B., Carrillo, J.A.C., Juanco, F.E., Fernández, A.J.G., Martínez, F.N., Fernández, A.O., 2016. Fundamentals of Machine Theory and Mechanisms. Springer.
Matsushita, O., Tanaka, M., Kanki, H., Kobayashi, M., Keogh, P., 2017. Vibrations of Rotating Machinery. Springer.
Midha, A., Erdman, A., Frohrib, D., 1979. A closed-form numerical algorithm for the periodic response of high-speed elastic linkages. ASME Journal of Mechanical Design 101 (1), 154–162.
Miedema, B., Mansour, W., 1976. Mechanical joints with clearance: a three-mode model. ASME Journal of Engineering for Industry 98 (4), 1319–1323.
Mihelj, M., Bajd, T., Ude, A., Lenarčič, J., Stanovnik, A., Munih, M., Rejc, J., Šlajpah, S., 2019. Robotics. Springer.
Mitchell, D.B., 1950. Tests on dynamic response of cam-follower systems. Mechanical Engineering 72, 467–471.
Nayfeh, A.H., Mook, D.T., 2008. Nonlinear Oscillations. John Wiley & Sons.

Norton, R.L., 2004. Design of Machinery: An Introduction to the Synthesis and Analysis of Mechanisms and Machines. McGraw-Hill Higher Education, Boston.

Oliver, J., Wysocki, D., Thompson, B., 1985. The synthesis of flexible linkages by balancing the tracer point quasi-static deflections using microprocessor and advanced materials technologies. Mechanism and Machine Theory 20 (2), 103–114.

Özgüven, H.N., Houser, D.R., 1988. Mathematical models used in gear dynamics—a review. Journal of Sound and Vibration 121 (3), 383–411.

Parker, R.G., 2000. A physical explanation for the effectiveness of planet phasing to suppress planetary gear vibration. Journal of Sound and Vibration 236 (4), 561–573.

Petyt, M., 2010. Introduction to Finite Element Vibration Analysis. Cambridge University Press.

Pfeiffer, F., Kunert, A., 1990. Rattling models from deterministic to stochastic processes. Nonlinear Dynamics 1 (1), 63–74.

Pratap, R., 2010. Getting Started with MATLAB: A Quick Introduction for Scientists and Engineers. Oxford University Press, New York.

Preumont, A., 2018. Vibration Control of Active Structures: An Introduction, vol. 246. Springer.

Rankine, W., 1869. Centrifugal whirling of shafts. Engineer 27, 249.

Rao, S.S., 2010. Mechanical Vibrations, 5th edition. Pearson.

Rivin, E.I., 2003. Passive Vibration Isolation. ASME Press.

Rothbart, H.A., 2004. Cam Design Handbook. McGraw-Hill Education.

Sauer, T., 2012. Numerical Analysis, 2nd edition. Pearson Education Inc., Upper Saddle River.

Schlegel, R.G., Mard, K.C., 1967. Transmission noise control-approaches in helicopter design. In: Mechanical Engineering, vol. 89. ASME, p. 55.

Smith, J.D., 2003. Gear Noise and Vibration. CRC Press.

Song, Y., Zhang, C., Yu, Y., 2001. Neural networks based active vibration control of flexible linkage mechanisms. ASME Journal of Mechanical Design 123 (2), 266–271.

Stoddart, D.A., 1953. Polydyne Cam Design. Penton Publishing Company.

Sun, Y., Zhang, C., Pan, F., Chen, S., Huang, Y., 2000. Dynamic model of a spur gear transmission system vibration. Chinese Journal of Mechanical Engineering 36 (8), 47–50.

Sunada, W., Dubowsky, S., 1983. On the dynamic analysis and behavior of industrial robotic manipulators with elastic members. ASME Journal of Mechanisms, Transmissions, and Automation in Design 105, 42.

Tang, Y., Chang, Z., Dong, X., Hu, Y., Yu, Z., 2013. Nonlinear dynamics and analysis of a four-bar linkage with clearance. Frontiers of Mechanical Engineering 8 (2), 160–168.

Tepper, F., Lowen, G., 1972. General theorems concerning full force balancing of planar linkages by internal mass redistribution. Journal of Engineering for Industry 94 (3), 789–796.

Tesar, D., Matthew, G.K., 1976. The Dynamic Synthesis, Analysis, and Design of Modeled Cam Systems. Lexington Books.

Theodossiades, S., Natsiavas, S., 2000. Non-linear dynamics of gear-pair systems with periodic stiffness and backlash. Journal of Sound and Vibration 229 (2), 287–310.

Thompson, B.S., Zuccaro, D., Gamache, D., Gandhi, M.V., 1983. An experimental and analytical study of the dynamic response of a linkage fabricated from a unidirectional fiber-reinforced composite laminate. ASME Journal of Mechanisms, Transmissions, and Automation in Design 105 (3), 526–533.

Thornton, W., Willmert, K., Khan, M., 1979. Mechanism optimization via optimality criterion techniques. ASME Journal of Mechanical Design 101, 392–397.

Tian, Q., Flores, P., Lankarani, M., 2018. A comprehensive survey of the analytical, numerical and experimental methodologies for dynamics of multibody mechanical systems with clearance or imperfect joints. Mechanism and Machine Theory 122, 1–57.

Tuplin, W., 1950. Gear-tooth stresses at high speed. Proceedings of the Institution of Mechanical Engineers 163 (1), 162–175.
Turner, M., 1956. Stiffness and deflection analysis of complex structures. Journal of the Aeronautical Sciences 23 (9), 805–823.
Volmer, J., 1976. Getriebetechnik Kurvengetriebe. Veb Verlag Technik, Berlin.
Wagg, D., Neild, S., 2016. Nonlinear Vibration with Control. Springer.
Wang, J., Li and, R., Peng, X., 2003. Survey of nonlinear vibration of gear transmission systems. ASME Applied Mechanics Review 56 (3), 309–329.
Wang, S., 2005. Theoretical and experimental investigations on dynamics of spur planetary gear transmissions based on planet phasing theory. PhD thesis, PhD Dissertation. Tianjin University.
Wang, S.Y., Song, Y.M., Shen, Z.G., Zhang, C., Yang, T.Q., Xu, W.D., 2005. Research on natural characteristics and loci veering of planetary gear transmissions. Journal of Vibration Engineering 18 (4), 412–417.
Weaver, W., Timoshenko, S.P., Young, D.H., 1990. Vibration Problems in Engineering. John Wiley & Sons.
Wei, S., Han, Q., Chu, F., 2016. Review on research of uncertain dynamics for gear system. Journal of Mechanical Engineering 52 (1), 1–19.
Wiederrich, J., Roth, B., 1978. Design of low vibration cam profiles. In: Cams and Cam Mechanisms.
Yan, H.S., Tsai, M.C., Hsu, M.H., 1996. Variable-speed method for improving motion characteristics of cam-follower systems. ASME Journal of Mechanical Design 118 (2), 250–258.
Yan, H., Yao, Y., Zou, H., 2001. Polydyne servo-cam design. Journal of the Chinese Society of Mechanical Engineers, Transactions of the Chinese Institute of Engineers, Series C/Chung-Kuo Chi Hsueh Kung Ch'eng Hsuebo Pao 22 (2), 99–104.
Yao, Y., Zhang, C., 1998. Theoretical explanation on the relation between residual vibration and motion continuity of cam mechanisms. Mechanical Science and Technology 17 (6).
Zhang, C., Grandin, H., 1982. Kinematic refinement technique in optimum design of flexible mechanisms. ASME Paper (82-DET-21).
Zhang, C., Grandin, H., 1983. Optimum design of flexible mechanisms. ASME Journal of Mechanisms, Transmissions, and Automation in Design 105 (2), 267–272.
Zhang, C., Huang, Y., Wang, Z., Chen, S., 1997. Analysis and Design of Elastic Linkages. China Machine Press, Beijing.
Zhang, C., Wang, S., Song, Y., Yang, T., 2005. Modification of parameters of planetary gear trains. Journal of Tianjin University 38 (4), 283–287.
Zhang, C., Yang, J., 2020. A History of Mechanical Engineering. Springer.
Zhang, C., Yang, Y., Zhou, X., Yao, Y., 1995. General purpose harmonic trapezoidal combinational cam curve with continuous jerk. In Chinese. Journal of Dalian Institute of Light Industry 14 (3), 8–15.

Index

D

N

O

P

R

S

T

V

W

Kolibri 360 GmbH
Ferdinand-Jühlke-Straße 7
99095 Erfurt, Deutschland
produktsicherheit@kolibri360.de